The Langevin and Generalised Langevin Approach to the Dynamics of Atomic, Polymeric and Colloidal Systems

The Langevin and Generalised Langevin Approach to the Dynamics of Atomic, Polymeric and Colloidal Systems

Ian Snook

Applied Physics
School of Applied Sciences
RMIT University
Melbourne, Australia

ELSEVIER

Amsterdam • Boston • Heidelberg • London • New York • Oxford
Paris • San Diego • San Francisco • Singapore • Sydney • Tokyo

Elsevier
Radarweg 29, PO Box 211, 1000 AE Amsterdam, The Netherlands
The Boulevard, Langford Lane, Kidlington, Oxford OX5 1GB, UK

First edition 2007

Library of Congress Cataloging-in-Publication Data
A catalog record for this book is available from the Library of Congress

British Library Cataloguing in Publication Data
A catalogue record for this book is available from the British Library

ISBN-13: 978-0-444-52129-3
ISBN-10: 0-444-52129-1

For information on all Elsevier publications
visit our website at books.elsevier.com

Printed and bound in the United Kingdom

Transferred to Digital Print 2010

I dedicate this book to my mother Joan, my wife Marie and my children Stuart, Graeme, Tamara and Simon

Contents

Preface

The stochastic description of the dynamics of a many-body system consisting of a single, large particle, suspended in a fluid made up of an enormous number of much smaller, molecular-sized particles was a major advance in Physics. This approach to describe Brownian motion was pioneered by Einstein [1] and Langevin [2], and their work marks the beginning of a new way to treat the dynamics of many-body systems where a dynamical variable of interest (in this case the velocity of the large particle) is singled out and some other aspects of the problem are treated by the theory of random processes. This treatment is in contrast with the more traditional, kinetic-theory approach of considering the explicit dynamics of all the particles in the system.

Since this early work these stochastic methods have been generalised and refined into a field of major significance. Many areas of research have been influenced by this Brownian approach and stochastic methods have been applied to problems in such diverse areas of physics, chemistry, biology and engineering [3–5]. One of the main uses of these methods in the theory of matter is to treat many-body systems in which different degrees of freedom operate on very different time scales. This is because this approach allows the separation of these very different degrees of freedom and the treatment of each of them by different methods. This separation is achieved by averaging over some variables, which results in equations of motion for only the chosen variables. This approach to singling out particular variables is, however, quite general and not restricted to situations where there is a clear distinction between "fast" and "slow" variables, and can be applied to any system whose dynamics is described by equations of motion such as Newton's equations. The equations resulting from this alternative approach are exact and equivalent to the starting equations from which they are derived and, thus, may be used to provide an exact, but alternative description of the time evolution of any dynamical variable. However, the form of these new equations of motion also enables certain variables to be treated approximately and simply, while allowing the remaining variables to be treated in greater detail. Many methods of this type have been developed under the heading of generalised Langevin methods (GLE) and have provided insight into the properties of many-body systems ranging from atomic liquids, atoms on surfaces and polymeric solutions to colloidal suspensions.

Theories of this type have been considered by many authors from a variety of different points of view ranging from atomic to macroscopic and many excellent treatises and review articles are available (see for example references to Chapters 2 and 3) that deal with specialist aspects of these methods, for example, time correlation functions of atomic and molecular systems. By contrast this treatise attempts to unify this approach to a large number of areas encountered in condensed-matter theory. Explicitly, his book is primarily intended as a fundamental treatment of the background and derivation of GLE's and Langevin equations of use in the interpretation of the dynamics and dynamical properties of many-body systems governed by Classical Mechanics. Such systems may be atomic, molecular, colloidal or polymeric. The attempt is to provide a reasonably rigorous and complete text, which starts from first principles and provides detailed derivations and discussions of these fundamentals. No attempt is made to cover all applications, but only a few typical applications are discussed in order to illustrate these fundamentals. However, the references given should enable the reader to get a reasonable idea of applications to various areas.

Discussions of the use of approximations are given which should help the reader to make use of approximate schemes to calculate and interpret time-correlation functions. Furthermore, we have included methods to construct numerical GLE algorithms to assist researchers to carry out computer simulations of the dynamical properties of atomic, colloidal and polymeric systems using "Brownian Dynamics" methods. To help in this, details are also given of relevant macroscopic hydrodynamics, which is needed to treat the dynamics of colloidal suspensions. In addition, some Fortran programs are included in Appendix O and references to where other computer programs and routines may be found in order to help the interested reader produce their own results and, hopefully, to aid some people in their research.

Finally, I would like to thank some of the many people who have helped me in this area over the years and attempted to educate me about specific aspects of the theory covered in this book. In no particular order I would like to thank, Bill van Megen, Peter Daivis, Kevin Gaylor, Bob Watts, Brendan O'Malley, Ed Smith, Peter Pusey, Heiner Versmold, Ubo Felderholf, Denis Evans, Rudolf Klein, Kurt Binder, Mike Towler and Ron Otterwill. I would also like to thank my young colleagues who helped me with some aspects of this book, Nicole Benedek, Manolo Per, Tanya Kairn, Mat McPhie, Stephen Williams and Rob Rees.

REFERENCES

1. A. Einstein, *Investigations on the Theory of Brownian Movement*, Dover, New York, USA, 1956.
2. P. Langevin, *Comp. Rendus*, **146**, 530 (1908).
3. S. Chandrasekhar, Stochastic problems in physics and astronomy, *Rev. Mod. Phys.*, **15,** 1–91 (1943).
4. N. Wax (Ed.), Selected papers on noise and stochastic processes, Dover, New York, 1954.
5. N.G. Van Kampen, *Stochastic processes in physics and chemistry*, North-Holland, Amsterdam, 1992.

Notation

The following notation, definitions and symbols have been used throughout this book.

A. POTENTIAL ENERGY FUNCTIONS

Hard Sphere Potential

$$u(r) = \infty, \quad r \le d$$
$$= 0, \quad r > d$$

Lennard-Jones 12-6 (or LJ12-6) potential

$$u_{ij}(r) = 4\varepsilon_{ij}\{(\sigma_{ij}/r)^{12} - (\sigma_{ij}/r)^{6}\}$$

or with a core,

$$u_{ij}(r) = 4\varepsilon_{ij}[\{\sigma_{ij}/(r-d_{ij})\}^{12} - \{\sigma_{ij}/(r-d_{ij})\}^{6}]$$

Chandler–Weeks–Andersen (or CWA) potential

$$u_{ij}(r) = 4\varepsilon_{ij}\{(\sigma_{ij}/r)^{12} - (r/\sigma_{ij})^{6}\} + \varepsilon, \mathrm{r} \le \mathrm{r}_{0ij}$$

$$= 0, \quad r > r_{0ij}$$

or with a core,

$$u_{ij}(r) = 4\varepsilon_{ij}[\{\sigma_{ij}/(r-d_{ij})\}^{12} - \{\sigma_{ij}/(r-d_{ij})\}^{6}] + \varepsilon_{ij}, \quad r \le r_{0ij}$$

$$= 0, \quad r > r_{0ij}$$

where ε_{ij} is the well depth, σ_{ij} the value of r for which $u(r) = 0$, r_0 is the position of the well, i.e., $u_{ij}(r_0) = -\varepsilon_{ij}$ and d_{ij} is the core diameter

Soft-sphere potential

$$u(r) = 4\varepsilon_{ij}(\sigma_{ij}/r)^{\mathrm{n}}$$

where n is an integer.

B. SYMBOLS USED

A, A_{ij}	a scalar
$\underline{A}$	a vector,
$\mathbf{A}_{\mathrm{ij}}$	a tensor
$\mathbf{A}$	an operator

C. OPERATIONS

$$\underline{A} \bullet \underline{B} = A_x B_x + A_y B_y + A_z B_z$$

$$\mathbf{A_{ij}} : \mathbf{B_{ij}} = \Sigma\Sigma A_{ij}\ B_{ji} = A_{ij}\ B_{ji}$$

– 1 –

Background, Mechanics and Statistical Mechanics

In order to fully appreciate how to calculate and use generalised Langevin equations (GLEs) it is first necessary to review the mechanics upon which these GLEs are based and the statistical mechanics which is used in order to calculate bulk properties from the information which these equations generate.

1.1 BACKGROUND

The instantaneous mechanical state of system described by classical mechanics requires only the specification of a set of positions and momenta of the particles making up the system and provided that these particles are "heavy enough" this classical mechanical approach will provide an accurate description of the physical state of a many-body system. In practice this applies to systems consisting of most atoms under normal physical conditions, except for hydrogen and helium. Then the common approach to describing the time evolution of this mechanical state of such a many-body system, its dynamics, is by use of a coupled set of differential equations, for example, Newton's equations of motion, which describes the detailed, individual dynamics of all the particles in the system. In the study of molecular systems this approach has led to the development of the widely used numerical technique called molecular dynamics (MD). This method has provided numerous insights into the behaviour of molecular systems and there is now an extensive literature on the method and its application.

However an alternative approach to describing dynamics is to use equations of motion that describe the dynamics of only some, selected particles moving in the presence of the other particles in the system which are now regarded as a background or bath whose detailed dynamics is not treated. Thus, we select out a typical particle or set of particles in which we are interested and find an equation which describes the dynamics of these chosen particles in the presence of the other particles. The classical example of this is the Langevin equation (LE) developed in a heuristic way by Paul Langevin to describe the Brownian motion of a large particle suspended in a fluid consisting of an enormously large number of lighter particles.

The LE is an equation of motion for the velocity $\underline{V}_B$ of a single Brownian (B) particle of mass M_B suspended in a "bath" consisting of an enormous number, N_b, of particles of much smaller mass m and is

$$M_B\left(\frac{d}{dt}\right)\underline{V}_B = -\zeta\underline{V}_B + \underline{F}_B^R(t) \tag{1.1}$$

where ζ is the friction factor, $-\zeta\underline{V}_B$ the drag force due to the bath particles and $\underline{F}_B^R(t)$ the random force due to random thermal motion of the bath particles. Thus, instead of a description of the dynamics of the system by writing down the coupled set of Newton equations for the total system of particles consisting of the B particle and all the N_b background particles, we write down an equation for the dynamics of only the B particle in a background of the small particles. The effect of the small particles on the dynamics of the B particle now only appears via the drag force $-\zeta\underline{V}_B$ and the random force , a description of their detailed dynamics no longer is needed nor, in fact, is possible in this treatment. Further, the drag force $-\zeta\underline{V}_B$ is calculated by the theory of macroscopic hydrodynamics and the random force is assumed to be a Gaussian random variable, that is, is treated stochastically and only its statistical properties are needed. Thus, we have reduced the description of the dynamics of an $N_b + 1$ body problem to that of a one-body problem. However, we have lost information as we now no longer are able using this approach to follow the dynamics of the N_b bath particles as we have averaged-out or coarse-grained over their motion.

This approach pioneered by Langevin[1] has been made formal, vastly extended, and has been shown to be applicable to any dynamical variable. The basic ideas used to generalise the traditional Langevin equation are:

1. First define the dynamical variables of interest, for example, the velocities of particles.
2. Write down a set of coupled equations of motion for these variables in operator form, for example, for the velocities of some particles of the system.
3. Rewrite these equations of motion so as to project out the variables in which we are interested by use of projection operators.

This projecting out averages over the motion of the other particles in the system and is a coarse graining of the equations of motion which provides a description only of the variables which are not averaged over, thus, losing information about the dynamics of the system. Furthermore, this averaging or projection leaves us with terms, "random forces" in the resulting equations of motion for the desired variables which we only have limited information about. These terms are deterministic in the sense that, if we were to go back to the full equations of motion we would know them exactly but this would defeat the purpose of deriving the projected equations of motion. Thus, the "random forces" must, of necessity, be treated in practice as stochastic or random variables about which we only know their statistical properties.

The result of this process is as in the LE approach, one equation for each dynamical variable chosen rather than many coupled equations with which we started. There may still, of course, be many coupled equations but many fewer than we started with and which explicitly involve many less variables than we started with. For example in the Mori–Zwanzig approach[2] we may derive a single, exact equation for the velocity of a typical particle of each species and not one equation for every particle in the system that would constitute the normal kinetic theory approach. However, it should be emphasised that the resulting equations are still exactly equivalent to the original coupled equations on which they are based and we have not eliminated the coupling of the dynamics of each particle. This coupling will be shown to be represented by a Kernel (or memory function) and a "random force" appearing in these equations. Thus, until we are able to calculate or approximate these two terms we have not achieved a solution to the problem of describing the dynamical properties of the system and their time evolution.

There are, however, advantages to this approach some of which are

1. There is one equation per "species" to be solved.
2. These equations are exact and entirely equivalent to the original equations of motion.
3. These equations may be readily used to construct approximate equations of motion.
4. Equations of motion for time-correlation functions, which can be used to calculate linear transport coefficients and scattering functions, may be directly derived from these basic equations.
5. The form of the memory function often gives us physical insight into the processes involved in the relaxation of a variable to equilibrium.

In order to carry out the above derivations we must first give an outline of the classical mechanical description of dynamics mentioned above and then establish the basis for deriving the generalised Langevin description. For completeness we will also provide an outline of how such dynamical information may be used to calculate the mechanical and non-mechanical properties of a many-body system. In subsequent chapters equations of the GLE type will be derived, basic applications given and numerical schemes outlined for solving them which are analogous to the MD method based on Newton's equations of motion.

1.2 THE MECHANICAL DESCRIPTION OF A SYSTEM OF PARTICLES

The classical mechanical description of the instantaneous mechanical state of system only requires the specification of the set of positions and momenta for all the N particles in the system. Then the common approach to describing the time evolution of these variables, the dynamics, of such a many-body system is by use of a coupled set of differential equations describing the detailed time evolution of the position and momentum of each particle. Usually this time evolution is described by Newton's equations of motion written in terms of Cartesian co-ordinates but other equations of motion may be used such as the Lagrange or Hamilton equations and various co-ordinate and momenta schemes used as appropriate. However, we will not give a detailed description of these aspects of mechanics as they are very well known.[3] It does, however, seem wise to comment on the description of a many-body system as given by this classical treatment, show how its output may be used to calculate the macroscopic properties of the system, to discuss the limitations it imposes on the calculation of these properties and what extensions are needed to this basic approach to treat some properties and some types of systems. These points need to be clarified before transforming this familiar Newtonian mechanical description of a system into the less familiar generalised Langevin one.

To reiterate as the classical mechanical state of a system at a particular time is specified by giving the value of the positions of the particles and their momenta (represented by the symbol Γ) then all mechanical properties may similarly be expressed in terms of these quantities.[3] For example, the mechanical energies of the system, its pressure and its pressure tensor may be expressed as a function of these particle positions and momenta. Since the total number of particles N, the volume V and the total energy U are fixed these equations describe an isolated or (NVU) system. The observable macroscopic, mechanical properties of the system may then be obtained by time averaging these instantaneous values of the property B over an appropriate time interval[3]

$$\langle B \rangle = \lim_{T\to\infty} \frac{1}{T} \int_{t_0}^{t_0+T} B(t)\,\mathrm{d}t \tag{1.2}$$

However, to find the non-mechanical properties of the system, such as the equilibrium thermodynamic and transport coefficients, we have to invoke other hypotheses as such properties cannot be derived from the postulates of classical mechanics alone.[3]

For example, in order to calculate equilibrium thermodynamic properties from the positions and momenta obtained from a solution of Newton's equations we may use Gibb's microcanonical, ensemble-averaging method and the Ergodic hypothesis.[3,4] This hypothesis states that we may replace the above-mentioned time averages over a dynamical trajectory by a probabilistic or ensemble average over all possible mechanical states of the system consistent with the constant values of N, V and U (a Microcanonical ensemble)

$$\langle B \rangle = \int B(\Gamma) f^{\mathrm{eq}}(\Gamma)\,\mathrm{d}\Gamma \tag{1.3}$$

where $f^{\mathrm{eq}}(\Gamma)$ is a probability function which characterises the equilibrium state of an (NVU) system (see Appendix A).

This method enables the calculation of such thermodynamic properties as the heat capacity and entropy to be made using the positions and momenta obtained from classical mechanics but without the need to modify the basic equations of motion.

Similarly, to calculate linear-transport coefficients such as the self-diffusion constant we may use the method of linear response theory and time-correlation functions pioneered by Kubo and Green.[5,6] This approach, essentially, allows the transport coefficients which control the rate of linear, dissipative processes to be related to fluctuations of variables occurring in a mechanical system at equilibrium which are related to time-correlation functions of mechanical variables defined by (see Appendix A)

$$\langle A(0)\, B(t) \rangle = \int A^{*}(0) B(t)\, f^{\mathrm{eq}}(\Gamma)\,\mathrm{d}\Gamma \tag{1.4}$$

Thus, we have to go beyond classical mechanics in order to find expressions for equilibrium and transport coefficients but they may still be calculated in terms of the basic quantities arising from this mechanical description. We do not need to alter the basic equations of motion.

However, a further problem encountered in using the traditional mechanical description of a many-body system at equilibrium is that Newton's (and Hamilton's) equations of motion describe an isolated (*NVU*) system which at equilibrium, may be represented by the Microcanonical ensemble as mentioned above. Now at equilibrium it is often found to be more useful to work with systems other than isolated systems, that is ones whose physical state is defined by constant values of system variables other than *N*, *V* and *U*. Experimentally, this is achieved by putting the system of interest in contact with an external system with which it can exchange energy and/or particles. For example, we may choose to use a closed system of constant volume in contact with an external thermostat or heat bath which maintains constant temperature in the system, such a system is an (*NVT*) system not an (*NVU*) one. Gibb's method for a microcanonical ensemble may be readily extended to calculate the equilibrium thermodynamic properties of such systems by deriving different distribution functions, that is, probability weightings in the ensemble averaging process.[3,4,7] However, once again in achieving this we still only use the properties obtained from the unmodified equations of mechanics but with the addition of terms representing other contributions to the energy, see Table A.1 of Appendix A. Of course, we also allow the energy to not be a constant of the motion. Also, even though in an experiment the interaction of the system with external reservoirs are needed to maintain constant values of variables other than *N*, *V* and *U* the explicit mechanical properties of the external reservoirs are not needed to calculate equilibrium properties. Only the values of parameters that are maintained constant, for example, temperature, *T* are added as constraints in determining the probabilities of occurrence of each state. Thus, eq. (1.3) is still applicable and only the form of f^{eq} differs from one ensemble to another. The form of f^{eq} for the commonly used ensembles is given in Table A.1 of Appendix A.

This Gibbsian method does not, however, give rise to an equation of motion appropriate to such a system. Recent work has, however, addressed this question and in what might loosely be thought of as using the Ergodic hypothesis in reverse these alternative approaches have sought to find generalised Newton's equations of motion which do maintain constant values of variables other than *N*, *V* and *U*, for example, constant temperature.[8–10] Then equilibrium properties may be obtained by time averaging as can be done in an (*NVU*) system instead of ensemble averaging. In order to illustrate how this may be achieved we will describe one method of maintaining constant temperature in Section 1.2.2 below. Once such equations are obtained the calculation of time-correlation functions and transport coefficients in an (*NVT*) system follows the same procedure as that in the (*NVU*) system.

However, an extension of time-correlation function methods, based on fluctuations in an equilibrium system, to calculate non-linear transport coefficients has proved to be frustratingly difficult and there appears to be no generalisation of this approach to use in such circumstances which does not contain theoretical difficulties. Now, in order to measure transport coefficients experimentally we normally set up a non-equilibrium steady state which is maintained in a time-independent but non-equilibrium state by a combination of some external field and the provision of some mechanism which adds or removes heat from the system. This is done by means of interaction of the system with external systems just as is used to achieve thermostated equilibrium systems. This experimental method allows the measurement of both linear and non-linear transport phenomena, for example, rheological behaviour of suspensions. Using this method transport properties may be calculated in terms of appropriate variables averaged over time in the steady state. For example, the coefficient of viscosity η may be measured by setting up an experimental system with a constant gradient du_x/dy in the x-component of the flow velocity, u_x in the y direction and is given by

$$\eta = \frac{\sigma}{(du_x/dy)} \tag{1.5}$$

where σ is the macroscopic shear stress.

Unfortunately, methods based on the direct application of Newton equations without an external force cannot be used to generate such non-equilibrium steady states as starting from an arbitrary (and in general non-equilibrium) state given sufficient time the state of such a system will eventually reach equilibrium. In order to achieve a non-equilibrium steady state one can set up a system analogous to that used in experiments described above, that is, one which is maintained in this state by means of walls and external reservoirs. However, this type of inhomogeneous system has proven to be difficult to use in computer simulation if one wishes to model a homogeneous system in a well-defined physical state.[9–14] This is because the effect of the boundaries (for example, non-uniform density) is large unless a very large system is used.[14] Of course, if one wishes to simulate such a system then there is no alternative but to simulate the whole system. Also the theory of non-equilibrium states in inhomogeneous systems is far from fully developed. Thus, other extensions of the Newton equations have been developed which can describe general non-equilibrium states and, in particular, non-equilibrium steady states which are based on modifying these equations of motion in such a way that homogeneous systems may be used. This is done by an extension of the methods used to generalise Newton's equations of motion to treat constant-temperature systems at equilibrium outlined above.[9,15] We will describe one particular approach of this type, the so-called SLLOD equations of motion and its generalisation which has proved to be very useful in the MD study of non-equilibrium phenomena and has been recently used to derive GLEs appropriate to non-equilibrium states.[16] Of course, once such states are set up properties may be obtained by time averaging, for example, in a system at steady state undergoing simple shear we would estimate the viscosity η by[9]

$$\eta = \frac{\langle P_{xy} \rangle}{(du_x/dy)} \tag{1.6}$$

where P_{xy} is now the xy component of the microscopically defined pressure tensor and $\langle \cdots \rangle$ represents time averaging. In computer-simulation studies this approach has the advantage that relatively small systems may be used and the system's state variables (e.g. defined by the temperature and density) are uniform throughout the system. This is in contrast to the simulation of systems with explicit use of walls to maintain the steady state.[11–14]

As with the constant-temperature equations of motion these new equations do not need to explicitly include a description of the dynamics of the external systems with which the system of interest is in contact and which are needed to maintain the non-equilibrium state.

Having outlined the main ideas behind the mechanical description of a system of particles we will now set up a convenient notation and mathematical methods to describe this mechanical state of a system of interacting particles and its time evolution.

1.2.1 Phase space and equations of motion

The classical state of the system of N particles at a particular time, t is described by the $3N$ momenta, $\underline{p}_i = m_i\underline{v}_i$ and $3N$ position co-ordinates $\underline{r}_i$ of the N particles. These $6N$ variables define a point in $6N$ dimensional phase space denoted by the vector,

$$\Gamma = (\underline{r}_1, \underline{r}_2, \ldots, \underline{r}_N, \underline{p}_1, \underline{p}_2, \ldots, \underline{p}_N)$$

Alternatively we may use the two $3N$ dimensional vectors defined by

$$\underline{p}^{(N)} = (\underline{p}_1, \underline{p}_2, \ldots, \underline{p}_N) \text{ and } \underline{r}^{(N)} = (\underline{r}_1, \underline{r}_2, \ldots, \underline{r}_N)$$

Thus, in systems described by classical mechanics all the mechanical properties are functions of Γ only and equilibrium properties may be expressed as time (or ensemble) averages and linear-transport coefficients and scattering functions as equilibrium time correlation functions, see Appendix A.

In order to describe the time evolution of the mechanical state of the system, that is, its dynamics we need to solve the equations of motion which describe the time evolution of the system in phase space. Such a solution gives us the time evolution of Γ and so we may think of the time evolution of the system as describe by the $2N$, 3-dimensional vectors $\{\underline{r}_1, \underline{r}_2, \ldots, \underline{r}_N, \underline{p}_1, \underline{p}_2, \ldots, \underline{p}_N\}$ or equivalently the time evolution of a single-point phase-space point Γ in $6N$ dimensional phase-space. In order to describe this dynamics we need to develop equations of motion.

1.2.2 In equilibrium

For a system of particles at equilibrium we will consider a set of N particles of masses m_i interacting according to a central force law $\underline{F}_i$ whose dynamics may be described by Newton's equations of motion

$$\frac{\underline{F}_i(t)}{m_i} = \left(\frac{\mathrm{d}}{\mathrm{d}t}\right)\underline{v}_i(t) = \left(\frac{\mathrm{d}}{\mathrm{d}t}\right)\underline{p}_i(t)/m_i \tag{1.7}$$

and

$$\left(\frac{\mathrm{d}}{\mathrm{d}t}\right)\underline{r}_i(t) = \underline{v}_i(t) = \frac{\underline{p}_i(t)}{m_i} \tag{1.8}$$

where $i = 1, \dots, N$

There is one vector eq. (1.7) for the velocity $\underline{v}_i$ or momentum $\underline{p}_i$ per particle and one vector eq. (1.8) for the position co-ordinate $\underline{r}_i$. The equation of motion of each particle i is coupled to all the other equations via the force $\underline{F}_i$. These equations are often solved numerically in order to generate a trajectory in phase-space, this is the method of (*NVU*) equilibrium molecular dynamics.[9]

One should be a little more precise about what we mean by saying that Newton's equations may be used to describe the dynamics of a system at equilibrium. We should state that using Newton's laws then:

1. Starting from an arbitrary (and in general non-equilibrium) state given sufficient time the state of such a system will eventually reach equilibrium.
2. Starting from an equilibrium state the system will pass through a series of equilibrium states whose probability of occurrence is given by $f_{eq}(\Gamma)$.
3. We cannot directly generate a non-equilibrium steady state.

As outlined above in order to describe other equilibrium thermodynamic systems we need to modify Newton's equations of motion in order to keep variables other than total energy U constant. This may be done in a variety of ways which may be readily implemented in MD simulations. For example constant temperature may be attained by ad-hoc scaling of velocities, the use of Gauss's principle of least constraint[9,10] or by the Nose–Hoover method.[19] In practice, all methods seem to work equally well at or close to equilibrium but there is still much research being carried out to find which methods are most appropriate far from equilibrium, this will be discussed in Section 1.2.3. We will now outline the use of Gauss's principle of least constraint and the iso-kinetic equations of motion as an example of such a technique.[8–10]

Gauss's principle of least constraint can be used to impose a constant value on some properties by adding constraints to Newton's equations of motion. This principle is used to modify Newton's equations to maintain constant values of some variables but in such a way that the trajectories followed deviate as little as possible, in a least-squares sense, from the unconstrained Newtonian trajectories. To achieve constant kinetic temperature a constraint is imposed to maintain constant kinetic energy which leads to the modified equations of motion,

$$\left(\frac{\mathrm{d}}{\mathrm{d}t}\right)\underline{r}_i = \frac{\underline{p}_i}{m_i} \tag{1.9}$$

$$\left(\frac{\mathrm{d}}{\mathrm{d}t}\right)\underline{p}_i = \underline{F}_i - \alpha\underline{p}_i$$

where α is a thermostating parameter given by

$$\alpha = \sum_{i=1}^{N}(\underline{p}_i \cdot \underline{F}_i/\mathrm{m}_i) / \sum_{i=1}^{N} \underline{p}_i^2/\mathrm{m}_i \tag{1.10}$$

and this factor ensures that the kinetic temperature remains constant. It should be noted that this method at equilibrium does not generate a canonical ensemble but rather an isokinetic ensemble.

Gauss's principle of least constraint can also be used to maintain constant values of other parameters, for example pressure, stress or entropy.[9,10] Several other methods are also available to maintain constant values of selected parameters using dynamical equations, for example the use of ad-hoc scaling of velocities,[17] the addition of a stochastic term[18] and the Nose–Hoover method[19] to maintain constant temperature and Anderson's method to maintain constant pressure.[18]

1.2.3 In a non-isolated system

As discussed in Section 1.2 non-equilibrium states such as a steady state can be maintained by the use of external sources which impose gradients in some properties and which balance the dissipative processes occurring in the system, for example, heating effects due to an imposed shear field. This may be done as it is experimentally by the direct modelling of a system of particles placed between walls which are also modelled explicitly, for example, moving walls that shear the system and are maintained at constant temperature. However, such methods can lead to very inhomogeneous physical states of the system if small numbers of particles are used. Thus, methods have been developed which involve modifying Newton's equations of motion by including terms representing fictitious fields as was described in Section 1.2.2 to maintain constant temperature in equilibrium. It has been shown that using field-dependent equations of motion leads to homogeneous systems which when combined with appropriate boundary conditions allow relatively small systems to be used to get accurate values of transport coefficients. One such method is a generalization the so-called SLLOD equations of motion developed for systems undergoing shear which are Newton's equations of motion modified to maintain non-equilibrium states and constant temperature, these equations are[9,15,16]

$$(\mathrm{d}/\mathrm{d}t)\underline{r}_i = \underline{p}_i/m_i + \underline{C}_i\underline{F}_e \tag{1.11}$$

$$(\mathrm{d}/\mathrm{d}t)\underline{p}_i = \underline{F}_i + \underline{D}_i\underline{F}_e - \alpha\underline{p}_i$$

where $\underline{C}_i$ and $\underline{D}_i$ describe the coupling between the particles and the (constant) applied field $\underline{F}_e$ that is required to generate the desired dissipative flux, and α is a "thermostating" parameter that ensures that the temperature or some other relevant variable remains constant. Under these equations of motion the dissipation (the rate of change of the internal energy due to the external field) is given by

$$J(\Gamma)\underline{F}_e = \sum_{i=1}^{N}\left[\underline{C}_i\cdot\underline{F}_i - \underline{D}_i\cdot\left(\frac{\underline{p}_i}{m_i}\right)\right]\underline{F}_e \tag{1.12}$$

where $J(\Gamma)$ represents the instantaneous value of the appropriate dissipative flux, which is a function of the phase point Γ.

For constant-temperature simulations at nonzero values of the field, these equations of motion generate an average rate of decrease in the volume of the phase space occupied by the phase-space distribution function that is directly proportional to the rate of heat removal by the thermostat. By contrast in equilibrium simulations this phase space compression does not occur.

The thermostating parameter α which is introduced to keep the temperature or some other variable constant has been derived by a variety of methods. The expression for α depends on the expression used to evaluate the temperature and/or what property is held constant. Two examples of expressions for α the first based on keeping the kinetic energy constant (the Gaussian Isokinetic) and the second may be used to keep the energy constant (the Gaussian Iso-energetic),[9,15]

$$\alpha_K = \frac{\sum_{i=1}^{N}(\underline{F}_i \cdot \underline{p}_i/m_i + \underline{F}_e \cdot \underline{D}_i \cdot \underline{p}_i/m_i)}{\sum \underline{p}_i^2/m_i} \tag{1.13}$$

and

$$\alpha_E = \underline{F}_e \cdot \frac{\sum_{i=1}^{N}(\underline{D}_i \cdot \underline{p}_i/m_i + \underline{F}_i \cdot \underline{C}_i)}{\sum_{i=1}^{N} \underline{p}_i^2/m_i} = \frac{-JV \cdot \underline{F}_e}{\sum_{i=1}^{N} \underline{p}_i^2/m_i} \tag{1.14}$$

Other methods of achieving constant temperature have been developed, for example, the Nose–Hoover method which lead to different expressions for α and to different algorithms for computing it.[15,19]

This and other related methods have been extensively used to describe systems in non-equilibrium states. There is, however, still much discussion as to whether some of the effects produced by these methods such as phase space compression are a characteristic of real systems or whether they are merely an artefact of the methods used to create a steady state in a system without boundaries.[20] In particular as the fields which appear in these equations of motion are not the external fields which one would apply to a physical system, they are introduced in order to constrain some of the system's properties to have constant values. However, such discussions will not concern us here as these methods can be shown to lead to the correct description of the measurable properties of a system even far from equilibrium.[15]

Thus, these methods and the numerical schemes based on them, the methods of non-equilibrium molecular dynamics (NEMD) lead to the correct values of correlation functions and transport properties in a non-equilibrium state.

1.2.4 Newton's equations in operator form

In order to derive equations of motion for particles in a background of other particles from the above formalisms we firstly introduce the linear, Hermitian, Liouville differential operator **L**

$$\mathbf{L} = -i\left\{\sum_{i=1}^{N}\left[\underline{v}_i \bullet \frac{\partial}{\partial \underline{r}_i} + \frac{\underline{F}_i}{m_i} \cdot \frac{\partial}{\partial \underline{v}_i}\right]\right\} \tag{1.15}$$

Note that **L** is only Hermitian if there is no phase-space compression and that i**L** is real.

We now rewrite Newton's equations (1.7) and (1.8) as the following equations (see Appendix A),

$$\left(\frac{\mathrm{d}}{\mathrm{d}t}\right)\underline{v}_i(t) = \underline{a}_i(t) = \underline{F}_i(t)/m_i = \mathrm{i}\mathbf{L}\,\underline{v}_i(t) \\ = \mathrm{i}\mathbf{L}\{\mathrm{e}^{\mathrm{i}\mathbf{L}t}\,\underline{v}_i(0)\} = \mathrm{i}\mathbf{L}\{\mathbf{G}(t)\underline{v}_i(0)\} \tag{1.16a}$$

$$\left(\frac{\mathrm{d}}{\mathrm{d}t}\right)\underline{r}_i(t) = \mathrm{v}_i(t) = \mathrm{i}\mathbf{L}\,\underline{r}_i(t) = \mathrm{i}\mathbf{L}\{\mathrm{e}^{\mathrm{i}\mathbf{L}t}\,\underline{r}_i(0)\} = \mathrm{i}\mathbf{L}\{\mathbf{G}(t)\underline{r}_i(0)\} \tag{1.16b}$$

where $\mathbf{G}(t) = \mathrm{e}^{i\mathbf{L}t}$ is called the propagator as it propagates a time-dependent variable forward in time, and $\underline{a}_i(t) = \underline{F}_i(t)/m_i$ is the acceleration of particle i.

These equations written in operator form are entirely equivalent to Newton's equations and if solved exactly will lead to precisely the same trajectories of each particle as the exact solution of Newton's equations.

1.2.5 The Liouville equation

A fundamental equation, which may also be used to describe the dynamics of a system of interacting particles, is the Liouville equation which is an equation of motion for the N-body distribution function which gives the probability of finding the system at a particular point Γ in phase space at a particular time t. This equation may be derived from the equations of motion of the system and will, thus, depend on the form of this equation of motion.

1.2.6 Liouville equation in an isolated system

In equilibrium since any function of the phase space vector may be written in terms of **L** as in eq. (1.16) and we may write such an equation for the N-body equilibrium distribution function (Γ,t) which is called the Liouville equation[2,9,15]

$$-\mathrm{i}\mathbf{L}f^N(\Gamma,t) = \mathrm{i}\mathbf{L}\{\mathrm{e}^{-\mathrm{i}\mathbf{L}t}f^N(\Gamma,0)\} \\ = (\partial/\partial t)f^N \tag{1.17}$$

Because of the fundamental roles played by the Liouville operator and the N-body distribution function the above equations are of central importance.

1.2.7 Expressions for equilibrium thermodynamic and linear transport properties

As any phase variable A (scalar, vector or tensor) in classical mechanics is a function of $\underline{r}_i$ and $\underline{v}_i$ only it may be expressed as $A(\mathbf{\Gamma},t)$ then

$$\left(\frac{\mathrm{d}}{\mathrm{d}t}\right)A(t) = \mathrm{i}\mathbf{L}\,A(t) = \mathrm{i}\mathbf{L}\{\mathbf{G}(t)A(0)\} \tag{1.18}$$

which by integration (see Appendix A) leads to

$$A(t) = e^{i\mathbf{L}t} A(0) = \mathbf{G}(t)A(0) \tag{1.19}$$

A time-correlation function may then be expressed as

$$\langle A(0)\,B(t)\rangle = \int f_{eq} A^*(0)\,B(t)\,d\Gamma = \int f_{eq} A^*(0)[e^{i\mathbf{L}t}\,B(0)]\,d\Gamma \tag{1.20}$$

An equilibrium time average, that is, an equilibrium ensemble average may be expressed as a zero-time value of a correlation function,

$$\langle A(0)B(0)\rangle = \int f_{eq}\, A^*(0)\,B(0)\,d\Gamma \tag{1.21}$$

Hence, all equilibrium thermodynamic, linear transport coefficients and scattering functions may be expressed in operator form also. A summary of the relations between correlation functions and many of these properties and derivations are given in Appendix A.

1.2.8 Liouville equation in a non-isolated system

In non-equilibrium states described by the equations of motion eq. (1.11) the form of the Liouville equation that is commonly used to describe the evolution of the N-particle distribution function for Hamiltonian systems is then invalid, and a more general form of the Liouville equation, including the effect of phase-space compression, must be used instead. The Liouville equation in this case takes the form[9,15,16]

$$\partial f/\partial t = -\partial/\partial\Gamma\cdot\dot{\Gamma}(f(\Gamma,t)) \tag{1.22a}$$

$$= -i\mathbf{L}_f\, f \tag{1.22b}$$

$$= -(i\mathbf{L}_p + (-\partial/\partial\Gamma\cdot\dot{\Gamma}))f \tag{1.22c}$$

which defines the f-Liouvillean, $\mathbf{L}_f$, the p-Liouvillean, $\mathbf{L}_p$, and the phase space compression factor $\partial/\partial\Gamma\cdot\dot{\Gamma} = \Lambda$. The f-Liouvillean propagates the distribution function and the p-Liouvillean propagates phase variables. A similarly modified form of the Liouville equation is also required for the case of a system in thermal contact with the environment (see Section 9:12 of McLennan's book[21] for an illuminating discussion). An interaction of this type is essential for the establishment of a steady state, because dissipated heat must be removed from the system in order to maintain a fixed temperature.

1.2.9 Non-equilibrium distribution function and correlation functions

Consider an equilibrium system to which a steady field is applied at time $t = 0$ which produces a non-equilibrium state. The time-dependent distribution function describing the system will be now be given by the solution to the Liouville equation,[22,23]

$$f(\Gamma,t) = f_{\mathrm{eq}}(\Gamma,0)\exp\left(-\int_0^t J(-\tau)F_{\mathrm{e}}\,\mathrm{d}\tau\right) \tag{1.23}$$

where $f_{\mathrm{eq}}(\Gamma,0)$ is the equilibrium distribution function of the original system, and $J(-t)F_{\mathrm{e}}$ is the dissipation within the system due to the external field, propagates backwards in time under the field-dependent, thermostated equations of motion (eq. (1.11)). This distribution function was originally derived by Yamada and Kawasaki[23] to describe the evolution of a system subject to field-dependent adiabatic equations of motion, but it was subsequently shown by Evans and Morriss[9,23] that the form of the distribution function is unchanged by including a thermostat in the equations of motion and allowing for phase-space compression in the Liouville equation.

The Kawasaki distribution function contains the integral of the dissipation explicitly, so it can never be independent of time, even if the system is in a steady-state. This can be seen by taking the time derivative of eq. (1.23):

$$\frac{\partial f(\Gamma,t)}{\partial t} = -\beta J(-t)F_{\mathrm{e}} f(\Gamma,t) \tag{1.24}$$

The non-equilibrium ensemble average of a phase variable A at time t is defined by

$$\langle A(t)\rangle = \int f(0)A(t)\,\mathrm{d}\Gamma \tag{1.25a}$$

$$= \int f(t)A\,\mathrm{d}\Gamma \tag{1.25b}$$

$$= \int f(\tau)A(t-\tau)\,\mathrm{d}\Gamma \tag{1.25c}$$

where $A \equiv A(0)$, and f is the non-equilibrium N-particle distribution function (1.23). The three definitions of the non-equilibrium ensemble average are the Heisenberg representation (1.25a), the Schrödinger representation (1.25b) and a more general *mixed* representation (1.25c). All three are equivalent due to the adjoint relation between the f- and p-Liouvillean operators.[9]

The system is said to be in a non-equilibrium steady-state if the ensemble averages of all phase variables become time-independent in the presence of the field. The time derivative of the ensemble average of an arbitrary phase variable[23] is

$$\mathrm{d}/\mathrm{d}t\langle \mathrm{A}(t)\rangle = -\beta F_{\mathrm{e}}\langle A(t)J(0)\rangle \tag{1.26}$$

where the Schrödinger equation and mixed eq. (1.25c) representations, and the time derivative of the Kawasaki distribution function (1.23) has been used. If the system displays mixing then all long time correlations between phase variables vanish. Therefore as $t \to \infty$,

$$\mathrm{d/dt}\langle \mathrm{A}(t)\rangle = -\beta F_{\mathrm{e}}\langle A(t)J(0)\rangle \to -\beta F_{\mathrm{e}}\langle A(t)\rangle\langle J(0)\rangle = 0 \tag{1.27}$$

because the average value of the dissipative flux in the original equilibrium system is zero, that is, $\langle J(0)\rangle=0$.

The non-equilibrium time-correlation function of two phase variables is defined as

$$\langle A(t+\tau)B^*(t)\rangle = \int f(\Gamma,0)A(t+\tau)B^*(t)\,\mathrm{d}\Gamma \tag{1.28a}$$

$$= \int f(\Gamma,t)A(\tau)B^*\,\mathrm{d}\Gamma \tag{1.28b}$$

where the first line is the Heisenberg representation definition, and the second line is a mixed representation definition. Similar to eq. (1.25), the non-equilibrium time-correlation function becomes independent of the time origin t as $t \to \infty$, if the system is mixing.

In the development of the generalized Langevin equation in non-equilibrium, it is necessary to know how the p-Liouvillean operator behaves inside the time-correlation function defined by eq. (1.28).

Using the mixed representation, we have

$$\begin{aligned}\langle \mathrm{i}\mathbf{L}_{\mathbf{p}}A(t+\tau)B^*(t)\rangle &= \int f(\Gamma,t)[\mathrm{i}\mathrm{L}A(t)]B^*\mathrm{d}\Gamma \\ &= \int f(\Gamma,t)[\dot{\Gamma}\cdot\partial A(\tau)]B^*\,\mathrm{d}\Gamma\end{aligned} \tag{1.29}$$

where $\mathrm{i}\mathbf{L}_{\mathbf{p}}$ is the p-Liouvillean defined earlier. Integrating eq. (1.29) by parts,

$$\begin{aligned}\langle \mathrm{i}\mathbf{L}_{\mathbf{p}}A(t+\tau)B^*(t)\rangle = &\left[f(\Gamma,t)\dot{\Gamma}A(t)B^*\right]_S \\ &-\int (\partial/\partial\Gamma\cdot(\dot{\Gamma}f(\Gamma,t)))\mathrm{A}(\tau)B^*\,\mathrm{d}\Gamma \\ &-\int f(\Gamma,t)\mathrm{A}(\tau)\dot{\Gamma}\cdot\partial\mathrm{B}^*/\partial\Gamma\,\mathrm{d}\Gamma\end{aligned} \tag{1.30}$$

The first term vanishes due to the distribution function going to zero at the boundary of the system and when the momentum of any particle tends to infinity. The second term is simplified by using the Liouville equation (1.22a) and the partial time derivative of the distribution function (1.24). The third term in eq. (1.30) is simplified because the equations of motion are real, that is, $\dot{\Gamma}^*=\dot{\Gamma}$.

Thus eq. (1.30) becomes

$$\begin{aligned}\langle \mathrm{i}\mathbf{L}_{\mathbf{P}} A(t+\tau)B^*(t)\rangle &= -\beta \mathrm{F}_e \langle A(t+\tau)B^*(t)J(0)\rangle \\ &\quad - \langle A(t+\tau)(\mathrm{i}\mathbf{L}B(t))^*\rangle\end{aligned} \tag{1.31}$$

If the system is mixing, then as $t \to \infty$, the p-Liouvillean becomes Hermitian in the correlation function defined by eq. (1.28), that is,

$$\langle \mathrm{i}\mathbf{L}_{\mathbf{P}} A(t+\tau)B^*(t)\rangle = -\langle A(t+\tau)(\mathrm{i}\mathbf{L}B(t))\rangle \tag{1.32}$$

The instantaneous and time-averaged values of variables in non-equilibrium states may also be computed from the solution of the equations of motion (eq. (1.11)) and from knowledge of non-equilibrium distribution functions. For example, we may calculate the values of the components of the pressure tensor P_{xij} from which we may calculate the shear viscosity even under conditions where the time correlation expression given in Appendix A is not valid.

This formalism is now sufficient to describe the properties of a system in non-equilibrium states which may be generated by use of the generalised SLLOD equations of motion.

1.2.10 Other approaches to non-equilibrium

The above formalist, based on the SLLOD-type equations of motion is not the only approach which may be used for treating non-equilibrium systems. Two other approaches, which have been used to derive Langevin type equations, are the non-equilibrium distribution function approach of Oppenheim and Levine[24] and the projector operator method of Ichiyanagi.[25] The former approach, based on the assumption of local thermodynamic equilibrium has been used to derive a Langevin and a Fokker–Planck equation for one Brownian particle in a non-equilibrium bath.[24] Ichiyanagi's study[25] uses projection operator methods in conjunction with the von Neumann equation to derive a GLE for quantum mechanical non-equilibrium steady-state systems. The results are largely formal and are not directly useful for NEMD simulations because of the way heat is assumed to be removed from the system.

1.2.11 Projection operators[2,9,26]

One of the purposes of the generalised Langevin approach is to reduce the 6N coupled equations of motion to one equation per species or to those appropriate to all the particles of a particular species. Thus, we will derive an equation for a "typical" particle of each species present or for that species that we are interested in and once we have written down the fundamental equations of motion for the all the particles in the system we want to "project out" the dynamical variables of interest. In practice this projecting out involves averaging (integration) over the degrees of freedom in which we are less interested and lead to a more coarse-grained description of the dynamics than that provided by the starting equations of motion. To do this we start with the full description given in the sections above and use a projection operator, **P**. The resulting equations in their exact form are entirely equivalent to the original coupled equations of motion from which they are derived however their use will be shown to offer many advantages in certain circumstances.

To introduce the concept of a projection operator we will define the general properties of such an operator **P**, that is[2,9,26]

$$\mathbf{P}A = \mathbf{P}_A\, A = A \tag{1.33}$$

$$\mathbf{P}_A \mathrm{P}_A\, A = A$$

So,

$$\mathbf{P}_A \mathbf{P}_A = \mathbf{P}_A^2 = \mathbf{P}_A \quad \text{(idempotency)} \tag{1.34}$$

We may also define the complement of **P** by $\mathbf{Q} = \mathbf{1} - \mathbf{P}$ so from eq. (1.33),

$$\mathbf{Q}A = (\mathbf{1} - \mathbf{P})A = \mathbf{1}A - \mathbf{P}A = A - A = 0 \tag{1.35}$$

For any projection operator we have, for a system with a time-independent Liouville operator, an exact relationship, the Dyson decomposition, which will enable us to write exact equation of motion for a particle (see Appendix B),

$$\mathbf{G}(t) = \mathrm{e}^{\mathrm{i}\mathbf{L}t} = \mathrm{e}^{\mathrm{i}\mathbf{QL}t} + \int_0^t \mathrm{e}^{\mathrm{i}\mathbf{L}(t-\tau)}\, \mathrm{i}\mathbf{PL}\, \mathrm{e}^{\mathrm{i}\mathbf{QL}\tau}\, \mathrm{d}\tau \tag{1.36}$$

We will make extensive use of operators and, in particular, projection operators in the formulation of GLEs. The actual form of the projection operator **P** will depend on the problem at hand and its characteristics. Thus, the form of **P** will be dependent on which variables are averaged over, for example, all particles but one and what the physical state of the system is, for example, at equilibrium or in a steady state. A majority of examples studied to date have, however, been for homogeneous systems.

1.3 SUMMARY

Since the equations introduced here will appear often in subsequent chapters it is useful to briefly summarize below the approach and some of the methods discussed in this chapter.

EQUILIBRIUM	NON-EQUILIBRIUM
Equations of Motion	
$\left(\frac{\mathrm{d}}{\mathrm{d}t}\right)\underline{p}_i(t)=\underline{F}_i(t),\qquad \left(\frac{\mathrm{d}}{\mathrm{d}t}\right)\underline{r}_i(t)=\frac{\underline{p}_i(t)}{m_i}\quad (i=1,\ldots,N)$	$\left(\frac{\mathrm{d}}{\mathrm{d}t}\right)\underline{p}_i=\underline{F}_i+\underline{D}_i\,\mathrm{F_e}-\alpha\underline{p}_i,$ $\left(\frac{\mathrm{d}}{\mathrm{d}t}\right)\underline{r}_i=\underline{p}_i/m_i+\underline{C}_iF_e\quad (i=1,\ldots,N)$
Liouville Equation	
$\left(\frac{\mathrm{d}}{\mathrm{d}t}\right)f(\Gamma,t)=\mathrm{i}\mathbf{L}f(\Gamma,t)$	$\left(\frac{\mathrm{d}}{\mathrm{d}t}\right)f(\Gamma,t)=-\partial/\partial\dot{\Gamma}\cdot(\dot{\Gamma}f(\Gamma,t))$ $=-\mathbf{L}_\mathrm{f}f(\Gamma,t)$ $=-(\mathrm{i}\mathbf{L_P}-\partial/\partial\Gamma\cdot\Gamma)f(\Gamma,t)$
Time-Dependent Distribution Function	
$f(\Gamma,t)=\exp(-\mathrm{i}\mathbf{L}t)\,f(\Gamma,0)$	$f(\Gamma,t)=f(\Gamma,0)\exp(-\beta\int_0^t \mathrm{J}(-\tau)\,F_\mathrm{e}\mathrm{d}\tau)$
Ensemble Averages	
$\langle A(0)\rangle=\int f_\mathrm{eq}(\Gamma)\,A(0)\,\mathrm{d}\Gamma$	$\langle A(t)\rangle=\int f(0)A(t)\,\mathrm{d}\Gamma$ $=\int f(t)A(0)\,\mathrm{d}\Gamma$ $=\int f(\tau)A(t-\tau)\,\mathrm{d}\Gamma$
Time-Correlation Functions	
$\langle A(0)B(0)\rangle=\int f_\mathrm{eq}(\Gamma)A^*(0)B(t)\,\mathrm{d}\Gamma$ $=\int f_\mathrm{eq}(\Gamma)A^*(0)\exp(\mathrm{i}\mathbf{L}t)\,B(0)\,\mathrm{d}\Gamma$	$\langle A(T+\tau)B^*(t)\rangle=\int f(\Gamma,0)\,A(t+\tau)B^*(t)\,\mathrm{d}\Gamma$ $=\int f(\Gamma,t)\,A(\tau)B^*(0)\,\mathrm{d}\Gamma$

1.4. CONCLUSIONS

In order to describe the classical state of a system at a particular time we must be able to specify the position and momentum of each particle in the system or equivalently to list the phase space co-ordinate of the system. The evolution of this mechanical state of the system, that is, its dynamics, requires an equation of motion. Newton's equations of motion are sufficient to describe the dynamics of an isolated system but modifications of these equations of motion have to be made in order to describe the state of more general systems such as the equilibrium state at constant temperature and non-equilibrium steady states.

In order to calculate the non-mechanical properties of a system we must introduce new physical laws such as the Ergodic Hypothesis, over and above those needed to define the equations of motion. However, to calculate either equilibrium thermodynamic properties or linear transport coefficients this is all that is needed beyond the output of classical mechanics.

However, further extensions of mechanics are needed to keep variables other than N, V and U constant at equilibrium and to treat non-equilibrium states. These involve modifying the equations of motion themselves.

One may write the classical equations of motion for a system of particles appropriate to different states in operator language. This leads to an equivalent operator form of these equations of motion, to different forms of Liouville's equation and to expressions for the properties of these many-body systems which we shall make use of in subsequent chapters.

REFERENCES

1. P. Langevin, *Comptes Rendus*, **146**, 530 (1908).
2. B.J. Berne and G.D. Harp, On the calculation of time correlation functions. In: *Advances in Chemical Physics*, XVII (Eds I. Prigogine and S.A. Rice), Wiley, New York, 1970, p. 63.
3. R.C. Tolman, *The Principles of Statistical Mechanics*, Dover, New York, 1979.
4. J.W. Gibbs, *Elementary Principles in Statistical Mechanics*, Dover, New York, 1960.
5. R. Kubo, M. Yokota and S. Nakagima, *J. Phys. Soc. Japan*, **12**, 1203 (1957).
6. M.S. Green, *J. Chem. Phys.*, **20**, 1281 (1952); **22**, 398 (1954).
7. See also Appendix A.
8. W.G. Hoover, A.J.C. Ladd and B. Moran, *Phys. Rev. Lett.*, **48**, 1818 (1982); D.J. Evans, *J. Chem. Phys.*, **78**, 3297 (1983); D.J. Evans, W.G. Hoover, B.H. Failor, B. Moran and A.J.C. Ladd, Phys. Rev. A, **28**, 1016 (1983).
9. D.J. Evans and G.P. Morriss, *Statistical Mechanics of Nonequilibrium Liquids*, Academic Press, New York, 1990.
10. D.J. Evans and G.P. Morriss, Non-equilibrium molecular dynamics, *Computer Phys. Rep.*, **1**, 297 (1984).
11. W.G. Hoover and W.T. Ashurst, Nonequilibrium molecular dynamics, *Adv. Theor. Chem.*, **1**, 1–51 (1975).
12. W.G. Hoover, Nonequilibrium molecular dynamics, *Ann. Rev. Phys. Chem.*, **34**, 103–127 (1983).
13. W.T. Ashurst and W.G. Hoover, *Phys. Rev. A*, **11**, 658 (1975).
14. S.Y. Liem, D. Brown and J.H.R. Clarke, *Phys. Rev. A*, **45**, 3706 (1992).
15. S.S. Sarman, D.J. Evans and P.T. Cummings, Recent developments in non-newtonian molecular dynamics, *Phys. Rep.*, **305**, 1–92 (1998).

16. M.G. McPhie, P.J. Daivis, I.K. Snook, J. Ennis and D.J. Evans, *Physica A*, **299**, 412 (2001).
17. L.V. Woodcock, *Chem. Phys. Lett.*, **10**, 257 (1970).
18. H.C. Anderson, *J. Chem. Phys.*, **72**, 2384 (1980).
19. S. Nose, *Mol. Phys.*, **52**, 255 (1984); W.G. Hoover, *Phys. Rev. A*, **31**, 1695 (1985).
20. M. Mareschal and B.L. Holian (Eds), *Microscopic Simulations of Complex Hydrodynamic Phenomena*, NATO ASI Series B, Physics, Vol. 292, Plenum Press, New York, 1992. See particularly the discussions commencing on p. 323.
21. J.A. McLennan, *Introduction to Nonequilibrium Statistical Mechanics*, Prentice-Hall, Englewood Cliffs, NJ, 1989.
22. T. Yamada and K. Kawasaki, *Prog. Theor. Phys.*, **38**, 1031 (1967).
23. G.P. Morriss and D.J. Evans, *Mol. Phys.*, **54**, 629 (1985).
24. J.-E. Shea and I. Oppenheim, *J. Phys. Chem.*, **100**, 19035 (1996); I. Oppenheim and R.D. Levine, *Physica A.*, **99**, 383 (1979).
25. M. Ichiyanagi, *J. Phys. Soc. Japan*, **62**, 1167 (1993).
26. B.J. Berne, Projection operator techniques in the theory of fluctuations in statistical mechanics. In: *Modern Theoretical Chemistry 6. Statistical Mechanics, Part B: Time-Dependent Processes* (Ed. B.J. Berne), Plenum Press, New York, 1977.

– 2 –

The Equation of Motion for a Typical Particle at Equilibrium: The Mori–Zwanzig Approach

In order to illustrate the generalised Langevin approach and to introduce the method of derivation we will choose one particle in a system at equilibrium (a typical or "tracer" particle) and derive an equation of motion for it. This is the approach pioneered by Mori[1] and Zwanzig.[2] We will also use velocity of this particle as an example to introduce these methods in a relatively easy way and subsequently generalise this method to the description of the time dependence of other dynamical variables and of time correlation functions.

2.1 THE PROJECTION OPERATOR

Following Berne[3,4] we will initially consider only the normalised velocity $\underline{u}_i$ of particle i. First let us define the symbol $\langle \cdots \rangle$ by

$$\langle AB \rangle = \int f_{\mathrm{eq}}\, A\, B\, \mathrm{d}\Gamma$$

and then define the normalised velocity by $\underline{u}_i$,

$$\underline{u}_i = \frac{\underline{v}_i}{\langle \underline{v}_i\, \underline{v}_i \rangle^{1/2}} \tag{2.1}$$

so

$$\langle \underline{u}_i\, \underline{u}_i \rangle = \frac{\langle \underline{v}_i\, \underline{v}_i \rangle}{\left(\langle \underline{v}_i\, \underline{v}_i \rangle^{1/2}\right)^2} = 1 \tag{2.2}$$

$$\langle \underline{u}_i\, \underline{u}_j \rangle = \frac{\langle \underline{v}_i\, \underline{v}_j \rangle}{\left(\langle \underline{v}_i\, \underline{v}_i \rangle^{1/2}\right)^2} = 0, \quad i \neq j$$

This normalised form of dynamical variable means that $\{\underline{u}_i\}$ forms an orthonormal set of vectors, which is convenient particularly when generalising to Quantum Mechanical variables,

$$\left(\frac{\mathrm{d}}{\mathrm{d}t}\right)\underline{u}_i = \left(\frac{\mathrm{d}}{\mathrm{d}t}\right)\left\{\frac{\underline{v}_i}{\langle \underline{v}_i\,\underline{v}_i\rangle^{1/2}}\right\} = \frac{\underline{a}_i}{\langle \underline{v}_i\,\underline{v}_i\rangle^{1/2}} = \underline{b}_i \tag{2.3}$$

where we have defined the normalised vector $\underline{b}_i$ by $\underline{a}_i/\langle \underline{v}_i \underline{v}_i\rangle^{1/2}$ (note that the normalised acceleration b has dimension t^{-1}).

In order to average over other variables we will define the projector

$$\begin{aligned}
\mathbf{P}\,A(\Gamma) &= \mathbf{P}_i\,A \\
&= \underline{u}_i\{\langle \underline{u}_i\,A\rangle\} \\
&= \underline{u}_i\left\{\int f_{\mathrm{eq}}\,\underline{u}_i\,A\,\mathrm{d}\Gamma\right\} \\
&= \underline{v}_i\left(\frac{\langle \underline{v}_i\,A\rangle}{\langle v_i^2\rangle}\right) \\
&= \left(\frac{\langle \underline{v}_i\,A\rangle}{\langle v_i^2\rangle}\right)\underline{v}_i \\
&= \left(\int f_{\mathrm{eq}}\,\underline{u}_i\,A\,\mathrm{d}\Gamma\right)\underline{u}_i
\end{aligned} \tag{2.4}$$

Here, $f_{\mathrm{eq}}(\Gamma)$ is the N-body equilibrium distribution function which involves all particles in the system even including the selected or "tracer" particle. We choose this to be the equilibrium distribution function appropriate to an NVT system which is given by

$$f_{\mathrm{eq}} = \frac{\mathrm{e}^{-H(\Gamma)/kT}}{\int \mathrm{e}^{-H(\Gamma)/kT}\,\mathrm{d}\Gamma} \tag{2.5}$$

where H is the Hamiltonian for the system. Thus, we have introduced a constraint on the system from the outset by use of this projection operator, i.e. we assume that the system is a homogeneous, equilibrium, NVT one.

In particular,

$$\begin{aligned}
\mathbf{P}\,\underline{u}_i &= \langle \underline{u}_i\,\underline{u}_i\rangle\underline{u}_i \\
&= \left(\frac{\langle \underline{v}_i\,\underline{v}_i\rangle}{\langle v_i^2\rangle}\right)\frac{\underline{v}_i}{\langle \underline{v}_i\,\underline{v}_i\rangle^{1/2}} \\
&= \frac{\underline{v}_i}{\langle \underline{v}_i\,\underline{v}_i\rangle^{1/2}} \\
&= \underline{u}_i
\end{aligned} \tag{2.6}$$

and

$$\begin{aligned}\mathbf{P}^2\,\underline{u}_i &= \mathbf{P}(\mathbf{P}\underline{u}_i)\\ &= (\mathbf{P}\underline{u}_i)\\ &= \underline{u}_i\end{aligned} \tag{2.7}$$

$$\begin{aligned}\mathbf{P}\underline{u}_j &= \langle \underline{u}_i\,\underline{u}_j\rangle \underline{u}_i \quad i \neq j\\ &= 0\end{aligned} \tag{2.8}$$

so, clearly, **P** is a projection operator.

The complement **Q** of **P** is given by

$$\mathbf{Q} = \mathbf{1} - \mathbf{P}, \quad \text{i.e. } \mathbf{Q} + \mathbf{P} = \mathbf{1} \tag{2.9}$$

$$\begin{aligned}\mathbf{Q}\underline{u}_i &= \mathbf{1}\underline{u}_i - \mathbf{P}\underline{u}_i\\ &= \underline{u}_i - \underline{u}_i\\ &= 0\end{aligned} \tag{2.10}$$

We could equally well use $\underline{v}_i$ and then define **P** by

$$\begin{aligned}\mathbf{P}_i' A(\Gamma) = \mathbf{P}' A &= \frac{\underline{v}_i \langle \underline{v}_i\, A\rangle}{\langle \underline{v}_i^2\rangle}\\ &= \frac{\underline{v}_i \int f_{\text{eq}}\, \underline{v}_i\, A\, \mathrm{d}\Gamma}{\int f_{\text{eq}}\, \underline{v}_i\, \underline{v}_i\, \mathrm{d}\Gamma}\\ &= \underline{v}_i \left(\frac{\langle \underline{v}_i\, A\rangle}{\langle v_i^2\rangle}\right)\end{aligned} \tag{2.11}$$

If we write $\underline{u}_i$ as the vector $|\underline{u}_i\rangle$, then $\langle \underline{u}_i\, \underline{u}_i\rangle$ is the scalar product and therefore we may write $\mathbf{P} = |\underline{u}_i\rangle\langle \underline{u}_i|$ and $\mathbf{P}' = |\underline{v}_i\rangle\langle \underline{v}_i| / \langle \underline{v}_i\, \underline{v}_i\rangle$. Thus, it is obvious that working with $|\underline{u}_i\rangle$ (i.e., $\underline{u}_i$) and **P** is entirely equivalent to using $|\underline{v}_i\rangle$ (i.e., $\underline{v}_i$) and **P′**. We have chosen to use the normalised velocity and the definition of **P** given in eq. (2.6) to agree with the notation used in references 3 and 4. The definition used in eq. (2.11) would, however, result in the same form of equations and conversion from one set to another is easy.

2.2 THE GENERALISED LANGEVIN EQUATION

We apply the Dyson decomposition (1.36) to $\underline{u}_i(t)$ and this leads to

$$\left(\frac{\mathrm{d}}{\mathrm{d}t}\right)\underline{u}_i(t) = \mathrm{e}^{\mathrm{i}\mathbf{QL}t}\mathbf{L}\underline{u}_i(0) + \int_0^t \mathrm{e}^{\mathrm{i}\mathbf{L}(t-\tau)}\mathrm{i}\mathbf{PL}\mathrm{e}^{\mathrm{i}\mathbf{QL}\tau}\left[\mathbf{Q}\,\mathrm{i}\mathbf{L}\underline{u}_i(0)\right]\mathrm{d}\tau \tag{2.12}$$

or

$$\left(\frac{\mathrm{d}}{\mathrm{d}t}\right)\underline{v}_i(t) = \mathrm{e}^{\mathrm{i}\mathbf{QL}t}\mathbf{L}\underline{v}_i(0) + \int_0^t \mathrm{e}^{\mathrm{i}\mathbf{L}(t-\tau)}\mathrm{i}\mathbf{PL}\mathrm{e}^{\mathrm{i}\mathbf{QL}t}\left[\mathbf{Q}\,\mathrm{i}\mathbf{L}\underline{v}_i(0)\right]\mathrm{d}\tau \tag{2.13}$$

We will now interpret the terms in (2.12) and (2.13) and first the term

$$\begin{aligned} \underline{b}_i^{\mathrm{R}}(t) &= \mathrm{e}^{\mathrm{i}\mathbf{QL}\,t}\mathrm{i}\mathbf{L}\underline{u}_i(0) \\ &= \mathrm{e}^{\mathrm{i}(1-\mathbf{P})\mathbf{L}\,t}\mathrm{i}\mathbf{L}\underline{u}_i(0) \end{aligned} \tag{2.14}$$

is the "Random Force" by analogy with the Random Force introduced by Langevin. Note, however, that in its current form it does not have the dimensions of force, in fact, its dimensions are t^{-1}. Also there is nothing random about this "force", rather it has often been approximated by a random term. This is often necessary in practice since we usually only know its first two moments (see below) and from an information theory point of view then the most reasonable approximation to this term is a Gaussian random variable.

Similarly, the other term in (2.12) may be written in a more understandable way (see Appendix B),

$$\begin{aligned} \int_0^t \mathrm{e}^{\mathrm{i}\mathbf{L}(t-\tau)}\,\mathrm{i}\mathbf{PL}\mathrm{e}^{\mathrm{i}\mathbf{QL}\tau}\left[\mathbf{Q}\,\mathrm{i}\mathbf{L}\underline{u}_i(0)\right]\mathrm{d}\tau &= -\int_0^t \underline{u}_i(t-\tau)\left\langle \mathrm{i}\mathbf{L}\underline{u}_j(0)\,\mathrm{e}^{\mathrm{i}\mathbf{QL}\tau}\,\mathrm{i}\mathbf{L}\underline{u}_i(0)\right\rangle \mathrm{d}\tau \\ &= -\int_0^t K_u(\tau)\underline{u}_i(t-\tau)\,\mathrm{d}\tau \end{aligned} \tag{2.15}$$

and the term

$$K_u(\tau) = \left\langle \mathrm{i}\mathbf{L}\underline{u}_i(0)\,\mathrm{e}^{\mathrm{i}\mathbf{QL}\tau}\,\mathrm{i}\mathbf{L}\underline{u}_i(0)\right\rangle \tag{2.16}$$

is usually called the memory function.

Thus, the equation of motion of a single particle or generalised Langevin equation (GLE) for particle i may be written as

$$\left(\frac{\mathrm{d}}{\mathrm{d}t}\right)\underline{u}_i(t) = -\int_0^t K_u(\tau)\underline{u}_i(t-\tau)\,\mathrm{d}\tau + \underline{b}_i^{\mathrm{R}}(t) \tag{2.17}$$

where

$$\underline{b}_i^{\mathrm{R}}(t) = \mathrm{e}^{\mathrm{i}\mathbf{QL}t}\,\mathrm{i}\mathbf{L}\underline{u}_i(0) \tag{2.18}$$

and

$$K_u(t) = \left\langle \mathrm{i}\mathbf{L}\underline{u}_i(0)\,\mathrm{e}^{\mathrm{i}\mathbf{QL}t}\mathrm{i}\mathbf{L}\underline{u}_i(0)\right\rangle \tag{2.19}$$

$$\begin{aligned}\langle \underline{b}_i^{\mathrm{R}}(0)\cdot \underline{b}_i^{\mathrm{R}}(t)\rangle &= \langle \mathrm{e}^{\mathrm{i}\mathbf{QL}0}\,\mathrm{i}\mathbf{L}u_i(0)\mathrm{e}^{\mathrm{i}\mathbf{QL}t}\,\mathrm{i}\mathbf{L}u_i(0)\rangle \\ &= \langle \mathrm{i}\mathbf{L}u_i(0)\mathrm{e}^{\mathrm{i}\mathbf{QL}t}\,\mathrm{i}\mathbf{L}u_i(0)\rangle \\ &= K_u(t)\end{aligned} \tag{2.20}$$

Thus, the memory function is the time correlation function of the random force (see Appendix B).

It may also be shown that the following relations hold (see Appendix B):

$$\langle \underline{b}_i^{\mathrm{R}}(0)\cdot \underline{b}_i^{\mathrm{R}}(t)\rangle = \frac{\langle \underline{a}_i^{\mathrm{R}}(0)\cdot \underline{a}_i^{\mathrm{R}}(t)\rangle}{\langle \underline{v}_i(0)\underline{v}_i(0)\rangle} = K_u(t) \tag{2.21}$$

$$\langle \underline{u}_i(0)\cdot \underline{b}_i^{\mathrm{R}}(t)\rangle = 0 \tag{2.22}$$

$$\langle \underline{b}_i^{\mathrm{R}}(t)\rangle = \langle \underline{b}_i^{\mathrm{R}}(0)\rangle = 0 \tag{2.23}$$

$$\mathbf{Q}\mathrm{i}\mathbf{L}\underline{u}_i(t) = (\mathbf{1}-\mathbf{P})\mathrm{i}\mathbf{L}\underline{u}_i(t) = \left(\frac{\mathrm{d}}{\mathrm{d}t}\right)[\underline{u}_i(t)] \tag{2.24}$$

Thus, the random force is uncorrelated with the velocity and all we know about this random force is its mean value (which is zero), its mean square value, that its time autocorrelation function is the memory function and that it is uncorrelated with $\underline{u}_i(t)$ (i.e., with $\underline{v}_i(t)$). Hence, a reasonable approximation to this random force is a Gaussian random variable with a mean of zero and a mean square value equal to the memory function for $\underline{u}_i(t)$. In fact, from an Information Theory perspective as we know only its first two moments then the least-biased estimate of $\underline{b}_i^{\mathrm{R}}(t)$ is to assume it is a Gaussian random variable (see Section 3.4).

Thus, the "random" term $\underline{b}_i^{\mathrm{R}}(t)$ in the GLE is not random rather it is uncorrelated with the chosen variable, the particle velocity $\underline{u}_i(t)$ and it is often said that it occupies the space orthogonal to that occupied by $\underline{u}_i(t)$. Thus, by averaging over certain of the N degrees of freedom of the system we have, in effect, lost some information. We could of course go back to the full description of the system and solve the N equations of motion and obtain all the dynamical information but if we wish to only follow the trajectory of one representative particle i using the GLE (2.17) then we can only know a much more limited amount of information than we started with using the description based on Newton's laws. This new, "coarse grained", equation of motion contains a force term about which we know only limited information, i.e. contained in eqs. (2.21)–(2.23) and we cannot therefore exactly determine this force.

This is reminiscent of our treatment of a non-isolated mechanical system in contact with its surroundings, if we wish to describe such a system in terms of mechanical equations of motion then we must create a super-system consisting of the system plus its surroundings. If not and we wish to only treat the system in terms of the particles contained in our system of interest then we treat the environment as providing a constraint on the system's properties and the surroundings only appear as constants in the equations describing the system's variables. In contrast to this case, however,

1. We are forced to use a stochastic description because of lack of detailed knowledge of $\underline{b}_i^{\mathrm{R}}(t)$ whereas the dynamical equations derived to maintain constant temperature, e.g. (1.9) are entirely deterministic.
2. We have eliminated all the particles of each species but one representative one.

However, we could define a projection operator to project out the effects of all the particles belonging to the surroundings and then we would obtain a stochastic equation which would represent the effect of the surroundings by a random force and a memory function. We will return to this idea in Chapters 5.

One of the important concepts in the theory of random processes is that of a Markov Process (MP), i.e. one in which the next state of the system only depends on its current state and not the history of how the system got to that state (see Section 9.4). Now the future state of a system described by Newton's equations of motion (or by the time dependent Schrodinger equation) is fully determined by the present state of the system and so the dynamics may, in this sense, be thought of as Markovian but are also deterministic. However, the GLE for the evolution of the velocity (momentum) and position depend on the memory function $K_{\mathrm{v}}(t)$ which is a function of the dynamical history of the particle prior to the current time, i.e. the GLE is non-Markovian. This aspect has been introduced by the averaging over all the velocities of the other particles in the system by use of the projection operator **P**. They are also stochastic due to the appearance of the random force. Thus, we have introduced both non-Markovian behaviour and random (stochastic) behaviour by our averaging processes.

2.3 THE GENERALISED LANGEVIN EQUATION IN TERMS OF THE VELOCITY

All the above has been written in terms of the normalized velocity $\underline{u}_i$ which is given by $\underline{u}_i = \underline{v}_i / \langle \underline{v}_i \cdot \underline{v}_i \rangle^{1/2}$, so we may write (2.17) as

$$\left(\frac{\mathrm{d}}{\mathrm{d}t}\right)\underline{v}_i(t) = -\int_0^t K_{\mathrm{v}}(\tau)\,\underline{v}_i(t-\tau)\mathrm{d}\tau + \underline{a}_i^{\mathrm{R}}(t) \tag{2.25}$$

where $\underline{a}_i^{\mathrm{R}}(t) = \underline{b}_i^{\mathrm{R}}(t)\,\langle \underline{v}_i \cdot \underline{v}_i \rangle^{1/2}$.

$$K_{\mathrm{v}}(t) = K_{\mathrm{u}}(t) = \frac{\langle \underline{a}_i^{\mathrm{R}}(0)\cdot\underline{a}_i^{\mathrm{R}}(t)\rangle}{\langle \underline{v}_i\cdot\underline{v}_i\rangle} \tag{2.26}$$

or

$$\langle \underline{a}_i^{\rm R}(0)\cdot\underline{a}_i^{\rm R}(t)\rangle = \langle \underline{v}_i\cdot\underline{v}_i\rangle K_{\rm v}(t) = \left(\frac{3k_{\rm B}T}{m_i}\right)K_{\rm v}(t)$$

and

$$\langle \underline{a}_i^{\rm R}(0)\rangle = \langle \underline{v}_i\cdot\underline{v}_i\rangle\langle \underline{b}_i^{\rm R}\rangle = 0 \tag{2.27}$$

Note that we have imposed the condition that $\langle \underline{v}_i\cdot\underline{v}_i\rangle = (3k_{\rm B}T/m_i)$ and that the dimension of a_i is Lt^{-2} and so it is an acceleration.

We may write an equation of motion analogous to Newton's second law by multiplying (2.25) by the mass of particle i,

$$m_i\left(\frac{\rm d}{{\rm d}t}\right)\underline{v}_i(t) = -m_i\int_0^t K_{\rm v}(\tau)\,\underline{v}_i(t-\tau)\,{\rm d}\tau + \underline{F}_i^{\rm R}(t) \tag{2.28}$$

where

$$\underline{F}_i^{\rm R}(t) = m_i\,\underline{b}_i^{\rm R}(t)\langle \underline{v}_i\,\underline{v}_i\rangle^{1/2} = m_i\,\underline{a}_i^{\rm R}(t) \tag{2.29}$$

$$\langle \underline{F}_i^{\rm R}(0)\cdot\underline{F}_i^{\rm R}(t)\rangle = m_i^2\langle \underline{a}_i^{\rm R}(0)\cdot\underline{a}_i^{\rm R}(t)\rangle$$

$$K_{\rm v}(t) = K_{\rm u}(t) = \frac{\langle \underline{a}_i^{\rm R}(0)\cdot\underline{a}_i^{\rm R}(t)\rangle}{\{3k_{\rm B}T/m_i\}} = \frac{\langle \underline{F}_i^{\rm R}(0)\cdot\underline{F}_i^{\rm R}(t)\rangle/(m_i^2)}{\{3k_{\rm B}T/m_i\}} = \frac{\langle \underline{F}_i^{\rm R}(0)\cdot\underline{F}_i^{\rm R}(t)\rangle}{\{3k_{\rm B}Tm_i\}}$$

or

$$\langle \underline{F}_i^{\rm R}(0)\cdot\underline{F}_i^{\rm R}(t)\rangle = (3k_{\rm B}Tm_i)K_{\rm v}(t) \tag{2.30}$$

and

$$\langle \underline{F}_i^{\rm R}(0)\rangle = \langle \underline{a}_i^{\rm R}(0)\rangle m_i^2 = 0 \tag{2.31}$$

Eq. (2.28) is the generalisation of the empirically derived Langevin equation (LE) (1.1) and is usually called a GLE. This type of equation is Newton's second law with force terms representing the effect of the background particles on the selected particle. In fact we have started with Newton's equation for $\underline{v}_i$ (one equation for each particle in the system) and derived from them one equation for the velocity $\underline{v}_i$ for a typical particle of each distinct species of particles for a system at equilibrium.

It is often stated that the forces in this GLE are of two types, i.e. a drag force which slows the particle down and a random force which compensates for this drag and provides energy from the bath to keep the temperature fixed at a constant value T, i.e. it is an equation of motion for a particle in a canonical ensemble. This is at first sight a curious thing as the original Newton's equations do not represent the dynamics of a system whose temperature is constant but one with a constant total energy, i.e. a microcanonical ensemble. But we have introduced the constraint of constant temperature in the derivation of the GLE by defining the projection operator **P** in terms of the equilibrium distribution function for an *NVT* system and by imposing the condition that $\langle \underline{v}_i \cdot \underline{v}_i \rangle = (3k_B T/m_i)$.

2.4 EQUATION OF MOTION FOR THE VELOCITY AUTOCORRELATION FUNCTION

An added advantage of this formalism is that by multiplying (2.25) by $\underline{v}_i(0)$ and then taking an equilibrium ensemble average we obtain an equation of motion for the velocity autocorrelation function $C_v(t)$ and for the normalised velocity autocorrelation function $c_v(t)$,

$$\frac{\mathrm{d}}{\mathrm{d}t} C_v(t) = -\int_0^t K_v(\tau) C_v(t-\tau)\,\mathrm{d}\tau \tag{2.32}$$

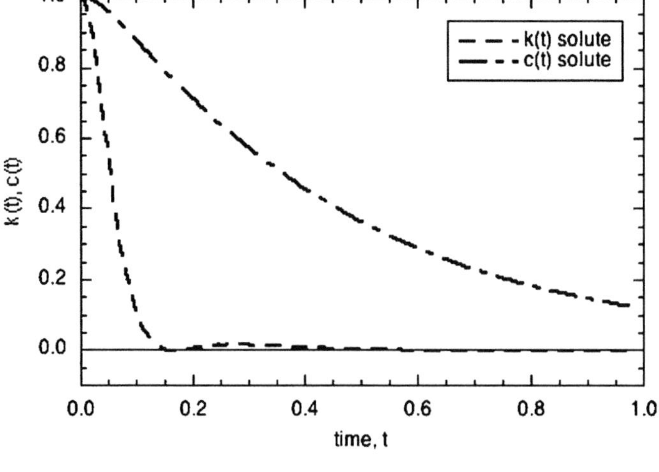

Figure 2.1 The normalised velocity autocorrelation function, $c(t)$ and corresponding memory function, $K(t)$ for the "solute" particles in a mixture of heavy "solute" and light "solvent" particles interacting via a CWA potential.

or

$$\frac{d}{dt}c_v(t) = -\int_0^t K_v(\tau)c_v(t-\tau)d\tau \tag{2.33}$$

where $C_v(t) = \langle \underline{v}_i(0)\cdot\underline{v}_i(t)\rangle$ is the velocity autocorrelation function and $c(t) = \langle \underline{v}_i(0)\cdot\underline{v}_i(t)\rangle / \langle \underline{v}_i(0)\cdot\underline{v}_i(0)\rangle$ is the normalised velocity autocorrelation function.

Thus, we have explicit equations of motion for time correlation functions from which $K_v(t)$ may be evaluated numerically by the use of EMD simulations or we may theoretically model the behaviour of $C_v(t)$ by modelling $K_v(t)$ which will be discussed in Chapter 3.

The velocity autocorrelation function $c_v(t)$ and the corresponding memory function $K_v(t)$ for a heavy particle in a bath of lighter particles all interacting via a CWA potential as calculated by the EMD method may be seen in Figure 2.1. This figure illustrates the general principle that a time correlation function is of much longer range in time than is its related memory function.

2.5 THE LANGEVIN EQUATION DERIVED FROM THE MORI APPROACH: THE BROWNIAN LIMIT

In order to provide a link between the GLE and the Brownian motion we will now give an approximate derivation of the LE starting from the GLE. In the Brownian limit, the ratio of the mass m of the background particles to that of the selected heavy B particle M_B, $\lambda = m/M_B$, becomes small, it is then convenient to divide the particles up into two subgroups because of the enormous difference in the time scales of motion of the B and of the bath particles. We will do this rigorously in Chapter 5, however we can get an idea of what the result will be by looking at the GLE (2.28) for a single "tracer" particle given above.

Now in the Brownian Limit when $\lambda = (m/M_B)^{1/2} \rightarrow 0$, i.e. $M_B \gg m$, then it is reasonable to assume that the memory function for the heavy particle is given by a delta function in time as the motion of the bath particles and the relaxation time of their dynamical variables is very, very much faster than that of the velocity of the B particle (see Appendix D for a more detailed discussion),

$$K_v(t) = \lambda_1\,\delta(t) \tag{2.34}$$

or

$$\tilde{K}_v(s) = \lambda_1 = \frac{\zeta}{M_B} = \gamma$$

where γ is the friction coefficient and ζ the friction factor defined as $\zeta = M_B\,\gamma$.

We will show later that this is equivalent to the use of the simplest truncation of the rigorous Mori continued fraction expression for the memory function (see Section 3.3). Writing the velocity of the chosen particle $\underline{v}_i$ as $\underline{V}_B$ and its mass as M_B and substituting (2.34) into eq. (2.28),

$$M_B\frac{d}{dt}\underline{V}_B(t) = -\zeta\underline{V}_B(t) + \underline{F}_B^R(t) \tag{2.35}$$

and from (2.30) and (2.31),

$$\langle \underline{F}_{\mathrm{B}}^{\mathrm{R}}(0) \rangle = 0 \tag{2.36}$$

$$\begin{aligned} \langle \underline{F}_{\mathrm{B}}^{\mathrm{R}}(0) \cdot \underline{F}_{\mathrm{B}}^{\mathrm{R}}(t) \rangle &= 3\zeta\, k_{\mathrm{B}} T\, \delta(t) \\ &= 3\gamma M_{\mathrm{B}}\, k_{\mathrm{B}} T\, \delta(t) \end{aligned} \tag{2.37}$$

In this case we may also approximate the friction coefficient by (see Appendixes D and E),

$$\gamma_{\mathrm{B}} \approx \left(\frac{1}{M_{\mathrm{B}} k_{\mathrm{B}} T} \right) \int_0^\infty \langle \underline{F}_{\mathrm{B}}(0) \cdot \underline{F}_{\mathrm{B}}(t) \rangle_0 \mathrm{d}t \tag{2.38}$$

where the subscript 0 means ensemble averaging over a system with Hamiltonian H_0 which is the Hamiltonian for the total system minus the terms pertaining to the momentum of the Brownian particle. This is of the form of the empirical LE.

As was remarked in Section 2.2 Newton's equations can be thought of as a deterministic, Markovian description of dynamics as the future dynamical state of the system is only dependent on the current state of the system. However, when we transform Newton's equations by projection (averaging) the exact GLE (2.28) is non-Markovian since we have to know the history of the system to calculate the memory function and it is also non-deterministic due to the presence of the random force of which we only have limited knowledge. Now we have approximated the full GLE by the LE (2.35) and obtained an equation which describes a Markov process (because of the delta function approximation to the memory function (2.34)) but which is now stochastic due to the random force term. In essence the assumption of a delta correlated memory function means that we are assuming that all the processes happening in the bath particles in the presence of the Brownian particle have relaxed to equilibrium before the Brownian particle has moved.

More discussions as to the usefulness of the equations of motion in the Brownian Limit will be made elsewhere.

2.6 GENERALISATION TO ANY SET OF DYNAMICAL VARIABLES

In Section 2.1 we used the velocity of a single particle $\underline{v}_i$ as an example of a dynamical variable and derived an equation of motion for this variable. Let us now consider any dynamical variable $A(t)$, i.e. any function of Γ and t and return to using the vector notation, i.e. we will write $A(t)$ as $|A(t)\rangle$ and define the projection operator $\mathbf{P}_A$ by[5,6]

$$\mathbf{P}_{\mathrm{A}} = | A(0) \rangle \langle A(0) | \langle A(0) A(0) \rangle^{-1} \tag{2.39}$$

where $\langle \cdots \rangle$ once again means ensemble averaging as in Section 2.1 but we will not assume that $|A(t)\rangle$ is normalised.

From (1.18) and (1.19),

$$|A(t)\rangle = e^{iLt}\,|A(0)\rangle$$

so

$$\begin{aligned}
&\frac{d}{dt}|A(t)\rangle \\
&= \frac{d}{dt}e^{iLt}\,|A(0)\rangle \\
&= e^{iLt}i\mathbf{L}\,|A(0)\rangle \\
&= e^{iLt}i\mathbf{L}\mathbf{1}\,|A(0)\rangle \\
&= e^{iLt}i\mathbf{L}(\mathbf{P}+\mathbf{Q})\,|A(0)\rangle
\end{aligned} \tag{2.40}$$

Using (2.39),

$$\begin{aligned}
&e^{iLt}i\mathbf{L}\,\mathbf{P}\,|A(0)\rangle \\
&= \langle A(0)\,|\,i\mathbf{L}\,|\,A(0)\rangle\,(\langle A(0)A(0)\rangle)^{-1}\,e^{iLt}\,|\,A(0)\rangle \\
&= i\Omega\,|\,A(t)\rangle
\end{aligned}$$

where we have defined the frequency Ω by

$$i\Omega = \langle A(0)\,|\,i\mathbf{L}\,|\,A(0)\rangle(\langle A(0)A(0)\rangle)^{-1} \tag{2.41}$$

Thus, using the Dyson decomposition (see Section 1.2.11 and (1.36)) of the operator e^{iLt} we have

$$\frac{d}{dt}|A(t)\rangle = i\Omega\,|\,A(t)\rangle + e^{iQLt}i\mathbf{QL}\,t + \int_0^t i\mathbf{LP}\,e^{iQL\tau}i\mathbf{QL}\,|\,A(t)\rangle\,d\tau \tag{2.42}$$

Define the random force by

$$\begin{aligned}
\underline{F}^{R}(t) &= e^{iQLt}i\mathbf{QL}\,|\,A(t)\rangle \\
&= \mathbf{Q}e^{iQLt}i\mathbf{QL}\,|\,A(t)\rangle
\end{aligned} \tag{2.43}$$

and, thus it may be easily seen that

$$\langle \underline{F}^{R}(t)A\rangle = 0 \tag{2.44}$$

Now, $i\mathbf{PL}\underline{F}^{R}(t) = i\mathbf{PLQ}\underline{F}^{R}(t) = \langle i\mathbf{PL}\underline{F}^{R}(t)A\rangle\langle AA\rangle^{-1}A(t)\rangle$ but since $\mathbf{Q}$ and $\mathbf{L}$ are Hermitian then,

$$\langle i\mathbf{LQ}\,\underline{F}^{R}(t)A\rangle = -\langle \underline{F}^{R}(t)i\mathbf{QL}\,A\rangle = \langle \underline{F}^{R}(t)\underline{F}^{R}(t)\rangle \tag{2.45}$$

and so,

$$\mathbf{iPL}\underline{F}^{\mathrm{R}}(t) = \langle \underline{F}^{\mathrm{R}}(t)\underline{F}^{\mathrm{R}}(t)\rangle \langle AA\rangle^{-1} \,|\, A(t)\rangle \tag{2.46}$$

So once again we may define a memory function,

$$K_A(t) = \langle \underline{F}^{\mathrm{R}}(t)\underline{F}^{\mathrm{R}}(t)\rangle \langle AA\rangle^{-1} \tag{2.47}$$

and using eqs. (2.42), (2.46) and (2.47) we have the equation of motion or GLE for the dynamical variable $|A(t)\rangle = \mathrm{A}(t)$,

$$\frac{\mathrm{d}A(t)}{\mathrm{d}t} = \mathrm{i}\Omega A(t) - \int_0^t K_A(\tau)A(t-\tau)\,\mathrm{d}\tau + \underline{F}^{\mathrm{R}}(t) \tag{2.48}$$

This is the generalisation of (2.25) and once again the random force is uncorrelated with the variable $A(t)$ ((2.44)) and all we know about it is that it is proportional to the memory function (from (2.47)) and its mean value is zero.

As before we may take the scalar product of (2.48) with $\langle A(0)|$, i.e. multiply $\mathrm{A}(t)$ by $A(0)$ on the right of (2.48) and ensemble average then we have an equation of motion, a GLE for the autocorrelation function $C_A(t) = \langle A(0)A(t)\rangle$,

$$\frac{\mathrm{d}C_A(t)}{\mathrm{d}t} = \mathrm{i}\Omega C_A(t) - \int_0^t K_A(\tau)C_A(t-\tau)\,\mathrm{d}\tau \tag{2.49}$$

Eqs. (2.48) and (2.49) may be readily generalised to the case of a set of M variables $\{A_1, A_2,\ldots, A_M\}$ and if we represent this set by a column matrix $\mathbf{A}$ and define the correlation matrix $\mathbf{C}(t) = \langle \mathbf{A}(t)\mathbf{A}^{+}(0)\rangle$ where $\mathbf{A}(t)$ is a column matrix with elements $\mathrm{e}^{i\mathbf{L}\,t}A(t)$ and $\mathbf{A}^{+}(t)$ is the Hermitian conjugate of $\mathbf{A}(t)$ and, thus, $\mathbf{C}(t)$ is an $M \times M$ matrix whose ijth element is $C_{ij}(t) = \langle A_i(t)A_j^{*}(t)\rangle$ then define the static susceptibility matrix $\boldsymbol{\chi}$ by

$$\mathbf{C}(0) = \langle \mathbf{AA}^{+}\rangle = \beta^{-1}\boldsymbol{\chi} \tag{2.50}$$

and a projection operator $\mathbf{P}$ by

$$\mathbf{P} = \langle \ldots\mathbf{A}^{+}\rangle\cdot\langle \mathbf{AA}^{+}\rangle\cdot\mathbf{A} = \beta^{-1}\langle \ldots\mathbf{A}^{+}\rangle\cdot\boldsymbol{\chi}\cdot\mathbf{A} \tag{2.51}$$

Following the procedure above used to derive (2.48) and (2.49) we obtain the GLE for the set $\mathbf{A}$,

$$\frac{\mathrm{d}\mathbf{A}(t)}{\mathrm{d}t} = +\mathrm{i}\boldsymbol{\Omega}\cdot\mathbf{A}(t) - \int_0^t K_A(\tau)\,\mathbf{A}(t-\tau)\,\mathrm{d}\tau + \mathbf{F}^{\mathrm{R}}(t) \tag{2.52}$$

and the equation of motion for the time autocorrelation function matrix $\mathbf{C_A}(t)$,

$$\frac{\mathrm{d}\mathbf{C_A}(t)}{\mathrm{d}t} = +\mathrm{i}\boldsymbol{\Omega}\cdot\mathbf{C_A}(t) - \int_0^t \mathbf{K_A}(\tau)\mathbf{C_A}(t-\tau)\mathrm{d}\tau \tag{2.53}$$

Thus, we may write equations of motion for any dynamical variable or set of dynamical variables.

2.7 MEMORY FUNCTIONS DERIVATION OF EXPRESSIONS FOR LINEAR TRANSPORT COEFFICIENTS

A simple argument may be used to show that the linear transport coefficients may be related to integrals over autocorrelation functions.[4,9,10] Returning to the GLE equations for any set of dynamical variables $\mathbf{A}(t)$ and their correlation functions $\mathbf{C}_\mathrm{A}(t) = \langle\mathbf{A}(0)\mathbf{A}(t)\rangle$, i.e. (2.52) and (2.53) if we average (2.52) over a non-equilibrium ensemble, then

$$\frac{\mathrm{d}\langle\mathbf{A}(t)\rangle_\mathrm{ne}}{\mathrm{d}t} = +\mathrm{i}\boldsymbol{\Omega}\cdot\langle\mathbf{A}(t)\rangle_\mathrm{ne} - \int_0^t \langle\mathbf{A}(t-\tau)\rangle_\mathrm{ne}\mathbf{K_A}(t)\mathrm{d}\tau + \langle\mathbf{F}^\mathrm{R}(t)\rangle_\mathrm{ne} \tag{2.54}$$

If we can assume that $\langle\mathbf{F}^\mathrm{R}(t)\rangle_\mathrm{ne} = 0$, e.g. in a state "close to equilibrium" then from (2.53) and (2.54) we have that the dynamics of non-equilibrium and the behaviour of $C_\mathrm{A}(t)$ are governed by the same equation and, thus, that the linear decay of a dynamical variable **A,** i.e. the dissipation in a non-equilibrium state is governed by the same processes as the decay of the equilibrium correlation function which is a fluctuation at equilibrium. This is Onsager's famous result relating dissipative processes in a non-equilibrium state to fluctuations in an equilibrium one.[10]

Furthermore, if we write these linear rate laws governing dissipation of currents as

$$\frac{\partial\langle\mathbf{A}(\mathrm{t})\rangle_\mathrm{ne}}{\partial t} = \mathbf{F}\cdot\langle\mathbf{A}(t)\rangle_\mathrm{ne} \tag{2.55}$$

and

$\mathbf{F} = [\mathrm{i}\mathbf{Z}+\mathbf{L}]\chi^{-1}$ where $\mathbf{L}$ are the Onsager coefficients. Then taking the limit $t \to \infty$ and making the heuristic argument that[10]

$$\int_0^\infty \langle\mathbf{A}(t-\tau)\rangle_\mathrm{ne}\mathbf{K_A}(\tau)\mathrm{d}\tau = \int_0^\infty \mathbf{K_A}(\tau)\mathrm{d}\tau\langle\mathbf{A}(t-\tau)\rangle_\mathrm{ne} \tag{2.56}$$

then (2.54) and (2.55) give,

$$\mathrm{L}_{ij} = \int_0^\infty \mathrm{K_{ij}}(s)\mathrm{d}s \tag{2.57}$$

which is the Green–Kubo relationship for L_{ij}.

The "derivation" given above[10] is not rigorous but serves to indicate that we may use the Mori–Zwanzig GLE to obtain expressions for the linear transport coefficients in terms of correlation functions in equilibrium. There are many more rigorous derivations of the time correlation function expressions for transport coefficients[11–16] (see Appendix A) and in the next section we carry out this derivation in a more formal way for shear viscosity using a method due to Evans and Morriss.[9] It is convenient to work in k-space and then take the limit as $k \to 0$ in order to get these expressions.

2.8 CORRELATION FUNCTION EXPRESSION FOR THE COEFFICIENT OF NEWTONIAN VISCOSITY

Now for any dynamical variable $A(t)$ we will write the GLE in the form

$$\frac{\mathrm{d}A(t)}{\mathrm{d}t} = -\int_0^t K_A(\tau)A(t-\tau)\mathrm{d}\tau + \underline{F}^{\mathrm{R}}(t) \tag{2.58}$$

where $K(t) = \langle \underline{F}^{\mathrm{R}*}(t)\, \underline{F}^{\mathrm{R}}(0)\rangle / \langle \underline{F}^{\mathrm{R}*}(0)\, \underline{F}^{\mathrm{R}}(0)\rangle$.

Defining the flux of momentum by

$$\underline{J}(\underline{r},t) = \rho(\underline{r},t)\underline{u}(\underline{r},t)$$

then

$$\begin{aligned} \underline{J}(\underline{k},t) &= \int \rho(\underline{k}-\underline{k}')\underline{u}(\underline{k}',t)\mathrm{d}\underline{k}' \\ &= i\underline{k}\cdot\sum m\underline{v}_i(t)\mathrm{e}^{\mathrm{i}\underline{k}\cdot\underline{r}i(t)} \\ &= \mathrm{i}\underline{k}\cdot P_{ij}(k,t) \end{aligned}$$

where the instantaneous density is given by

$$\rho(\underline{r},t) = \sum_{i=1}^{N} m\delta(\underline{r}-\underline{r}_i(t))$$

Thus,

$$\begin{aligned} \rho(\underline{k},t) &= \int \sum_{i=1}^{N} m\delta(\underline{r}-\underline{r}_i(t))\mathrm{e}^{\mathrm{i}\underline{k}\cdot\underline{r}}\mathrm{d}\underline{r} \\ &= \sum_{i=1}^{N} m\, e^{\mathrm{i}\underline{k}\cdot\underline{r}i(t)} \end{aligned}$$

Let us consider the particular case in which we want the flux of the x component of momentum in the y direction, then

$$\begin{aligned}J_x(k_y,t) &= \sum_{i=1}^{N} mv_{xi}(t)\mathrm{e}^{\mathrm{i}kyi(t)}\\ &= \sum_{i=1}^{N} p_{xi}(t)\mathrm{e}^{\mathrm{i}kyi(t)}\\ &= \mathrm{i}k_y P_{yx}(k_y,t)\end{aligned}$$

and obviously $J_x = J_x(k_y,t) \to P_{yx}(t)$ as $k_y \to 0$.

Now we have

$$\begin{aligned}\langle J_x(k_y)J_x(k_y)^*\rangle &= \left\langle \sum_{i=1}^{N} p_{xi}(0)\mathrm{e}^{\mathrm{i}kyi(0)} \sum_{j=1}^{N} p_{xi}(0)\mathrm{e}^{-\mathrm{i}kyj(0)} \right\rangle\\ &= N\langle p_{x1}(0)^2\rangle + N(N-1)\langle p_{x1}(0)p_{x2}(0)\mathrm{e}^{\mathrm{i}k(y1(0)-y2(0))}\rangle\\ &= Nmk_\mathrm{B}T\end{aligned}$$

and $\mathbf{P}P_{yx}(k_y) = \langle P_{yx}(k_y)\, J_x(-k_y)\rangle / \langle |J_x(k_y)|^2\rangle = 0$

$$\begin{aligned}F_x(0) &= \mathrm{i}\mathbf{QLJ}_x\\ &= (\mathbf{1}-\mathbf{P})\mathrm{i}k_y P_{yx}(k_y)\\ &= \mathrm{i}k_y P_{yx}(k_y)\end{aligned}$$

and

$$F_x(t) = \mathrm{e}^{\mathrm{i}\mathbf{QL}t}\mathrm{i}k_y P_{yx}(k)$$

Using the Dyson decomposition (1.36) of the operator $\mathrm{e}^{\mathrm{i}\mathbf{QL}t}$,

$$\mathrm{e}^{\mathrm{i}\mathbf{L}t} = \mathrm{e}^{\mathrm{i}\mathbf{QL}t} + \int_0^t \mathrm{e}^{\mathrm{i}\mathbf{L}(t-s)}\mathbf{P}\mathrm{i}\mathbf{L}\mathrm{e}^{\mathrm{i}\mathbf{QL}s}\,\mathrm{d}s$$

and combining this with the above,

$$\begin{aligned}\mathbf{P}\mathrm{i}\mathbf{L}B &= \frac{\langle J_x(k_y,t)^*\,\mathrm{i}\mathbf{L}B\rangle J_x(k_y,t)}{(Nmk_\mathrm{B}T)}\\ &= \frac{-\langle B(\mathrm{i}\mathbf{L}J_x(k_y,t))\rangle J_x(k_y,t)}{(Nmk_\mathrm{B}T)}\end{aligned}$$

Hence, we see that $e^{iQLt} \to e^{iLt}$ as $k_y \to 0$ and, thus,

$$K(t) = \frac{\langle \underline{F}^{R*}(t)\underline{F}^{R}(0)\rangle}{\langle \underline{F}^{R*}(0)\underline{F}^{R*}(0)\rangle} \xrightarrow[k_y \to 0]{} \frac{k_y^2\langle P_{yx}(k_y,t)P_{yx}(-k_y,0)\rangle}{(Nmk_BT)}$$

and substituting this into (2.58) and ignoring the random term we have,

$$\begin{aligned}
&\lim_{k_y\to 0}\frac{dJ_x(k_y,t)}{dt}\\
&= \frac{-k_y^2}{(Nmk_BT)}\int_0^t \langle P_{yx}(k_y,s)P_{yx}(-k_y,0)\rangle J_x(k_y,t-s)ds\\
&+ik_yP_{yx}(k_y,t)
\end{aligned}$$

and as $dJ_x(k_y,t)/dt = ik_yP_{yx}(k_y,t)$ and $-ik_y\,J_x(k_y,t-s)/\rho = \dot{\gamma}(k)$, i.e. $J_x(k_y,t-s) = -\dot{\gamma}(k)\rho/(ik_y)$ where $\rho = \sigma\, mN/V$ so,

$$\begin{aligned}
ik_yP_{yx}(k_y,t) &= \frac{-k_y^2}{(Nmk_BT)}\int_0^t \langle P_{yx}(k_y,s)P_{yx}(-k_y,0)\rangle \left(\frac{-\dot{\gamma}(k)Nm}{(-ik_y V)ds}\right)\\
&+ik_yP_{yx}(k_y,t)\\
&= -ik_y/(k_BTV)\int_0^t \langle P_{yx}(k_y,s)P_{yx}(-k_y,0)\rangle \dot{\gamma}(k)ds\\
&+ik_yP_{yx}(k_y,t)
\end{aligned}$$

So as $k_y \to 0$,

$$P_{yx}(k_y,t) = \frac{-1}{(k_BTV)}\int_0^t \langle P_{yx}(k_y,s)P_{yx}(-k_y,0)\rangle \dot{\gamma}(k)ds + P_{yx}(k_y,t)$$

i.e.,

$$P_{yx}V = \frac{-1}{(k_BT)}\dot{\gamma}\int_0^t \langle P_{yx}(s)P_{yx}(0)\rangle ds$$

or

$$\begin{aligned}
\frac{-P_{yx}}{\dot{\gamma}} &= \eta\\
&= \frac{1}{(k_BTV)}\int_0^\infty \langle P_{yx}(s)P_{yx}(0)\rangle ds \qquad (2.59)
\end{aligned}$$

This equation, eq. (2.59), is the Green–Kubo formula relating the Newtonian shear viscosity coefficient η to an integral over the stress autocorrelation function, $\langle P_{yx}(s)\, P_{yx}(0)\rangle$.

An example of a stress autocorrelation function, $\langle P_{yx}(s)\, P_{yx}(0)\rangle$ obtained from an MD simulation of a system which is 4% by number of heavy "colloid particles" in a background of light "solvent" particles interacting via a CWA potential at a reduced temperature $k_BT/\varepsilon = 1.0$ and reduced density $\rho\sigma^3 = 0.705$ is shown in Figure 2.2 and the integral over this function, $\int_0^\infty \langle P_{yx}(s)\cdot P_{yx}(0)\rangle$ may be seen in Figure 2.3.

A similar procedure may be used to derive expressions for other transport coefficients in terms of time correlation functions, some of these expressions will be found in Appendix A.

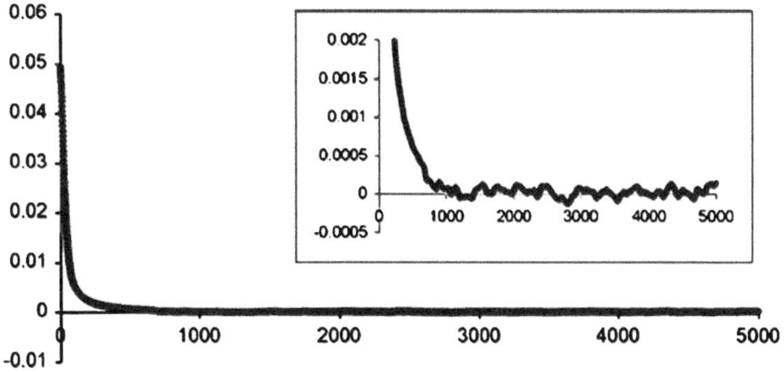

Figure 2.2 The stress autocorrelation function (inset shows detail) from an NEMD simulation of a system which is 4% by number of heavy "colloid particles" in a background of light "solvent" particles interacting via a CWA potential at a reduced temperature $k_BT/\varepsilon = 1.0$ and reduced density $\rho\sigma^3 = 0.705$. (Figure courtesy of Dr. T. Kairn, RMIT, Applied Physics.)

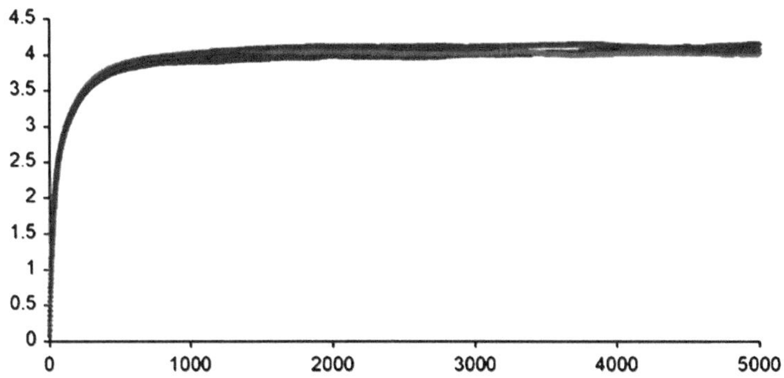

Figure 2.3 The normalised integral of stress correlation function (three sets of data) for the same system as shown in Figure 2.2. (Figure courtesy of Dr. T. Kairn, RMIT, Applied Physics.)

2.9 SUMMARY

As the ideas introduced in this chapter, such as the use of projection operators, are central concepts we summarise below in diagrammatic form the ideas involved in deriving a GLE.

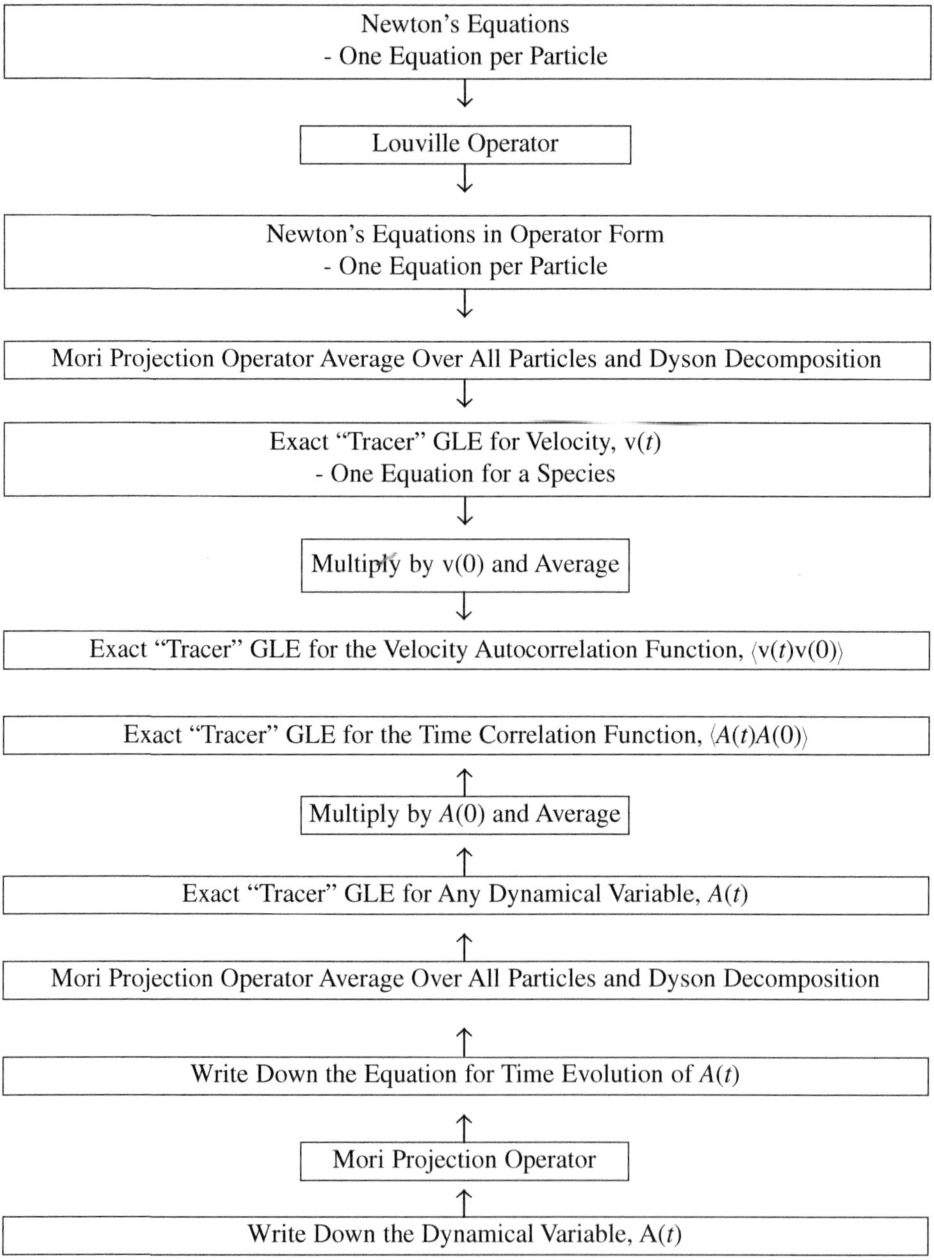

2.10 CONCLUSIONS

In order to illustrate the methods of deriving a GLE we have discussed the method of Mori and Zwanzig for deriving a GLE for the velocity and velocity autocorrelation function of a single particle in a many-body system. This enabled the introduction of the central method used to do this, i.e. the use of projection operators to average over some variables and the use of Dyson decomposition. The ideas of random forces and memory functions arise from this treatment and the GLE for the velocity may be written in terms of these quantities. The averaging performed means that we do not any longer have a completely deterministic description of the dynamics of the system since we cannot completely determine the random force which leads to a stochastic description. The appearance of a memory function means that the equation of motion is also non-Markovian. The form of the projection operator used to perform the averaging means that the system is regarded as an *NVT* equilibrium one. Also, the GLE may also be transformed into an equation of motion for the velocity autocorrelation function.

We subsequently generalised this method to the description of the time dependence of dynamical variable or set of dynamical variables and of other time correlation functions. This allowed a non-rigorous derivation of the Onsager principle that dissipative processes in a non-equilibrium system which is close to equilibrium are related to fluctuations in a system at equilibrium. This was explicitly shown for the coefficient of Newtonian shear viscosity, η by a more rigorous procedure and the Green–Kubo formulae for η obtained.

As in the rest of this book, in this chapter we have based our dynamics on the classical mechanical definition of the dynamical state of a system whose dynamics is described by classical equations of motion. The generalisation of the equations obtained here to systems which must be described by Quantum Mechanics is possible and, indeed, has been achieved.[3,4] For this reason we used the notation that a dynamical variable $A(t)$ is written as $|A(t)\rangle$ and that, e.g., an ensemble average may be written as $\langle A(0)A(0)\rangle$ and a time autocorrelation function as $\langle A(0)A(t)\rangle$. This notation may be re-interpreted as being for quantum mechanical state vectors and expectation values and, thus, once the dynamical equations become those of quantum mechanics the techniques used here may be adapted to quantum systems. The remaining problems are, of course, ones of applying the correct rules to manipulate symbols and then to interpret the results obtained.[3,4]

REFERENCES

1. H. Mori, *Prog. Theoret. Phys.*, **33**, 423 (1965); **34**, 399 (1965).
2. R. Zwanzig, *Lectures in Theoretical Physics, Volume 3*, Interscience, New York, 1961, pp. 135–141.
3. B.J. Berne, Time-dependent properties of condensed media. In: *Physical Chemistry, An Advanced Treatise, Vol. VIIIB, Liquid State* (ed. D. Henderson), Academic Press, New York, 1971, pp. 539–715.
4. B.J. Berne and G.D. Harp, On the calculation of time correlation functions. In: *Advances in Chemical Physics, XVII* (eds. I. Prigogine and S.A. Rice), Wiley, New York, 1970, pp. 63–227.

5. B.J. Berne, Projection operator techniques in the theory of fluctuations. In: *Modern Theoretical Chemistry 6. Statistical Mechanics, Part B, Time-Dependent Processes* (ed. B.J. Berne), Plenum, New York, 1977, pp. 233–257.
6. J-P. Hansen, Correlation functions and their relationship with experiment. In: *Microscopic Structure and Dynamics of Liquids* (Eds. J. Dupuy and A.J. Dianoux), Plenum, New York, 1977, pp. 3–68.
7. J.-P. Hansen and I. McDonald, *Theory of Simple Liquids, 2nd Edition*, Academic Press, New York, 1986.
8. P. Schofield, Theory of time-dependent correlations in simple classical liquids. In:*Specialist Periodical Report, Statistical Mechanics, Vol. 2* (Ed. K. Singer), The Chemical Society, London, 1975, pp. 1–54.
9. D.J. Evans and G.P. Morriss, *Statistical Mechanics of Nonequilibrium Liquids*, Academic Press, New York, 1990.
10. T. Keyes, Principles of mode-mode coupling theory. In: *Modern Theoretical Chemistry 6. Statistical Mechanics, Part B, Time-Dependent Processes* (Ed. B.J. Berne), Plenum, New York, 1977, pp. 259–309.
11. L. Onsager, *Ann. Rev. Phys. Chem.*, **16**, 67 (1965).
12. M.S. Green, *J. Chem. Phys,* **20**, 1281 (1952); **22**, 398 (1954).
13. R. Kubo, M. Yokota and S. Nakagima, J. Phys. Soc. Japan, **12**, 1203 (1957).
14. H. Mori, Phys. Rev., **112**, 1829 (1958).
15. J.A McLennan, Phys. Rev. A, **115**, 1405 (1959).
16. E. Helfand, Phys. Rev., **119**, 1 (1960).

– 3 –

Approximate Methods to Calculate Correlation Functions and Mori-Zwanzig Memory Functions

Since the GLE based on Newton's equations of motion for any dynamical variable $A(t)$ (e.g. velocity) or time correlation function, $\langle A(0)B(t)\rangle$ is exact and, therefore, is entirely equivalent to using these equations to calculate them, these new equations cannot, in general, be solved exactly. Thus, we will detail some schemes that are available for generating approximations to correlation functions and what general information may be obtained from the GLE. An ideal approach would be to identify all the physical processes which govern the time evolution of the dynamical variable, find their time evolution (e.g. by use of a kinetic theory) and then use this information to calculate the memory function $K(t)$ and the random force $\underline{F}^{R}(t)$. Then the GLE for the time evolution of the variable and/or for the desired time correlation functions could be solved, even if only numerically. This scheme is not possible in general, and thus we will outline some general characteristics of time correlation functions in order to guide us in finding mathematically or physically motivated approximations.

3.1 TAYLOR SERIES EXPANSION

Since we may expand the propagator e^{iLt} in a power series in time, this can be used as a basis for methods to approximately solve the GLE, that is,

$$e^{iLt} = 1 + \sum_{k=1}^{\infty} \left(\frac{1}{k!}\right) (iL)^k t^k \tag{3.1}$$

It should noted that eq. (3.1) is in fact a formal definition of e^{iLt}, and that it illustrates that all the time dependence of e^{iLt} is contained in the terms t^k.[1–5]

This means that time correlation functions may be similarly expanded, for example, for the velocity autocorrelation function we have

$$\langle \underline{v}_i(0)\cdot\underline{v}_i(t)\rangle = \frac{3k_B T}{m_i} - \frac{1}{2}\langle \underline{a}_i(0)\cdot\underline{a}_i(0)\rangle t^2 + \frac{1}{24}\left\langle \left(\frac{d}{dt}\right)\underline{a}_i(0)\cdot\left(\frac{d}{dt}\right)\underline{a}_i(0)\right\rangle t^4 + O(t^6) \tag{3.2}$$

and for the corresponding memory function $K_v(t)$ we have by substituting eq. (3.2) into eq. (2.32),

$$K_v(t) = \frac{\langle \underline{a}_i(0)\cdot\underline{a}_i(0)\rangle}{\langle \underline{v}_i(0)\cdot\underline{v}_i(0)\rangle} - \frac{1}{2}\left[\frac{\langle (\mathrm{d}/\mathrm{d}t)\underline{a}_i(0)\cdot(\mathrm{d}/\mathrm{d}t)\underline{a}_i(0)\rangle}{\langle \underline{a}_i(0)\cdot\underline{a}_i(0)\rangle} - \langle \underline{a}_i(0)\cdot\underline{a}_i(0)\rangle\langle \underline{v}_i(0)\cdot\underline{v}_i(0)\rangle\right]t^2 + \cdots \quad (3.3)$$

It should be noted that the coefficients in these expansions are evaluated at zero time and may therefore be calculated by the methods of equilibrium statistical mechanics. However, the higher order terms become very complicated rapidly and these expansions are really only useful at short time. They are very useful for helping to find the parameters appearing in approximate forms of the memory function, as we shall see below, and to develop numerical GLE algorithms (see Chapter 6).

For the general case, $C_A(t) = \langle A(0)A(t)\rangle = \langle A(0)\,\mathrm{e}^{\mathrm{i}\mathbf{L}t}\,A(0)\rangle$ and its corresponding memory function $K_A(t)$ we may use the Taylor series expansion of $\mathrm{e}^{\mathrm{i}\mathbf{L}t}$ and obtain

$$\begin{aligned} C_A(t) &= \left\langle A(0)\left\{1+\sum_{k=0}^{\infty}\left(\frac{1}{k!}\right)(\mathrm{i}\mathbf{L})^k\,t^k\right\}A(0)\right\rangle \\ &= \langle A(0)A(0)\rangle + \sum_{k=0}^{\infty}\left(\frac{1}{k!}\right)(-1)^k\,\langle A(0)(\mathrm{i}\mathbf{L})^k A(0)\rangle t^k \\ &= \langle A(0)A(0)\rangle + \sum_{k=0}^{\infty}\left(\frac{1}{2k!}\right)(-1)^k\,\langle A(0)(\mathrm{i}\mathbf{L})^{2k} A(0)\rangle\, t^{2k} \\ &= \sum_{k=0}^{\infty}\left(\frac{1}{2k!}\right)(-1)^k\,\gamma_{2k}\,t^{2k} \end{aligned} \quad (3.4)$$

where γ_{2k}, which are the moments of $C_A(t)$, are given by

$$\begin{aligned} \gamma_{2k} &= \langle A(0)(\mathrm{i}\mathbf{L})^{2k} A(0)\rangle \\ &= \langle (\mathrm{i}\mathbf{L})^k A(0)(\mathrm{i}\mathbf{L})^k A(0)\rangle \qquad (\mathbf{L}\ \text{is Hermitian}) \\ &= \langle A(0)^{(k)}\, A(0)^{(k)}\rangle \end{aligned} \quad (3.5)$$

where $A(0)^{(k)}$ is the kth derivative of $A(0)$ (i.e. $(\mathrm{i}\mathbf{L})^k A(0)$) and the coefficients of the terms of odd powers in t are zero by time reversal symmetry.

Thus, substituting eq. (3.4) into eq. (2.49) we have

$$K_A(t) = \sum_{k=0}^{\infty}\left[\frac{(-1)^k}{(2k)!}\right]\mu_{2k}\,t^{2k} \quad (3.6)$$

where the moments (or cumulants) μ_{2k} are given by[1,2]

$$\mu_{2k} = \langle B_k(0)\,B_k(0)\rangle \quad (3.7)$$

where

$$B_k(0) = [\mathrm{i}(\mathbf{I}-\mathbf{P})\mathbf{L}]^k\ \mathrm{i}\mathbf{L}A(0)$$

From this it may be seen that[1,2] that the cumulants μ_{2k} are related to the moments γ_{2k},

$$\begin{aligned} \mu_0 &= \gamma_2 \\ \mu_2 &= \gamma_4 - \gamma_2^2 \\ \mu_2 &= \gamma_6 - 2\gamma_4\gamma_2 + \gamma_2^2 \end{aligned} \tag{3.8}$$

3.2 SPECTRA

Following Berne,[2] we define the power spectra or spectral functions $G_A(\omega)$ and $L_A(\omega)$ by the Weiner–Kintchine theorem,

$$\begin{aligned} C_A(t) &= \int_{-\infty}^{\infty} \exp(-\mathrm{i}\omega t) G_A(\omega)\,\mathrm{d}\omega \\ G_A(\omega) &= (2\pi)^{-1/2} \int_{-\infty}^{\infty} \exp(\mathrm{i}\omega t) G_A(t)\,\mathrm{d}t \\ K_A(t) &= \int_{-\infty}^{\infty} \exp(-\mathrm{i}\omega t) L_A(\omega)\,\mathrm{d}\omega \end{aligned} \tag{3.9}$$

and

$$L_A(\omega) = (2\pi)^{-1/2} \int_{-\infty}^{\infty} \exp(\mathrm{i}\omega t) L_A(\mathrm{t})\mathrm{d}t$$

Thus, $G(\omega)$ and $L(\omega)$ are the power spectrum of the autocorrelation function and its memory function, respectively. Berne also defines them as the probability distribution function of which $C_A(t)$ and $K_A(t)$ are the characteristic functions.

We may note that the often used Laplace transform is related to the Fourier transform used above by the Hilbert transform,

$$\begin{aligned} \tilde{C}_A(s) &= \int_{-\infty}^{\infty} C_A(t)\mathrm{e}^{zt}\,\mathrm{d}t \\ &= -\mathrm{i}\int_{-\infty}^{\infty} C_A(\omega)(\omega - z)^{-1}\,\mathrm{d}\omega \end{aligned} \tag{3.10}$$

3.3 MORI'S CONTINUED FRACTION METHOD

One of the most useful methods is the method owing to Mori,[6] which begins by defining a heirachy of memory functions by[1–6]

$$\frac{\mathrm{d}}{\mathrm{d}t}K_n(t) = -\int_0^t K_{n+1}(\tau)K_n(t-\tau)\mathrm{d}\tau, \quad n = 0,\cdots,N \tag{3.11}$$

where

$$K_0(t) = C_A(t) \text{ and } K_1(t) = K_A(t).$$

Now taking the Laplace transform of eq. (3.11) with $n = 0$,

$$\tilde{K}_0(s) = \frac{K_0(0)}{(s+\tilde{K}_1(s))} \tag{3.12}$$

but from eq. (3.11) by Laplace transform with $n - 1$,

$$\tilde{K}_1(s) = \frac{K_1(0)}{(s+\tilde{K}_2(s))} \tag{3.13}$$

So substituting eq. (3.12) into eq. (3.13),

$$\tilde{K}_0(s) = K_0(0)/(s+K_1(0))/(s+\tilde{K}_2(s))$$

and, furthermore, by Laplace transform of eq. (3.11) for $n = 2$,

$$\tilde{K}_2(s) = K_2(0)/(s+\tilde{K}_3(s))$$

So,

$$\tilde{K}_0(s) = K_0(0)/(s+K_1(0))/(s+K_2(0))/(s+\tilde{K}_3(s))$$

Thus, in general,

$$\tilde{K}_0(s) = K_0(0)/(s+K_1(0))/(s+K_2(0))/(s+K_3(0))/\cdots/(s+\tilde{K}_n(s)) \tag{3.14}$$

where $n = 1$ to N.

This is the Mori continued fraction expression for $C(s) = K_0(s)$, that is,

$$C_A(s) = C_A(0)/(s+K_1(0))/(s+K_2(0))/(s+K_3(0))/\cdots/(s+\tilde{K}_n(s)) \tag{3.15}$$

where $n = 1$ to N and, in fact, it may be readily seen that there is such an expression for each $K_n(s)$.

In particular for $\tilde{K}_1(s)$,

$$\begin{aligned}&\tilde{K}_1(s)\\&=\tilde{K}_A(s)\\&=K_A(0)/(s+K_2(0))/(s+K_3(0))/(s+K_4(0))/\cdots/(s+\tilde{K}_n(s))\end{aligned} \tag{3.16}$$

where $n = 2$ to N.

The Mori continued fraction formulae are exact but to use them to calculate K_n one requires the knowledge of all the higher order memory functions $K_{n+1}, K_{n+2}, \ldots$. Thus, to use this scheme, we must truncate the continued fraction in some manner. Many schemes have been developed to construct approximate memory functions which are based on the Mori continued fraction method or rely on it in some way.[1–13]

One method of truncating the hierarchy is to assume that the kth-order memory function is "white noise",

$$\tilde{K}_k(s) = \lambda_k$$

or

$$K_k(t) = \lambda_k\,\delta(t)$$

Then,

$$\tilde{C}_A(s) = C_A(0)/(s+K_1(0))/(s+K_2(0))/(s+K_3(0))/\cdots/(s+\lambda_k) \tag{3.17}$$

and

$$\tilde{K}_A(s) = K_A(0)/(s+K_2(0))/(s+K_3(0))/(s+K_4(0))/\cdots/(s+\lambda_k)$$

Some particular cases are[1,2,4]

1. $k = 1$:

$$\begin{aligned}K_A(t) &= \lambda_1\,\delta(t)\\C_A(t) &= C_A(0)\mathrm{e}^{-\lambda_1 t}\end{aligned} \tag{3.18}$$

2. $k = 2$:

$$K_A(t) = K_A(0)\mathrm{e}^{-\lambda_2 t} \tag{3.19}$$

$$\text{If}\quad 4C_A(0)\,\lambda_1^{-2} < 1$$

$$c_A(t) = \frac{[z_2\,e^{-\lambda_1 t} - z_1\,e^{-\lambda_2 t}]}{(z_2 - z_1)}$$

where $z_1 z_2 = C_A(0)$, $z_1 + z_2 = \lambda_2$

or

$$\text{if} \quad 4C_A(0)\lambda_1^{-2} > 1$$

$$c_A(t) = \exp(-|t|/(2\lambda_2^{-1}))[\cos(\omega' t) + (2\omega'\lambda_2^{-1})^{-1}\sin(\omega' t)] \tag{3.20}$$

where $(\omega')^2 = C_A(0) - (2\lambda_2^{-1})^{-2.}$

To use these truncated formulae we need to find

1. $K_m(0)$, $m = 0,....,k$.
2. $\lambda_{k.}$

The first task may be done by using eq. (3.15) in conjunction with the time series expansion of $C_A(t)$ to find $K_m(t)$ or using the more formal definitions of $K_m(t)$,[1,2,7–13] which define them in terms of operators $\mathbf{L}^m$ that are related to $\mathbf{L}$. The second task may be accomplished in a variety of ways, for example by using the Green–Kubo expressions for the transport coefficients and fitting to known values of these coefficients.

3.4 USE OF INFORMATION THEORY

Often, all we know about a memory function are the first few moments $K_k(0)$ of the memory function and it is interesting to ask the question what we can infer about the form of $K(t)$ from this information? To do this, Berne[14] uses the power spectrum of the memory function, $L(\omega)$, defined in eq. (3.9) as a probability function with the characteristic function $K_A(t)$, and invokes information theory to find the best form of $K_A(t)$ consistent with limited information about $K_A(t)$.[14–17]

Now as we know the Taylor series expansion of $K_A(t)$ from eq. (3.6), we also know the moments of $L(\omega)$ and they are

$$\langle \omega^{2n+1} \rangle = 0$$

$$\langle \omega^0 \rangle = 1$$

$$\langle \omega^2 \rangle = \mu_2/\mu_0$$

and

$$\langle \omega^4 \rangle = \mu_4/\mu_0 \tag{3.21}$$

Berne and Harp[14] then define the information measure or entropy of the distribution $L(\omega)$ by

$$S[L(\omega)] = -\int_{-\infty}^{\infty} L(\omega) \ln L(\omega)\, \mathrm{d}\omega \tag{3.22}$$

According to information theory, the form of $L(\omega)$ is optimum which maximises $S[L(\omega)]$ subject to the known moments. For example, if we only know the first two moments,

$$\langle \omega^0 \rangle = 1 = \int_{-\infty}^{\infty} L(\omega)\, \mathrm{d}\omega$$

$$\langle \omega^2 \rangle = \mu_2 = \int_{-\infty}^{\infty} \omega^2 L(\omega)\, \mathrm{d}\omega$$

then we must find the form of $L(\omega)$ for which

$$\delta S[L(\omega)] = -\delta \int_{-\infty}^{\infty} L(\omega) \ln L(\omega)\, \mathrm{d}\omega = 0 \tag{3.23}$$

$$\delta \int_{-\infty}^{\infty} L(\omega)\, \mathrm{d}\omega = 0$$

and

$$\delta \int_{-\infty}^{\infty} \omega^2 L(\omega)\, \mathrm{d}\omega = 0$$

Berne and Harp[14] used Lagrange multipliers to solve this problem and found that

$$L(\omega) = [2\pi\mu_2]^{-1/2} \exp[-\omega^2/2\mu_2]$$

and then from eq. (3.9) we have

$$k_A(t) = k_G(t) = \exp\left[\frac{-\mu_2 t^2}{2}\right] \tag{3.24}$$

Thus, the optimum choice in this case is a Gaussian function of time.

Appendix O contains a Fortran computer program, VOLSOL, which will calculate a time correlation function $C_A(t)$ once a form is chosen for $K_A(t)$. To see how the form of $K_A(t)$ influences that of $C_A(t)$, it is a useful exercise to use the simple forms given above, for example the exponential form in eq. (3.19) or Gaussian form given above in eq. (3.24) to calculate $C_A(t)$ for various values of the parameters contained in $K_A(t)$.

3.5 PERTURBATION THEORIES

The most successful theories of the equilibrium behaviour of simple liquids are based on perturbation theory,[5] where the interatomic potential energy function, ϕ, is divided up into a short-ranged, harshly repulsive part, ϕ_R, which largely determines the structure of a dense fluid and a longer ranged, slowly varying perturbation potential, ϕ_p. Expansions of thermodynamic properties, usually the Helmholtz free energy, are then made about the properties of the system with interatomic potential energy function, ϕ_R, and the properties of this system are expressed in terms of those of a hard sphere system of some effective hard sphere diameter.[5,18–25]

Attempts have been made to extend these theories to nonequilibrium, which usually involves attempting to find a perturbation theory for $K_A(t)$. Early attempts at formulating such a theory were made by Singwi and Sjolander,[19] Corngold and Duderstadt,[20] Gaskell,[21] Mitra[22] and Schofield and Trainin,[23] the details of these studies being given in Schofield.[4]

All these theories make many approximations but produce rather similar results in that the equation obtained for $K_v(t)$ is the sum of two parts, that is one for binary hard-core collisions and one due to the slowly varying long-range part of the potential. All give similar results for the velocity autocorrelation function for simple fluids. However, none of these theories achieve the same level of rigor or accuracy as obtained by similar equilibrium theories and some of the assumptions made appear to be hard to check or justify. One of the ways of treating the effect of the long-range part of the potential is by means of mode–mode coupling theory (see Section 3.6), which essentially uses an expansion in terms of macroscopic hydrodynamic modes. Apart from the problem of whether such an expansion is accurate there is also the problem that such modes take time to form in an atomic fluid and thus their contribution must be damped at short times and should not simply be added to the contribution from the short-time "binary-collisions". Also, the hard repulsive interactions will give long-range hydrodynamic contributions as shown by molecular-dynamics (MD) simulations of systems with purely repulsive potentials, which must be included. Finally, $C_v(t)$ for systems with only repulsive and attractive contributions also often show backscattering at intermediate time delays due to the elastic solid-like response of the atoms surrounding a moving atom (see Section 3.7. This effect is contained neither in the usual binary collision nor in mode–mode coupling schemes.

Another approach to the problem is to use a division of the potential as in equilibrium theories and to make expansions of a Liouville operator as done below, to obtain perturbation expansions more analogous to those used in perturbation theories of equilibrium properties.[24,25]

A method which gives an expansion of the autocorrelation function by use of an expansion of the propagator $e^{-\mathbf{L}t}$ is based on this division of the potential.[24] Combining an expansion of the potential energy function Φ as $\Phi = \Phi^{R} + \alpha\Phi^{P}$ and then using the Liouville operator in eq. (3.25) we get

$$\begin{aligned}\mathbf{L} &= -\left\{\sum_{i=1}^{N}[\underline{v}_i\cdot\partial/\partial\underline{r}_i + \underline{F}_i/m_i\cdot\partial/\partial\underline{v}_i]\right\} \\ &= \mathbf{L}^{R} + \alpha\mathbf{L}^{P}\end{aligned} \tag{3.25}$$

where

$$\mathbf{L}^{R} = -\left\{\sum_{i=1}^{N}[\underline{v}_i\cdot\partial/\partial\underline{r}_i + \underline{F}_i^{R}/m_i\cdot\partial/\partial\underline{v}_i]\right\}$$

and

$$\mathbf{L}^{P} = -\left\{\sum_{i=1}^{N}[\underline{F}_i^{P}/m_i\cdot\partial/\partial\underline{v}_i]\right\}$$

Now since the operators $\mathbf{L}^{R}$ and $\mathbf{L}^{P}$ do not commute, to make an expansion of the propagator $\exp(-\mathbf{L}t)$, an operator $\mathbf{T}(t)$ is defined by

$$\exp(-\mathbf{L}t) = [\exp(-\mathbf{L}^{R}t)\mathbf{T}(t)\exp(-\mathbf{L}^{P}t)] \tag{3.26}$$

The form of $\mathbf{T}(t)$ may be found by differentiating both sides of eq. (3.26) with respect to t and equating the results gives

$$\frac{d\mathbf{T}(t)}{dt} = \alpha\exp(t\mathbf{L}^{R})[\exp(-t\mathbf{L}^{R})\mathbf{T}(t), \mathbf{L}^{P}] \tag{3.27}$$

where $[\mathbf{A},\mathbf{B}] = \mathbf{AB}-\mathbf{BA}$ and then

$$\begin{aligned}\mathbf{T}(t) &= 1 + \alpha\int_0^t \exp(s\mathbf{L}^{R})[\exp(-s\mathbf{L}^{R})\mathbf{T}(s), \mathbf{L}^{P}]\,ds \\ &= 1 + \alpha\mathbf{T}_1(t) + \alpha^2\mathbf{T}_2(t) + \cdots\end{aligned} \tag{3.28}$$

where

$$\mathbf{T}_1(t) = \int_0^t \{\mathbf{L}^{P} - \exp(s\mathbf{L}^{R})\mathbf{L}^{P}\exp(-s\mathbf{L}^{R})\}\,ds \tag{3.29}$$

This approach may be shown to be equivalent to the one introduced by Feynman in Quantum Mechanics.[24]

Thus,

$$\exp(-\mathbf{L}t) = \exp(-\mathbf{L}^{\mathrm{R}}t)\{1 - \alpha\mathbf{L}^{\mathrm{P}}t + \alpha\mathbf{T}_1(t) + \cdots\} \tag{3.30}$$

We may also expand any dynamical variable $A(0)$ as

$$A(0) = A^{\mathrm{R}}(0) + \alpha A^{\mathrm{P}}(0) \tag{3.31}$$

and using $A(\Gamma,t) = \exp(-\mathbf{L}t)\,A(\Gamma,0)$ we have from eq. (3.30),

$$\begin{aligned} C_A(t) &= \langle A(0)A(t)\rangle \\ &= \langle A^{\mathrm{R}}(0)A^{\mathrm{R}}(t)\rangle_{\mathrm{R}} - \alpha\beta\langle A^{\mathrm{R}}(0)A^{\mathrm{R}}(t)(\Phi^{\mathrm{P}} - \langle\Phi^{\mathrm{P}}\rangle_{\mathrm{R}})\rangle_{\mathrm{R}} - \alpha t\langle A^{\mathrm{R}}(0)\exp(-t\mathbf{L}^{\mathrm{R}}) \\ &\quad \times \mathbf{L}^{\mathrm{P}} A^{\mathrm{R}}(0)\rangle_{\mathrm{R}} + \alpha\langle A^{\mathrm{R}}(0)\exp(-t\mathbf{L}^{\mathrm{R}})\mathbf{T}_1(t)A^{\mathrm{R}}(0)\rangle_{\mathrm{R}} \\ &\quad + \alpha(\langle A^{\mathrm{R}}(0)A^{\mathrm{P}}(t)\rangle_{\mathrm{R}} + \langle A^{\mathrm{P}}(0)A^{\mathrm{R}}(t)\rangle_{\mathrm{R}}) \end{aligned} \tag{3.32}$$

where $\beta = 1/(k_{\mathrm{B}}T)$.

This result is difficult to use without introducing approximations of unknown accuracy, especially for the term involving $\mathbf{T}_1$ which can probably only be accurately carried out by numerical methods such as the EMD method. However, another possible approach might be to use path integrals to evaluate it. It should be noted that the theory is not applicable to a potential containing a hard sphere part, that is ϕ_{R} must be differentiable.

A second approach which seems more promising is to expand $K_A(t)$ in Laplace space, that is an expansion of $\tilde{K}_A(s)$.[25] This was done by using the Mori's continued fraction outlined in Section 3.3, which involves the calculation of equilibrium ensemble averages that can be expanded by equilibrium perturbation theory, that is

$$K_i(0) = A_i(0) + \alpha B_i(0) \tag{3.33}$$

and then we may write

$$\tilde{K}_A(s) = \tilde{L}_{\mathrm{R}}(s) + \tilde{\alpha}M(s) + \cdots \tag{3.34}$$

where $\tilde{L}_R(s)$ is the Laplace transform of the memory function of the reference system and $\tilde{M}(s)$ is the Laplace transform of the first-order correction. Using eq. (3.15) we have to first order in α,

$$\tilde{C}_A(s) = \tilde{C}_A^{\mathrm{R}}(s)[1 + \alpha/A_0(B_0 - \tilde{C}_A^{\mathrm{R}}(s)\tilde{M}_1(s))] \tag{3.35}$$

where $\tilde{C}_A^{\mathrm{R}}(s) = A_0/(s + \tilde{L}_{\mathrm{R}}(s))$ is the Laplace transform of the correlation function of $C_A(t)$ for the reference system.

The problem remaining is to find an expression for $\tilde{\mathrm{M}}_1(s)$. This may be done by making a power series expansion in s or by finding a continued-fraction expression for it. Both these schemes give formal expressions for $\tilde{M}_1(s)$, but a truncation is needed to obtain a practical scheme. One such truncation leads to[25]

$$\tilde{M}_1(s) = (B_1/A_1)\tilde{L}_1(s) \tag{3.36}$$

and

$$\tilde{C}_A(s) = \tilde{C}_A^{R}(s) + \alpha/(A_0A_1)[A_1B_0 - B_1(A_0^2 + s\tilde{C}_R(s))]\tilde{C}_R(s) \tag{3.37}$$

By taking the inverse Laplace transform of eq. (3.37) we have

$$C_A(t) = C_A^{R}(t) + \alpha/A_0\left[B_0\, C_A^{R}(t) - B_1/A_1\int_0^t C_A^{R}(s)C_A^{R'}(t-\tau)\,d\tau\right] \tag{3.38}$$

where $C_A^{R'}(t) = dC_A^{R}/dt$.

This is probably a more tractable method than that obtained using the expansion of the propagator but is still not applicable to a potential containing a hard sphere part. However, all quantities involved can easily be evaluated by an MD simulation using the reference potential only.

3.6 MODE COUPLING THEORY

Starting from the equation eq. (2.53),[18]

$$d\mathbf{C}_A(t)/dt = -i\mathbf{\Omega}\cdot\mathbf{C}_A(t) + \int_0^t \mathbf{C}_A(t-\tau)\mathbf{K}_A(\tau)\,ds \tag{3.39}$$

and taking the Laplace transform gives

$$\tilde{\mathbf{C}}_{\mathbf{A}}(k,z) = \mathbf{C}_A(\underline{k})/[z\mathbf{I} - ik\mathbf{B}(\underline{k}) + k^2\mathbf{D}(\underline{k},z)] \tag{3.40}$$

Now write

$$d\mathbf{A}_k(t)/dt = ik\mathbf{y}_k(t) \tag{3.41}$$

$$\Omega_{xk} = k\mathbf{B}(\underline{k})$$

and

$$\mathbf{K}_{xk}(t) = k^2\mathbf{D}(k,t)$$

In mode–mode coupling theory[26–38], the decay of $k^2\mathbf{D}(\underline{k},t)$ is taken to be via hydrodynamic modes which was first considered by Fixman[26] and formalised by Kadanoff and Swift[27] and Kawasaki,[28] and the basic idea[29,37] is to write $\mathbf{y}_k(t)$ in terms of sums and products of the hydrodynamic normal modes $\mathbf{x}'_k$,

$$\mathbf{y}_{k'}(t) = \sum[\mathbf{a}_2(k,k_2)\mathbf{x}'_{k1}(t)\mathbf{x}'_{k2}(t) + \mathbf{a}_3(k,k_2,k_3)\mathbf{x}'_{k1}(t)\mathbf{x}'_{k2}(t)\mathbf{x}'_{k3}(t) + \cdots \tag{3.42}$$

As to whether such an expansion is exact depends on the completeness of the hydrodynamic basis used and, in practice, the expansion is truncated at relatively low order and cannot be valid at times which are short on an atomic scale. The original theory was applied to the anomalous behaviour of transport coefficients near the critical point but then extensively applied to study the long-time tails in $C_A(t)$[31–37] and also to describe the dynamics of simple liquids.[38–40] It has subsequently been refined and extensively and successfully applied to describe the dynamics of supercooled liquids and glasses.[41]

3.7 MACROSCOPIC HYDRODYNAMIC THEORY

Another approach to the calculation of the memory function corresponding to the velocity autocorrelation function is to use models based on the motion of a macroscopic body in a fluid as described by hydrodynamics. Now, as the velocity autocorrelation function from the LE in eq. (2.35) is an exponential then following the treatment of Zwanzig and Bixon[42] we may write[42,43]

$$C_v(t) = \langle v_x(0)v_x(t)\rangle = 2/\pi \,\mathrm{Re}\int_0^\infty \cos(\omega t)\, m\langle v_x^2\rangle/(-\mathrm{i}\omega m + \zeta)\,\mathrm{d}\omega \tag{3.43}$$

where ζ is a friction constant. This suggests rewriting the general relationship between the memory function and $C_v(t)$ in a similar form:

$$\langle v_x(0)v_x(t)\rangle = 2/\pi \,\mathrm{Re}\int_0^\infty \cos(\omega t)\, m\langle v_x^2\rangle/(-\mathrm{i}\omega m + \zeta(\omega))\,\mathrm{d}\omega \tag{3.44}$$

where we have written the Fourier transform of the memory function as a generalised, frequency-dependent friction coefficient,

$$K(\omega) = \zeta(\omega) = \int_0^\infty K(t)\mathrm{e}^{\mathrm{i}\omega t}\,\mathrm{d}t \tag{3.45}$$

Thus, the LE assumes the memory function is a delta function or equivalently that the friction coefficient is a constant which by eq. (3.18) means that $C_v(t)$ is a exponentially decaying function of time. However, although this is accurate for a large Brownian particle suspended in a fluid (see Section 5.5) this does not describe the limit when the particle is of the same size and mass as those of the suspending fluid at liquid densities as can be seen by inspecting some particular results for a simple liquid.[1–5] Under these later circumstances, it may be seen that the main features of $\langle v_x(0)v_x(t)\rangle$ are

1. Power series behaviour in even powers of t at short time (except for a system with an impulsive force where it is a power series containing odd powders of time also).
2. A negative "backscatter" region at intermediate times.
3. A long-time algebraic tail of the form $-At^{-3/2}$ at long times.[35,44–46]

Such behaviour may be seen in Figure 3.1 which shows the normalised velocity autocorrelation function $c_v(t) = Z(t) = \langle \underline{v}(0)\bullet\underline{v}(t)\rangle/\langle \underline{v}(0)\bullet\underline{v}(0)\rangle$ versus delay time t for a stable hard sphere fluid as a function of volume fraction, $\phi = \pi/2\ (N/V)\ d^3$ calculated by the MD method.

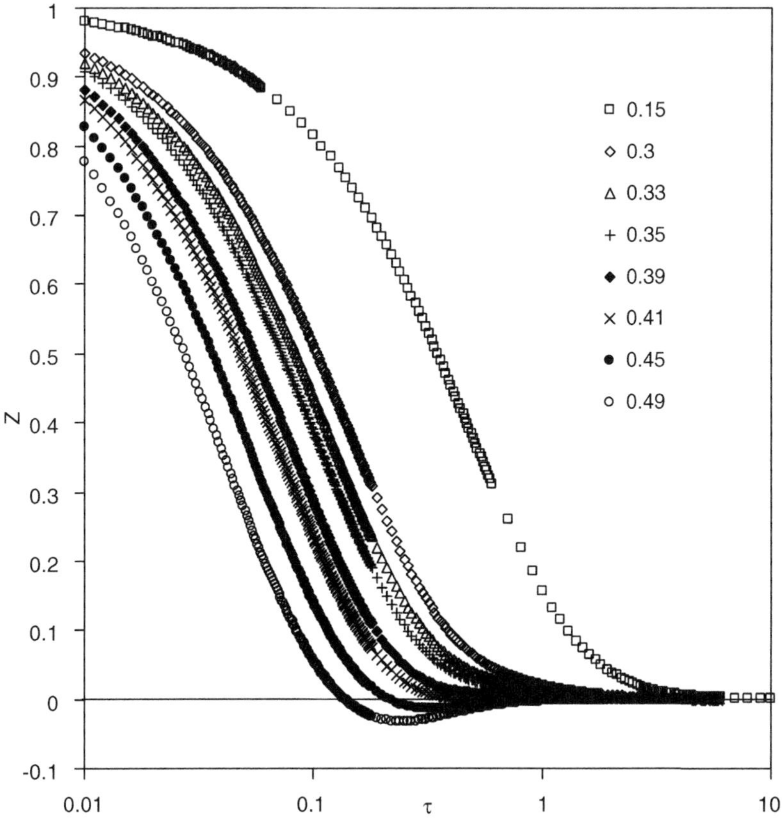

Figure 3.1 Normalised velocity autocorrelation function $c_v(\tau) = Z = \langle \underline{v}(0)\bullet\underline{v}(\tau)\rangle/\langle \underline{v}(0)\bullet v(0)\rangle$ versus delay time τ for a stable hard sphere fluid as a function of volume fraction, $\phi = \pi/2\ (N/V)\ d^3$ (courtesy of Dr Stephen Williams, ANU, Canberra, Australia).

To highlight the long-time tail, Figure 3.2 shows the normalised velocity autocorrelation function at long delay times for selected "tracer particles" in a mixture where the tracer particles have either the same mass (shown as •) or four times the mass (shown as Δ) of the other particles. The interaction potential of all the particles in the system was the same CWA pair potential.[46] This clearly shows the long-time tail for both cases and illustrates that the onset of this tail is at longer delay times as the mass ratio increases and that the magnitude of the effect is larger for larger mass ratios.[46]

Thus, a more complete theory of $\zeta(\omega)$, that is of $K_v(t)$, is needed to obtain all of these features. To achieve this, Zwanzig and Bixon[42] first noted that in fact Stokes' law (derived in 1851) for the nonsteady flow of a sphere for stick boundary conditions is[42]

$$\zeta(\omega) = 6\pi\eta_f a - 2\pi a^3 \, i\omega\rho_f/3 - 6\pi a^2 \, i(\omega\rho_f\eta)^{1/2} \tag{3.46}$$

and for slip boundary conditions it is[42]

$$\zeta(\omega) = 4\pi\eta_f a - 2\pi a^3 \, i\omega\rho_f/3 - 8\pi a^2 \, i(\omega\rho_f\eta_f)^{1/2}/(3 + ai(i\omega\rho_f\eta_f)^{1/2}) \tag{3.47}$$

where ρ_f and η_f are the density and coefficient of Newtonian viscosity of the fluid, respectively.

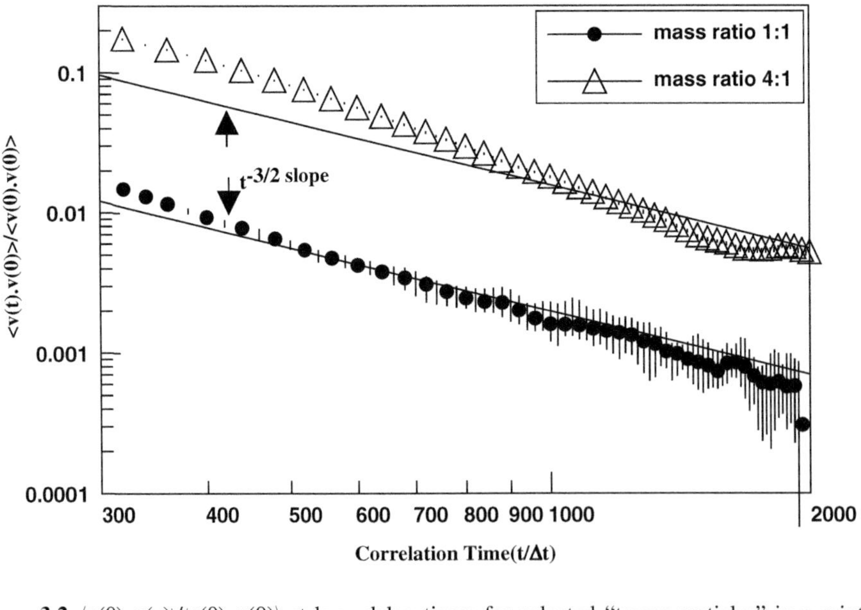

Figure 3.2 $\langle \underline{v}(0)\bullet\underline{v}(\tau)\rangle/\langle \underline{v}(0)\bullet\underline{v}(0)\rangle$ at long delay times for selected "tracer particles" in a mixture where the tracer particles have the same mass, • or four times the mass, Δ of the other particles (Figure courtesy of Dr. McDonough, CSIRO, Melbourne, Australia).

This expression in eq. (3.46) for $\zeta(\omega)$ can be converted into the time domain,[47–49] which was first done by Boussinesq in 1903,[48,50] where the force $F_H(t)$ on a sphere may be expressed as

$$F_H(t) = -6\pi\eta_f a v(t) - 2/3\pi\rho_f a^3\,(\mathrm{d}/\mathrm{d}t)v(t) - 6(\pi\eta_f\,\rho_f)^{1/2} a^2 \int_{-\infty}^{t} (t-s)^{-1/2}\,(\mathrm{d}v(s)/\mathrm{d}s)\mathrm{d}s \tag{3.48}$$

in which ρ_f is the fluid density and η_f its viscosity. This leads to the equation of motion[47,49] for $C_V(t)$,

$$M_{\text{eff}}\,\mathrm{d}C_v(t)/\mathrm{d}t = -6\pi\eta_f\,aC_v(t) - 6a^2\,k_BT(\pi\eta_f\,\rho_f/tM^2)^{1/2} - (6a^2\,k_BT/M)(\pi\eta_f\,\rho_f)^{1/2}\int_0^t (t-s)^{1/2}\,(\mathrm{d}C_v(s)/\mathrm{d}t)\,\mathrm{d}s$$

where $M_{\text{eff}} = M + 2/3\pi\rho_f\,a^3$ is the effective mass of the particle.

Thus, it is this nonsteady flow which leads to the long-time tail, $-At^{-3/2}$ in the velocity autocorrelation function[35,44–46] as first discovered for atomic systems by Alder and Wainwright[35] using MD simulations. So, the normal assumption that $\zeta(\omega)$ is a constant is not even correct for a macroscopic sphere moving in an incompressible, viscous fluid because this assumes steady motion.

However, the forms of $\zeta(\omega)$ given in eqs. (3.46) and (3.47) do not lead to the features 1 and 2 of $\langle v_x(0)v_x(t)\rangle$ listed above. So, Zwanzig and Bixon[42] allowed for the finite velocity of sound, C (i.e. allowing for compressibility) and viscoelasticity using a Maxwell single relaxation time approximation in the fluid surrounding the tracer particle to obtain[42,43]

$$\zeta_{\text{ZB}}(\omega) = [\eta k_t^2/k_l^2 - 2\eta_v](4/3\pi a)k_l a\,h_1(k_l a)A_L + (4/3\pi a)2\eta k_t a\,h_1(k_t a)A_N \tag{3.49}$$

where $k_1 = \omega^2/C_1^2$, $k_t = \omega^2/C_t^2$, C_l and C_t are the complex frequency-dependent sound velocities in the longitudinal and transverse directions given by

$$C_l^2 = C^2 - \mathrm{i}\omega\upsilon_l \text{ and } C_t^2 = -\mathrm{i}\omega\upsilon_t$$

where $\upsilon_1 = \eta_f/\rho_f$, $\upsilon_t = (4/3\eta_f + \eta_{fv})/\rho_f$, where η_{fv} is the bulk viscosity of the fluid.

Here $h_l(k_t a)$ is spherical Hankel function and A_N and A_L are coefficients determined by the boundary conditions. This extremely ingenious solution unfortunately does not allow $\zeta_{\text{ZB}}(\omega)$, $K_v(t)$ nor $\langle v_x(0)v_x(t)\rangle$ to be obtained analytically. However, $\langle v_x(0)v_x(t)\rangle$ may be obtained by numerical solution of eq. (3.44) using eq. (3.49), and if reasonable values of the parameters appearing in this equation are used, it gives results which, except at short times, agree very well with Rahman's MD results for dense Lennard–Jones liquid.[42] In particular, this model gives the negative, backscatter region in $\langle v_x(0)v_x(t)\rangle$ at intermediate times, indicating that to obtain this feature from such a model we need to include the effects of short-time compressibility and elastic properties of the fluid. The incorrect short-time exponential behaviour for nonhard sphere systems could be corrected by allowing for the proper short-time molecular behaviour to be incorporated into the expression for $\zeta(\omega)$.

The importance of this model is that it explains the main features of the velocity autocorrelation function for atomic fluids and leads to methods, which may be termed generalised hydrodynamics, which have been extensively used to interpret the results of experimental and computer simulation data on time correlation of many different kinds. In extensions of this approach,[51–54] generalised friction coefficients and generalised wavelength and/or frequency-dependent viscosities have been extensively developed and used in interpreting experimental data particularly for scattering data.[5,51,54]

3.8 MEMORY FUNCTIONS CALCULATED BY THE MOLECULAR-DYNAMICS METHOD

There have been a number of studies on liquids using MD methods, which have calculated time correlation functions, transport properties such as self-diffusion and viscosity coefficients, and scattering functions. Good references to this work may be found in the literature, such as Refs. 1–5, 55,56 . In many of these studies, time correlation functions have been used to calculate memory functions; one such velocity autocorrelation function and the memory function obtained from it may be seen in Figure 2.1. Some MD studies have also been made to test approximate memory functions because the MD method provides accurate numerical data for well-defined systems.[1–5,55,56]

Short-ranged approximations to $K_v(t)$, such as the exponential memory (eq. (3.19)) and the Gaussian form (eq. (3.24)) have been shown to give qualitatively reasonable results for dense fluids. These simple forms may be adjusted so that they give correct value of the lowest moments, although some treatments also use values of transport coefficients to obtain the parameters appearing in these expressions for $K_v(t)$. They generally describe the main features of the initial decay of $C_v(t)$ reasonably well and produce a backscatter region for intermediate times at high densities in agreement with MD results on the same system. However, they do not give exact agreement with these results and do not give a long-time tail in $C_v(t)$.

Thus, attempts have been made to improve these simple functions while retaining their desirable features. For example, Levesque and Verlet, in a pioneering study of self-diffusion in the LJ126 fluid, found that the following form of $K_v(t)$ gave good agreement with MD results:[55]

$$\begin{aligned} K_{\mathrm{LV}}(t) &= K_1(0)\exp(-B_0 t^2) + A_0 t^4 \exp(-\alpha_0 t) \\ &= K_{\mathrm{G}}(t) + A_0 t^4 \exp(-\alpha_0 t) \end{aligned} \tag{3.50}$$

Another form often used has been reported by Lee and Chung,[56] which is similar to eq. (3.50), that is,

$$K_{\mathrm{LC}}(t) = K_{\mathrm{G}}(t) + Bt^4 \exp(-\beta^2 t^2) \tag{3.51}$$

However, although both eqs. (3.50) and (3.51) are of longer range in time than the simple exponential or Gaussian memory $K_G(t)$ and are able to give a very accurate description of the MD data, neither gives the long-time tail $-At^{-3/2}$ in $C_V(t)$. Interestingly, although many theories predict long-time, algebraic tails in autocorrelation functions other than $C_V(t)$[34] there does not seem to be definitive evidence of such tails from MD simulations.[37]

Once again these and other explicit forms of $K_A(t)$ may be used in conjunction with the program VOLSOL in Appendix O to calculate $C_A(t)$ from memory functions, such as that given by eq. (3.50) or eq. (3.51).

3.9 CONCLUSIONS

The first essential step required in using GLEs to calculate the time evolution of $A(t)$ or $\langle A(0)A(t)\rangle$ is to obtain the memory function $K_A(t)$. This may be done by use of the MD method as pioneered by Alder,[35–37] Verlet and Levesque[55] and Rahman.[57] However, it is then superfluous to use a GLE except where one wishes to extend $K_A(t)$ beyond the maximum time available in an MD simulation by the use of some model.[55]

Alternatively, many schemes have been devised to generate approximate memory functions in order to understand transport and scattering data in terms of the physical processes responsible for them and construct approximate BD schemes. Methods that have been used to accomplish this include,

1. Time series expansions.
2. The Mori continued fraction method.
3. The use of information theory.
4. Perturbation theories.
5. Mode coupling theory.
6. Schemes based on macroscopic hydrodynamics and its generalisation, molecular hydrodynamics or generalised hydrodynamics.

In subsequent chapters we will use some of these methods and also refer to results obtained using some of them.

REFERENCES

1. B.J. Berne, Time-dependent properties of condensed media. In: *Physical Chemistry, An Advanced Treatise, Vol. VIIIB, Liquid State*, (Ed. D. Henderson), Academic Press, New York, 1971, pp. 539–715.
2. B.J. Berne and G.D. Harp, On the calculation of time correlation functions. In: *Advances in Chemical Physics*, Vol. XVII (Eds I. Prigogine and S.A. Rice), Wiley, New York, 1970, pp. 63–227.
3. B.J. Berne, Projection operator techniques in the theory of fluctuations in statistical mechanics. In: *Modern Theoretical Chemistry 6. Statistical Mechanics, Part B: Time-Dependent Processes* (Ed. B.J. Berne), Plenum Press, New York, 1977, pp. 233–257.

4. P. Schofield, Theory of time–dependent correlations in simple classical liquids. In: *Specialist Periodical Reports, Statistical Mechanics*, Vol. 2 (Ed. K. Singer), The Chem. Soc., London, 1975, pp. 1–54.
5. J–P. Hansen, Correlation functions and their relationship with experiment. In: *Microscopic Structure And Dynamics of Liquids* (Eds J. Dupuy and A.J. Dianoux), Plenum, New York, 1977, pp. 3–68; J.-P. Hansen and I.R. McDonald, *The Theory of Simple Liquids*, 2nd edn., Academic Press, London, 1991, pp. 193–363.
6. H. Mori, *Prog. Theory Phys.*, **34**, 399 (1965).
7. C.D Boley, *Ann. Phys.*, **86**, 91 (1974).
8. J.J. Duderstadt and A.Z. Akcasu, *Phys. Rev. A*, **1**, 905 (1970).
9. J.P. Boon and P. Deguent, *Phys. Rev. A*, **2**, 2542 (1970).
10. M.S. Jhon and J.S. Dahler, *J. Chem. Phys.*, **68**, 812 (1978).
11. M.S. Jhon and J.S. Dahler, *J. Chem. Phys.*, **68**, 812 (1978).
12. L.L. Lee, *Physica A*, **100**, 205 (1980).
13. S. Toxvaerd, *J. Chem. Phys.*, **81**, 5131 (1984).
14. B.J. Berne and G.D. Harp, On the calculation of time correlation functions. In: *Advances in Chemical Physics*, Vol. XVII (Eds I. Prigogine and S.A. Rice), Wiley, New York, 1970, pp. 118–120.
15. D.M. Heyes and J. Powles, *Mol. Phys.*, **71**, 781 (1990).
16. D.M. Heyes, J. Powles and J.C. Gil Montera, *Mol. Phys.*, **78**, 229 (1993).
17. R.K. Sharma, K. Tankeshwar and K.N. Pathak, *J. Phys. Condens. Matter*, **7**, 537 (1995).
18. P. Schofield, Theory of time-dependent correlations in simple classical liquids. In: *Specialist Periodical Reports, Statistical Mechanics*, Vol. 2 (Ed. K. Singer), The Chem. Soc., London, 1975, pp. 15–25.
19. K.S. Singwi and A. Sjolander, *Phys. Rev.*, **167**, 152 (1968).
20. N. Corngold and J.J. Duderstadt, *Phys. Rev.* A, **2**, 836 (1970).
21. T.G. Gaskell, *J. Phys. C*, **4**, 1466 (1971); M.J. Baker and T.G. Gaskell, *J. Phys. C*, **5**, 353 (1972).
22. S.K. Mitra, *Phys. Lett.*, **38A**, 471 (1972); S.K. Mitra, N. Dass and N.C. Varshneya, *Phys. Rev. A*, **6**, 1214 (1972); S.K. Mitra, *J. Phys. C*, **6**, 801 (1973).
23. Method discussed by P. Schofield on pages 24 and 25 in reference 18 above.
24. I.K. Snook and R.O. Watts, *Mol. Phys.*, **33**, 413–441 (1976) and K.E. Gubbins, "Thermal Transport coefficients for simple dense fluids", Specialist Periodical Report, Statistical Mechanics, Vol.1, Ed. K. Singer, The Chem. soc. London, 1972, pp. 238–241.
25. I.K. Snook and R.O. Watts, *Mol. Phys.*, **33**, 443–452 (1977).
26. M. Fixman, *J. Chem. Phys.* **36**, 310 (1962); **47**, 2808 (1967).
27. L. Kadanoff and J. Swift, *Phys. Rev.*, **165**, 310 (1968); **166**, 89 (1968).
28. K. Kawasaki, *Prog. Theor. Phys. (Japan)*, **39**, 1133 (1968); **40**, 11, 706, 930; *Ann. Phys. (NY)* **61**, 1 (1970).
29. P. Schofield, Theory of time-dependent correlations in simple classical liquids. In: *Specialist Periodical Reports, Statistical Mechanics*, Vol. 2 (Ed. K. Singer), The Chem. Soc., London, 1975, pp. 26–33.
30. T. Keyes, Principles of mode-mode coupling theory. In: *Modern Theoretical Chemistry 6. Statistical Mechanics, Part B, Time-Dependent Processes* (Ed. B.J. Berne), Plenum, New York, 1977, pp. 259–309.
31. M.H. Ernst, E.H. Hauge and J.M.J. van Leewen, *Phys. Rev. Lett.*, **25**, 1254 (1970).
32. M.H. Ernst and J.R. Dorfman, *Physics*, **61**, 157 (1972).
33. P. Reibois, *J. Stat. Phys.*, **2**, 21 (1970).
34. Y. Pomeau and P. Resibois, *Time dependent correlation functions and mode-mode coupling theories*, *Phys. Rep.*, **19**, 63–139 (1975).

35. B.J. Alder and T.E. Wainwright, *Phys. Rev. A*, **1**, 18 (1970).
36. B.J. Alder, Computer results on transport properties, In: *Stochastic Processes in Nonequilibrium Systems*, (Eds L. Garrido, P. Seglar and P.J. Shepherd), Lecture Notes in Physics, Vol. 84, Springer, Berlin, 1978, p 168.
37. B.J. Alder, Molecular Dynamics Simulations, *Molecular-Dynamics Simulation of Statistical-Mechanical Systems, Proceedings of the International School of Physics, Enrico Fermi*, Course XCVII (Eds G. Ciccotti and W.G. Hoover), North-Holland, Amsterdam, 1986, p. 66.
38. L. Sjogren and A. Sjolander, *J. Phys. C: Solid State Phys.*, **12**, 4369 (1979).
39. L. Sjogren, *J. Phys. C: Solid State Phys.*, **13**, 705 (1980).
40. M. Canales and J.A. Padro, *J. Phys.: Condens. Matter*, **9**, 11009 (1997).
41. W. Gotze and L. Sjogren, *Relaxation processes in supercooled liquids*, *Rep. Prog. Phys.*, **55**, 241−376 (1992); U. Bengtzelius, W. Gotze and A Sjolander, *J. Phys. C: Solid State Phys.*, **17**, 5915 (1984); E. Leutheusser, *Phys. Rev. A*, **29**, 2765 (1984); W. Gotze, *J. Phys.: Condens. Matter*, **11**, A1 (1999).
42. R. Zwanzig and M. Bixon, *Phys. Rev. A*, **2**, 2005 (1970).
43. H. Metiu, D.W. Oxtoby and K.F. Freed, *Phys. Rev. A*, **15**, 361 (1977).
44. D. Levesque and W.T. Ashurst, Phys. Rev. Lett., **33**, 277 (1974).
45. J.J. Erpenbeck and W.W. Wood, *Phys. Rev. A*, **26**, 1648 (1982); **32**, 412 (1985); J.J. Erpenbeck, *Phys. Rev. A*, **39**, 4718 (1989).
46. A. McDonough, S.P. Russo and I.K. Snook, *Phys. Rev. E*, **63**, 26109 (2001).
47. T.S. Chow and J.J. Hermans, *J. Chem. Phys.*, **56**, 3150 (1972).
48. J. Boussinesq, *Theorie Analytique de la Chaleur*, Vol. II, Gauthiers-Villars, Paris, 1927, p. 224.
49. A. Widom, *Phys. Rev. A*, **3**, 1394 (1971).
50. L.D. Landau and E.M. Lifshitz, *Fluid Mechanics*, Addison-Wesley, Reading, MA, 1959, pp. 96−99
51. J.R.D. Copley and S.W. Lovesey, The dynamic properties on monatomic fluids, *Rep. Prog. Phys.*, **38**, 461 (1975).
52. P. Schofield, Theory of time-dependent correlations in simple classical liquids. In: *Specialist Periodical Reports, Statistical Mechanics*, Vol. 2 (Ed. K. Singer), The Chem. Soc., London, 1975, pp. 33−45.
53. J.P. Boon and S.Yip, *Molecular Hydrodynamics*, McGraw-Hill, New York, 1980.
54. W.E. Alley, B.J. Alder and S. Yip, *Phys. Rev. A*, **27**, 3174 (1983).
55. D. Levesque and L. Verlet, *Phys. Rev. A*, **2**, 2514 (1970).
56. L.L. Lee and T.H. Chung, *J. Chem. Phys.*, **77**, 4650 (1982).
57. A. Rahman, *Phys. Rev. A*, **136** 405 (1964)

– 4 –

The Generalised Langevin Equation in Non-Equilibrium

As discussed in the previous chapters the Langevin equation (LE) describing Brownian motion of a colloidal particle in a solvent can be regarded as a special case in which the dynamical variable or phase variable of interest is the momentum of the colloidal particle, and the classical LE is obtained by selecting the momentum of a particle as the relevant phase variable and taking the Brownian limit for a system in equilibrium (see Section 2.5 and Chapter 5). Generalization of this technique to the description of non-equilibrium states has received relatively little attention. For example, Kawasaki and Gunton[1] and Zwanzig[2] have used projection operator methods to develop general theories of non-linear constitutive relations, and Ichiyanagi[3] used a similar approach to study quantum mechanical non-equilibrium steady-state systems. However, the results of these investigations are largely formal and they are not directly applicable to non-equilibrium molecular dynamics computer simulations because of differences in the way that heat is assumed to be removed from the system. Application of the generalised Langevin equation (GLE) to the investigation of atomic and Brownian motion in a strongly shearing fluid remains largely unexplored.

The purpose of this chapter is to derive a GLE for a specific class of non-equilibrium systems. The non-equilibrium systems considered are those generated by applying a constant external driving field and a spatially homogeneous non-holonomic constraint thermostat at time $t = 0$, to a classical fluid that is initially at equilibrium. This is typical of systems encountered in non-equilibrium molecular dynamics simulations that are used to obtain transport properties such as the shear viscosity, thermal conductivity and diffusion coefficient as discussed in Chapter 1 (see also Ref. [4]). The equations of motion discussed in Chapter 1 that generate our non-equilibrium steady states differ in crucial ways from those that have been used in previous work on non-equilibrium GLEs. They are non-Hamiltonian and they do not conserve phase volume. This requires us to use a form of the Liouville equation that explicitly accounts for phase space compression, and it also results in a difference between the Liouville operator for phase variables and the Liouville operator for the distribution function (see Section 1.2.8). An interesting consequence is that the distribution function is never time independent, even in the steady state, although the time averages of phase variables and correlation functions do become constant in the long time limit.

In our derivation we will follow the method of Berne,[5] which in turn closely parallels the original development of Mori.[6]

4.1 DERIVATION OF GENERALISED LANGEVIN EQUATION IN NON-EQUILIBRIUM

It is necessary for the derivation of the GLE given in Chapter 2 to define a projection operator, which will separate all phase variables into parts that are either correlated or uncorrelated with the phase variable of interest, A(t). Therefore, let us define a projection operator by

$$\mathbf{P}B(t+\tau) = \left\{\langle B(t+\tau)A^*(t)\rangle / \langle A(t)A^*(t)\rangle\right\}A(t) \tag{4.1}$$

where $\tau > 0$, and we assume that A is defined in such a way that its average value is equal to zero. This property is required to ensure that the definition of the time reversal operator is satisfied in the case of variables that are odd in time (this is discussed in more detail in the Appendix G). The function $\mathbf{P}B(t+\tau)$ is a measure of the correlation between $B(t+\tau)$ and $A(t)$. The complementary projection operator is defined by

$$\mathbf{Q}B(t+\tau) \equiv (\mathbf{1}-\mathbf{P})B(t+\tau) \tag{4.2}$$

where $\mathbf{1}$ is the identity operator. The complementary projection operator projects a phase variable onto the subspace of phase space which is orthogonal to $A(t)$, so we have $\langle \mathbf{Q}B(\mathrm{t}+\tau)\,A(t)\rangle = 0$. As in Chapter 2 the variables occupying this subspace are said to be random variables, in the sense that they are uncorrelated with $A(t)$. The projection operator and the complementary projection operator are both Hermitian in the correlation function in which they are defined.

We have chosen a very simple projection operator, containing only one variable for the purpose of this investigation. In addition, we only seek the equation of motion for that variable. More complicated projection operators are required when, for example, a complete description of the behaviour of the system in terms of the hydrodynamic variables is required. We refer the reader to Ref. 7 for more detail.

The GLE describes the development of an arbitrary phase variable, A, in terms independent of the $6N$ coordinates of the system, so that the only terms remaining are those which are linear in $A(t)$ and the so-called random terms.

To obtain the GLE we must first separate the time evolution of $A(t+\tau)$ into parts correlated and uncorrelated with $A(t)$, that is, upon using the non-equilibrium p-Liouvillean and non-equilibrium p-propagator,[4]

$$\begin{aligned}(\mathrm{d}/\mathrm{d}t)A(t+\tau) &= \exp(\mathrm{i}\mathbf{L}\tau)\mathrm{i}\mathbf{L}A(t)\\ &= \exp(\mathrm{i}\mathbf{L}\tau)(\mathbf{P}+\mathbf{Q})\mathrm{i}\mathbf{L}A(t)\end{aligned} \tag{4.3}$$

The first term is used to define the *frequency*, that is,

$$\begin{aligned}\exp(\mathrm{i}\mathbf{L}\tau)\mathbf{P}\mathrm{i}\mathbf{L}A(t) &= \exp(\mathrm{i}\mathbf{L}\tau)\left\{\langle \mathrm{i}\mathbf{L}A(t)A^*(t)\rangle/\langle A(t)A^*(t)\rangle\right\}A(t)\\ &= \left\{\langle \mathrm{i}\mathbf{L}A(t)A^*(t)\rangle/\langle A(t)A^*(t)\rangle \exp(\mathrm{i}\mathbf{L}\tau)\right\}A(t)\\ &= \mathrm{i}\Omega(t)A(t+\tau)\end{aligned} \tag{4.4}$$

and $\Omega(t)$ is the so-called *frequency* and is proportional to the correlation between $A(t)$ and $A^*(t)$. In equilibrium systems it is identically zero for the single-variable case. This follows from anti-symmetry of the equilibrium correlation function $\langle \dot{A}A \rangle$ about the time origin. In general, non-equilibrium correlation functions are symmetric about $t=0$ (the time at which the field is applied) but not about $\tau=0$ (the time origin of the correlation function). However, the frequency still approaches zero at sufficiently large values of t. In the single-variable case, this follows directly from eq. (1.32) when we take $\tau=0$. This issue is discussed in more detail in Appendix G.

Using the Dyson decomposition[8] (see Appendix B) to decompose the p-propagator, $\exp(i\mathbf{L}\tau)$, into a part that only occupies $\mathbf{Q}$-space, and a part that allows the development of correlations between the phase variable of interest and A, the second term in eq. (4.3) can be written as

$$\exp(i\mathbf{L}\tau)\mathbf{Q}i\mathbf{L}A(t) = e^{\mathbf{Q}i\mathbf{L}\tau}\mathbf{Q}i\mathbf{L}A(t) + \int_0^\tau e^{i\mathbf{L}(\tau-s)}\,\mathbf{P}i\mathbf{L}e^{\mathbf{Q}i\mathbf{L}s}\,\mathbf{Q}i\mathbf{L}A(t)\,ds \tag{4.5}$$

The term $e^{\mathbf{Q}i\mathbf{L}\tau}\mathbf{Q}i\mathbf{L}A(t)$ always occupies the space of phase variables uncorrelated with $A(t)$, and therefore is never correlated with $A(t)$. We follow the usual convention and call this phase variable the "random force", although it is not rigorously a random variable as discussed in Chapter 2:

$$R(t+\tau) = \exp(\mathbf{Q}i\mathbf{L}\tau)\mathbf{Q}i\mathbf{L}A(t) \tag{4.6}$$

An important property of the random force is that

$$R = \mathbf{Q}R \tag{4.7}$$

for all time which is a reiteration that the random force exists solely in the subspace orthogonal to $A(t)$.

Using the definition of the random force, eq. (4.4) can be written in a more compact form as

$$\exp(i\mathbf{L}\tau)\mathbf{Q}i\mathbf{L}A(t) = R(t+\tau) + \int_0^\tau e^{i\mathbf{L}(\tau-s)}\,\mathbf{P}i\mathbf{L}R(t+s)\,ds \tag{4.8}$$

Consider the integrand in eq. (4.8):

$$\begin{aligned} e^{i\mathbf{L}(\tau-s)}\mathbf{P}i\mathbf{L}R(t+s) &= e^{i\mathbf{L}(\tau-s)}\left\{\langle i\mathbf{L}R(t+s)A^*(t)\rangle / \langle A(t)A^*(t)\rangle\right\}A(t) \\ &= \left\{\langle i\mathbf{L}R(t+s)A^*(t)\rangle / \langle A(t)A^*(t)\rangle\right\}e^{i\mathbf{L}(\tau-s)}A(t) \\ &= \left\{\langle i\mathbf{L}R(t+s)A^*(t)\rangle / \langle A(t)A^*(t)\rangle\right\}A(t+\tau-s) \end{aligned} \tag{4.9}$$

The second line follows from the fact that the ensemble-averaged terms are independent of the phase and are unaffected by the propagator. The time-correlation function in the numerator can be simplified. Using eq. (1.31) the numerator can be written as

$$\langle \mathrm{i}\mathbf{L}R(t+s)A^*(t)\rangle = -\langle R(t+s)(\mathrm{i}\mathbf{L}A(t))^*\rangle + \beta F_\mathrm{e}\langle R(t+s)A^*(t)J(0)\rangle \tag{4.10}$$

If we use eq. (4.7) and the Hermitian property of the complementary projection operator in the correlation function in which it is defined, the first term in eq. (4.10) can be written as

$$\langle R(t+s)(\mathrm{i}\mathbf{L}A(t))^*\rangle = \langle R(t+s)(\mathbf{Q}\mathrm{i}\mathbf{L}A(t))^*\rangle = \langle R(t+s)(R(t))^*\rangle \tag{4.11}$$

Thus, we have a non-stationary autocorrelation function of the random force as part of the integrand. This is used to define the memory function by

$$K_\mathrm{A}(t+s) = \langle R(t+s)(R(t))^*\rangle / \langle A(t)A^*(t)\rangle \tag{4.12}$$

Thus, combining eqs. (4.3), (4.4), (4.8)–(4.10) and (4.12), we find that the evolution of an arbitrary phase variable is given by

$$\begin{aligned}(\mathrm{d}/\mathrm{d}t)A(t+\tau) = {} & \mathrm{i}\Omega(t)A(t+\tau) - \int_0^\tau K_\mathrm{A}(t+s)A(t+\tau-s)\,\mathrm{d}s + R(t+\tau)\\ & + \beta F_\mathrm{e}\int_0^\tau \{\langle R(t+s)A^*(t)J(0)\rangle / \langle A(t)A^*(t)\rangle\} A(t+\tau-s)\,\mathrm{d}s\end{aligned} \tag{4.13}$$

This is the GLE for an arbitrary phase variable in a non-equilibrium system, created by a steady external field applied to an equilibrium system some time t before. In the zero-field limit, this equation reduces to the usual equilibrium GLE of Mori[6] derived in Chapter 2.

If the system is mixing, then as $t \to \infty$,

$$\langle R(t+s)A^*(t)J(0)\rangle \to \langle R(t+s)A^*(t)\rangle\langle J(0)\rangle = 0 \tag{4.14}$$

and the non-equilibrium GLE (4.13) takes the same form as the equilibrium GLE. The steady-state GLE is then given by the $t \to \infty$ limit of eq. (4.13),

$$(\mathrm{d}/\mathrm{d}t)A(t+\tau) = \mathrm{i}\Omega(t)A(t+\tau) - \int_0^\tau K_\mathrm{A}(t+s)A(t+\tau-s)\,\mathrm{d}s + R(t+\tau) \tag{4.15}$$

Thus, eq. (4.14) is the GLE for the variable $A(t)$ for a system in a non-equilibrium steady state. We stress that although the form of the steady-state GLE is the same as that of the equilibrium GLE, it is fundamentally different in the sense that the time evolution is generated by propagators that include the field and the thermostat terms in the equations of motion.

When coupling between different macroscopic variables is possible, it is necessary to replace the variable A by a set of variables, usually a set of linearly independent, relevant or "slow" variables, which we may represent as a column vector (see Section 2.6),

$$\mathbf{A} = (A_1, ..., A_M) \tag{4.16}$$

The expression for the action of the projection operator on an arbitrary phase variable B is then given by

$$\mathbf{P}B(t+\tau) = \beta\langle B(t+\tau)\mathbf{A}^+(t)\rangle \cdot \boldsymbol{\chi}^{-1}(t) \cdot \mathbf{A}(t) \tag{4.17}$$

where $\mathbf{A}^+$ represents the Hermitian conjugate (complex conjugate of the transpose) of the vector $\mathbf{A}$, $\beta = k_B T$ and $\boldsymbol{\chi}^{-1}$ represents the inverse of the second-rank susceptibility matrix defined by

$$\boldsymbol{\chi}(t) = \beta\langle \mathbf{A}(t)\mathbf{A}^+(t)\rangle \tag{4.18}$$

In this case, the GLE can be written as

$$(\mathrm{d}/\mathrm{d}t)\,\mathbf{A}(t+\tau) = \mathrm{i}\Omega(t) \cdot \mathbf{A}(t+\tau) - \int_0^\tau K_A(t+s) \cdot \mathbf{A}(t+\tau-s)\,\mathrm{d}s + \mathbf{R}(t+\tau) \tag{4.19}$$

where the frequency matrix is defined as

$$\Omega(t) = \beta\langle \mathbf{L}\mathbf{A}(t)\mathbf{A}^+(t)\rangle \cdot \boldsymbol{\chi}^{-1}(t) \tag{4.20}$$

the random force vector is given by the generalised form of eq. (4.6),

$$\mathbf{R}(t+\tau) = \exp(\mathbf{Q}\mathrm{i}\mathbf{L}\tau)\mathbf{Q}\mathrm{i}\mathbf{L}\mathbf{A}(t) \tag{4.21}$$

and the memory function matrix $\mathbf{K}(t + s)$ is given by

$$\mathbf{K}(t+s) = \beta\langle R(t+s)\mathbf{R}^+(t)\rangle \cdot \boldsymbol{\chi}^{-1}(t) \tag{4.22}$$

This equation may be used to derive expressions for the time evolution of a dynamical variable or a non-equilibrium time-correlation function. In order to illustrate its use we will now derive the analogous equation in a shear field to the LE, which applies at equilibrium.

4.2 LANGEVIN EQUATION FOR A SINGLE BROWNIAN PARTICLE IN A SHEARING FLUID

We wish to express the equation of motion of a single, massive Brownian particle in a shearing fluid in terms of a "friction tensor" plus a random noise term that accounts for the effect of the solvent, and compare the resulting equation with the well-known equilibrium Langevin equation (LE) (1.1) and the usual generalisation including the effect of shear (see Section 6.5).

We will begin by discussing the general case in which the fluid consists of a set of particles of unspecified mass, with dynamics generated by the so-called SLLOD equations of motion, which are designed to generate shear flow.[4] These equations are a specific case of the SLOD equations (1.11), given by

$$\begin{aligned}(\mathrm{d}/\mathrm{d}t)\underline{r}_i &= \frac{\underline{p}_i}{m_i} + \underline{\mathrm{i}}\dot{\gamma}\, y_i \\ (\mathrm{d}/\mathrm{d}t)\underline{p}_i &= \underline{F}_i + \underline{\mathrm{i}}\dot{\gamma}\, p_{iy} - \alpha \underline{p}_i\end{aligned} \tag{4.23}$$

where the strain rate $\dot{\gamma}$ takes the role of the external field F_e. We wish to derive an equation of motion for one of these particles. It is well known that the self-diffusion coefficient becomes a second-rank tensor in shearing atomic and colloidal fluids, implying a coupling between different Cartesian components of the momentum.[9] This would be the motivation for introducing a second-rank friction tensor in this example.

The minimal set of slow variables required in this case therefore consists of the three components of the momentum of the chosen particle, i.e.

$$\mathbf{A} = (p_x, p_y, p_z) \tag{4.24}$$

In the description we have chosen, all elements of **A** have the same time-reversal symmetry. This means that all elements of the frequency matrix, $\Omega(t)$, are equal to 0 (see Appendix G), as in equilibrium systems. The equation of motion of any single particle within the system in a shearing steady state may therefore be written as

$$(\mathrm{d}/\mathrm{d}t)\underline{p}(t+\tau) = -\int_0^{\tau} \underline{K}_p(t+s)\cdot \underline{p}(t+\tau-s)\,\mathrm{d}s + \underline{R}(t+\tau) \tag{4.25}$$

The spatial symmetry of the system (invariance under a rotation of π in the x–y plane) imposes the following restrictions on the elements of the memory matrix: $K_{xy} = K_{yx}$; $K_{xz} = K_{zx} = K_{yz} = K_{zy} = 0$. The equation of motion derived above is exact, but the detailed properties of the "random force" are as yet, unknown.

In general, the GLE can be used to describe a non-Markovian process. To arrive at the simple LE from the GLE, a Markovian approximation is made. This is equivalent to the approximation (see Chapter 2 and Appendix D)

$$K_{ij}(t) = \zeta_{ij}\delta(t) \tag{4.26}$$

The physical meaning of this approximation is that the correlation function of the random force has decayed before there is any significant change in the momentum of a large mesoscopic particle within the system. This approximation becomes exact in the Brownian limit, when the ratio of the mass of the mesoscopic particle to the mass of a typical suspension particle tends to infinity (see Chapter 5). Substitution of eq. (4.26) into the equation of motion for a particle within the system (4.25) gives the simple LE,

$$(\mathrm{d}/\mathrm{d}t)\underline{p}(t) = -\zeta \underline{p}(t) + \underline{R}(t) \tag{4.27}$$

An expression for ζ for a spherical particle of radius a may be obtained from Stokes' law, $\zeta = k_B T/(6\pi\eta_f a)M_B$, which is appropriate for an infinite, isotropic, homogeneous medium of shear viscosity η_f (the zero shear, equilibrium limit).

We have shown that the GLE has the same form in non-equilibrium steady-state systems (4.15), as in equilibrium systems. Thus, there is some justification for making a Markovian approximation again for the equation of motion of the mesoscopic particle relative to the local velocity field within the fluid. In the case of a system undergoing planar shear the local velocity field is simply $\nabla\underline{u} = \dot{\gamma}_{ji}$. The LE is given by eq. (4.27), where $\underline{p}$ represents the thermal momentum of the particle. The thermal momentum is the actual momentum minus the local streaming velocity multiplied by the mass of the mesoscopic particle, i.e.

$$\underline{p} = M_B(\underline{r} - \dot{\gamma}\, y\underline{i}) \tag{4.28}$$

Substitution of eq. (4.28) and its time derivative into eq. (4.27) yields an LE for the motion of the mesoscopic particle of mass M_B,

$$M_B(\mathrm{d}^2/\mathrm{d}t^2)\underline{r}(t) = -\zeta M_B \cdot [(\mathrm{d}/\mathrm{d}t)\underline{r}(t) - \dot{\gamma} y\underline{i}] + M_B\dot{\gamma}\,\mathrm{d}y(t)/\mathrm{d}t\underline{i} + \underline{R}(t) \tag{4.29}$$

The first term is the standard term representing drag relative to the local streaming velocity, the second is a force in the x-direction due to the particle's movement across the streamlines and the third term is the random force.

The standard generalisation of the simple LE for particles in a fluid undergoing planar shear[10] obtained by merely adding a shear term to the equilibrium LE is (see Section 6.5)

$$M_B(\mathrm{d}^2/\mathrm{d}t^2)\,\underline{r}(t) = -\zeta M_B((\mathrm{d}/\mathrm{d}t)\underline{r}(t) - \dot{\gamma} y\,\underline{i}) + \underline{R}(t) \tag{4.30}$$

Assuming that the friction in eq. (4.29) is isotropic and given by Stokes' law, we can perform a simple comparison of these two expressions. The additional term in eq. (4.29) will be negligible compared to the frictional term if

$$M_B\dot{\gamma}/(6\pi\eta a) \ll 1 \tag{4.31}$$

or in terms of the density of the particle,

$$\dot{\gamma} \ll 9/2\eta/(\rho a^2) \tag{4.32}$$

For a neutrally buoyant particle of radius $a = 0.1$ μm in water ($\eta_f = 1$ mPa s), the right-hand side is equal to 4.5×10^8 s^{-1}, and for a 1 μm particle of the same density in air ($\eta_f = 2 \times 10^{-5}$ Pa s), it is 9×10^4 sc^{-1}. Thus, we see that the extra term can normally be neglected, as is commonly done.

The mean square displacements under shear can be calculated by solving the GLE for the thermal momentum(4.27). Because this equation is identical in form to the equilibrium LE, the solution is identical, but in this case, we obtain the mean square thermal displacements rather than the mean square total displacements. However, it is a relatively simple matter to obtain the mean square total displacements from the mean square thermal displacements. We define the thermal displacement in the x-direction by

$$\Delta q_x(t) = \int_0^t p_x(s)/M_B \, ds \tag{4.33}$$

The mean square thermal displacement in the x-direction is given by the well-known solution of eq. (4.27), assuming isotropic, constant friction,

$$\langle \Delta q_x^2(t) \rangle = (2k_BT/M_B\zeta)(t + 1/\zeta(e^{-\zeta t} - 1)) \tag{4.34}$$

To obtain the mean square total displacement, we use the definition of the thermal displacement on the left-hand side,

$$\Delta q_x(t) = \Delta x(t) - \dot{\gamma} \int_0^t y(s) ds \tag{4.35}$$

then ensemble average and rearrange to obtain the mean square total displacement,

$$\langle \Delta x^2(t) \rangle = \langle \Delta q_x^2(t) \rangle + \dot{\gamma}^2 \int_0^t \int_0^t \langle y(u) y(s) \rangle ds \, du \tag{4.36}$$

Evaluating the integral of the y-position correlation function using similar methods to Foister and van de Ven,[10] we obtain

$$\int_0^t \int_0^t \langle y(u) y(s) \rangle ds \, du = (k_BT/M_B\zeta^4)[2(\zeta t)^3/3 - (\zeta t)^2 + 2 - 2e^{-\zeta t} - 2(\zeta t)e^{-\zeta t}] \tag{4.37}$$

The diffusive limit of the shear rate-dependent part of the mean square total displacement is given by

$$\tfrac{2}{3}\dot{\gamma}^2 D_0\, t^3 \tag{4.38}$$

and the short-time limit is

$$\dot{\gamma}^2\,(k_B T/4M_B)t^4 \tag{4.39}$$

Remarkably, both of these limits are identical to the corresponding limits of the solution of eq. (4.30).[10] The full solution of eq. (4.27) is, however, much simpler than the full solution of eq. (4.30).[11]

A generalisation of the simple LE describing a single Brownian particle in a shearing fluid is obtained as a special case of our GLE. Our modified form of the LE equation (2.35) includes a force term due to the conversion of streaming motion into thermal motion (or vice versa) as particles cross streamlines. This term apparently has little effect on the overall mean square displacements of the particles, but its inclusion leads to a more self-consistent treatment of Brownian motion in shear, in which the thermal velocities are the variables of interest, just as they are in the equilibrium case. The solution of the corresponding LE is consequently simpler, and may be useful in further generalisations similar to that of Miyazaki and Bedeaux,[12] who investigated a non-Markovian generalisation of the LE for a Brownian particle in a shearing fluid, and allowed for anisotropic friction.

4.3 CONCLUSIONS

We have presented a derivation of the GLE for a system[13] that is subjected to a constant external field, explicitly taking into account the effect of the coupling to a thermostat or heatbath. This coupling is essential in order to establish a steady state. To do this we use the SLODD-type equations of non-equilibrium statistical mechanics (eq. (1.11)) as our basic dynamical equations of motion.

We find that the steady-state field-dependent GLE is identical in form to the equilibrium GLE, except that all the propagators must now include the field and thermostat terms. However, the derivation of a GLE that correctly describes the transient behaviour would appear to be substantially more difficult due to the presence of non-Hermitian terms that only approach zero in the steady state, and the absence of time-reversal symmetry in non-equilibrium correlation functions except in the long time (steady-state) limit.

REFERENCES

1. K. Kawasaki and J.D. Gunton, *Phys. Rev. A*, **8**, 2048 (1993).
2. R. Zwanzig, *Prog. Theor. Phys., Suppl.*, **64**, 74 (1978).
3. M. Ichiyanagi, *J. Phys. Soc. Japan*, **62**, 1167 (1993).
4. D.J. Evans and G.P. Morriss, *Statistical Mechanics of Non-equilibrium Liquids*, Academic Press, London, 1990.

5. B.J. Berne, *Projection Operator Techniques in the Theory of Fluctuations*. In: *Modern Theoretical Chemistry, Statistical Mechanics 6. Part B: Time-Dependent Processes* (Ed. B.J. Berne), Plenum Press, New York, 1977, Chapter 5, pp. 233–257; B.J. Berne and R. Pecora, *Dynamic Light Scattering*, Wiley, New York, 1976.
6. H. Mori, *Prog. Theor. Phys.*, **33**, 423 (1965); **34**, 399 (1965).
7. J.-E. Shea and I. Oppenheim, *J. Phys. Chem.*, **100**, 19035 (1996).
8. F.J. Dyson, *Phys. Rev.*, **75**, 486 (1949).
9. S. Sharman, D.J. Evans and P.T. Cummings, *J. Chem. Phys.*, **95**, 8675 (1991).
10. R.T. Foister and T.G.M. van der Ven, *J. Fluid Mech.*, **96**, 105 (1980).
11. M. San Miguel and J.M. Sancho, *Physica A*, **99**, 357 (1979).
12. K. Miyazaki and D. Bedeaux, *Physica A*, **217**, 53 (1995).
13. M.G. McPhie, P.J. Daivis, I.K. Snook, J. Ennis and D.J. Evans, *Physica A*, **299**, 412 (2001).

– 5 –

The Langevin Equation and the Brownian Limit

In this chapter we will discuss the approach to the Brownian limit starting with the microscopic Liouville equation. Here we will follow a more rigorous approach to the derivation of the equations for Brownian motion than that given in Section 2.5 which was pioneered by Mazur, Oppenheim, Deutch and Albers. The three key papers[1–3] of these authors are seminal papers in this area and because of their importance I will stay very close to their work, methods and notation. The latter is important so that these papers may be referred to in full for more details and to correct any errors that I have made. Also, this approach is easier to generalise to a system of many interacting Brownian particles than this previous method as we shall see in Section 5.2. There are several ways of deriving the equations appropriate to the Brownian limit including the treatments of Zwanzig[4] and Murphy and Aguire,[5] some of these will be found in the references at the end of this chapter and will be discussed in Section 5.5.

It should be emphasised that although we use the Deutch–Oppenheimer[2,3] approach to derive the Langevin equation in the Brownian limit their method is very general. Thus, their method may be used for other purposes, e.g. to separate a system into system and surroundings or to treat two or more component systems which may not be approximated as being in "The Brownian Limit" (see Chapter 7).

For the Brownian approach it is convenient to divide the system of N_t particles into a sub-system consisting of N bath or background particles of masses m_i and another of n particles of mass M_i, i.e. into the bath (solvent, background or suspension medium) and the particles of the suspended medium respectively. It is usually assumed that $M_i \gg m_i$ and $N \gg n$, i.e. we have a system of n heavy particles suspended in a bath or medium of much more numerous lighter particles. However, there are other conditions on the validity of the Langevin equation which will be discussed in Section 5.5.

In the Brownian case the momenta of the N particles of the bath are given by $\underline{p}_i$ and their positions by $\underline{r}_i$ and the momenta of the n suspended particles are given by $\underline{P}_i$ and their positions by $\underline{R}_i$.

Let us write,

$$\underline{p}^{(N)} = (\underline{p}_1, \underline{p}_2, \ldots, \underline{p}_N) \text{ and } \underline{r}^{(N)} = (\underline{r}_1, \underline{r}_2, \ldots, \underline{r}_N)$$

and

$$\underline{P}^{(n)} = (\underline{P}_1, \underline{P}_2, ..., \underline{P}_n) \quad \text{and} \quad \underline{R}^{(n)} = (\underline{R}_1, \underline{R}_2, ..., \underline{R}_n) \tag{5.1}$$

i.e.

$$\underline{v}_i = \underline{p}_i / m_i$$
$$\underline{V}_i = \underline{P}_i / M_i$$

If we define,

$$\mathbf{\Gamma}_0 = (\underline{p}^{(N)}, \underline{r}^{(N)}) = (\underline{p}_1, \underline{p}_2, ..., \underline{p}_N, \underline{r}_1, \underline{r}_2, ..., \underline{r}_N) \tag{5.2}$$

$$\mathbf{\Gamma}_1 = (\underline{P}^{(n)}, \underline{R}^{(n)}) = (\underline{P}_1, \underline{P}_2, ..., \underline{P}_n, \underline{R}_1, \underline{R}_2, ..., \underline{R}_n) \tag{5.3}$$

then

$$\mathbf{\Gamma} = \mathbf{\Gamma}_0 + \mathbf{\Gamma}_1$$

We start by considering the case of a single Brownian, B particle in a bath of small particles.

5.1 A DILUTE SUSPENSION – ONE LARGE PARTICLE IN A BACKGROUND[2]

Consider a system consisting of N bath particles of mass m_k and one particle of mass M_B, the Brownian particle which has momentum $\underline{P}_B$ and position vector $\underline{R}_B$.

The Hamiltonian of the system is given by

$$H = H_0 + H_B \tag{5.4}$$

where

$$H_0 = \sum_{k=1}^{N} \underline{p}_k \cdot \underline{p}_k / 2m_k + U(\underline{r}^{(N)}) + V(\underline{r}^{(N)}, \underline{R}_B) \tag{5.5.a}$$

or if all the m_k's are the same,

$$= \underline{p}^{(N)} \cdot \underline{p}^{(N)} / 2\mathrm{m} + U(\underline{r}^{(N)}) + V(\underline{r}^{(N)}, \underline{R}_B) \tag{5.5b}$$

and

$$H_B = \underline{P}_B^2 / 2M_B \tag{5.6}$$

where $\underline{P}_B = M_B \underline{V}_B$.

The potential energies U and V are as follows:

1. $U(\underline{r}^{(N)})$ the short-ranged intermolecular potential energy of interaction of the bath particles, and
2. $V(\underline{r}^{(N)}, \underline{R}_B)$ is the interaction potential between the bath particles and the Brownian particle which will be assumed to be given by

$$V(\underline{r}^{(N)}, \underline{R}_B) = \sum_{i=1}^{N} v(| r_i - \underline{R}_B |) \tag{5.7}$$

Also H_0 is a function of $(\underline{r}^{(N)}, \underline{p}^{(N)}, \underline{R}_B)$ not of $(\underline{P}_B)$ so,

$$H_0 = H_0(\underline{r}^{(N)}, \underline{p}^{(N)}, \underline{R}_B)$$

The Louville operator can thus be written as

$$\mathrm{i}\mathbf{L} = \mathrm{i}\mathbf{L}_0 + \mathrm{i}\mathbf{L}_B \tag{5.8}$$

where

$$\begin{aligned} \mathrm{i}\mathbf{L}_0 &= \sum_{k=1}^{N} [\underline{p}_k / m_k \cdot \underline{\nabla}_{rk} + F' \cdot \underline{\nabla}_{pk}] \\ &= \sum_{k=1}^{N} [\underline{p}_k / m_k \cdot \underline{\nabla}_{rk} - \underline{\nabla}_{rk} (U + V) \cdot \underline{\nabla}_{pk}] \end{aligned} \tag{5.9}$$

$$F' = -\underline{\nabla}_{rk} [U(\underline{r}^{(N)}) + V(\underline{r}^{(N)}, \underline{R}_B)]$$

$$\mathrm{i}\mathbf{L}_B = [\underline{P}/M_B \cdot \underline{\nabla}_R + \underline{F}_B \cdot \underline{\nabla}_P]$$

$$= [\underline{P}/M_B \cdot \underline{\nabla}_R - \underline{\nabla}_R V \cdot \underline{\nabla}_P] \tag{5.10}$$

and

$$\underline{F}_B = -\underline{\nabla}_R V(\underline{r}^{(N)}, \underline{R}_B)$$

Note as

$$\underline{V}_B = \underline{P}_B / M_B$$
$$\mathrm{i}\mathbf{L}_0 \ \underline{V}_B = 0$$
$$(\mathrm{i}\mathbf{L}_0 H_0) = 0$$

and

$$\mathrm{i}\mathbf{L}_B \underline{V}_B = \underline{F}_B / M_B = 1/M_B\,(\mathrm{d}/\mathrm{d}t)\underline{V}_B \tag{5.11}$$

Now, noting that unlike Albers, Deutch and Oppenheim I have used a subscript 0 on the averaging for reasons given below,

$$\begin{aligned}\langle A(\boldsymbol{\Gamma})\rangle_0 &= \int f_0^{\mathrm{eq}}(\underline{p}^{(N)},\underline{r}^{(N)},\underline{R}_B)A(\underline{p}^{(N)},\underline{r}^{(N)},\underline{R}_B,\underline{P}_B)\mathrm{d}\underline{p}^{(N)}\,\mathrm{d}\underline{r}^{(N)}\\ &= \int f_0^{\mathrm{eq}}(\boldsymbol{\Gamma}_0,\underline{R}_B)A(\boldsymbol{\Gamma})\mathrm{d}\boldsymbol{\Gamma}_0\\ &= D(\underline{R}_B,\underline{P}_B)\end{aligned} \tag{5.12}$$

where

$$\begin{aligned}f_0^{\mathrm{eq}}(\underline{p}^{(N)},\underline{r}^{(N)},\underline{R}_B) &= f_0^{\mathrm{eq}}(\boldsymbol{\Gamma}_0,\underline{R}_B)\\ &= \mathrm{e}^{-H_0/kT}\Big/\Big(\int \mathrm{e}^{-H_0/kT}\,\mathrm{d}\boldsymbol{\Gamma}_0\Big)\end{aligned}$$

and introduce the projection operator defined by,

$$\begin{aligned}\mathbf{P}A(\boldsymbol{\Gamma}) &= \mathbf{P}A(\boldsymbol{\Gamma}_0\boldsymbol{\Gamma}_1)\\ &= \int f_0^{\mathrm{eq}}(\boldsymbol{\Gamma}_1)A(\boldsymbol{\Gamma})\mathrm{d}\boldsymbol{\Gamma}_0\\ &= B(\boldsymbol{\Gamma}_1)\\ &= \langle A(\boldsymbol{\Gamma}_0\boldsymbol{\Gamma}_1)\rangle_0\end{aligned} \tag{5.13}$$

This operator will average over the phase space relevant to the N bath particles only. However, there is a further interesting feature of this operator in that even though it averages over all the bath particles' phase space, the weighting function for this averaging is not the N-body equilibrium distribution function appropriate to the system of bath particles by themselves. It is the N-body distribution function for the bath particles in the field of the fixed Brownian particle. In the Mori method (Chapter 2) the projection operator and the distribution functions refer to all particles, i.e. the selected ("tracer" or "representative") particle and all the other particles in the system.

So the notation $\langle A\rangle_0$ is defined by

$$\langle A(\boldsymbol{\Gamma})\rangle_0 = \int f_0^{\mathrm{eq}}(\boldsymbol{\Gamma}_0,\underline{R}_B)A(\boldsymbol{\Gamma})\,\mathrm{d}\boldsymbol{\Gamma}_0$$

where

$$f_0^{\mathrm{eq}}(\boldsymbol{\Gamma}_0,\underline{R}_B) = \mathrm{e}^{-H_0/kT}\Big/\Big(\int \mathrm{e}^{-H_0/kT}\,\mathrm{d}\boldsymbol{\Gamma}_0\Big)$$

whereas in the Mori method we have

$$\langle A(\mathbf{\Gamma})\rangle = \int f^{\mathrm{eq}}(\mathbf{\Gamma})A(\mathbf{\Gamma})\,\mathrm{d}\mathbf{\Gamma}$$

where

$$f^{\mathrm{eq}}(\mathbf{\Gamma}) = \mathrm{e}^{-H/kT}\Big/\left(\int \mathrm{e}^{-H/kT}\,\mathrm{d}\mathbf{\Gamma}\right)$$

and H is the Hamiltonian for the total system.

You could imagine another definition, i.e.

$$\langle A(\mathbf{\Gamma})\rangle_0 = \int f_0^{\mathrm{eq}}(\mathbf{\Gamma}_0)A(\mathbf{\Gamma})\,\mathrm{d}\mathbf{\Gamma}_0$$

where $f_0^{\mathrm{eq}}(\Gamma) = \mathrm{e}^{-Hb/kT}/\left(\int \mathrm{e}^{-Hb/kT}\mathrm{d}\Gamma_0\right)$ and H_b is the Hamiltonian of the bath system only,

$$H_\mathrm{b} = \sum_{k=1}^{N} \underline{p}_k\cdot\underline{p}_k/2m_k + U(\underline{r}^{(N)})$$

However, in this case if A were a function of Γ_1 only then $\langle A(\Gamma)\rangle_0 = A$.

Returning to the projection operator of Albers *et al.*[2] for the particular case $A = \underline{V}_\mathrm{B}$ we have

$$\begin{aligned}\mathbf{P}\underline{V}_\mathrm{B} &= \mathbf{P}\underline{V}_\mathrm{B}(\mathbf{\Gamma}_1)\\ &= \int f_0^{\mathrm{eq}}\,\underline{V}_\mathrm{B}\,\mathrm{d}\mathbf{\Gamma}_0\\ &= \underline{V}_\mathrm{B}\int f_0^{\mathrm{eq}}\,\mathrm{d}\mathbf{\Gamma}_0\end{aligned} \tag{5.14}$$

as $\underline{V}_\mathrm{B}$ is a function of $\underline{P}_\mathrm{B}$ only which is not in phase space Γ_0.

Also we have

$$\begin{aligned}\mathbf{P}\mathrm{i}\mathbf{L}_\mathrm{B}\,\underline{V}_\mathrm{B} &= \int f_0^{\mathrm{eq}}\,\mathrm{i}\mathbf{L}_\mathrm{B}\,\underline{V}_\mathrm{B}\,\mathrm{d}\mathbf{\Gamma}_0\\ &= \int f_0^{\mathrm{eq}}\,(\underline{F}_\mathrm{B}/M_\mathrm{B})\,\mathrm{d}\mathbf{\Gamma}_0\\ &= \langle\underline{F}_\mathrm{B}\rangle_0/M_\mathrm{B}\\ &= 0 \quad \text{for an isotropic system}\end{aligned} \tag{5.15}$$

5.1.1 Exact equations of motion for *A*(*t*)

Starting with the operator identity eq. (1.36),

$$\mathbf{G}(t) = \mathrm{e}^{\mathrm{i}\mathbf{L}t} = \mathrm{e}^{\mathrm{i}(\mathbf{1}-\mathbf{P})Lt} + \int_0^t \mathrm{e}^{\mathrm{i}\mathbf{L}(t-\tau)}\,\mathrm{i}\mathbf{P}\mathbf{L}\,\mathrm{e}^{\mathrm{i}(\mathbf{1}-\mathbf{P})\mathbf{L}\tau}\,\mathrm{d}\tau \tag{5.16}$$

and noting that

$$\begin{aligned}\mathbf{PiL}_0(\cdots) &= 0\\ \mathbf{PiL_B}\,A(0) &= (\underline{P}_B/M_B)\cdot\underline{\nabla}_R\,A(0)\end{aligned} \tag{5.17}$$

the latter because $\langle \underline{F}\rangle_0 = 0$ for an isotropic fluid.

To obtain an equation of motion for any arbitrary function $A(t)$ of $\underline{R}_B$ and $\underline{P}_B$ we use the operator identity (eq. (5.16)) which leads upon operation on $\mathbf{(1–P)iL}\,A(0)$ to the following exact equation of motion of the GLE type (see Appendix H),

$$dA(t)/dt = e^{i\mathbf{L}t}(\underline{P}_B/M_B)\cdot\nabla_R\,A(0) + K^+(t) + \int_0^t e^{i\mathbf{L}(t-\tau)}\langle i\mathbf{L}_B\,K^+(\tau)\rangle_0\,d\tau \tag{5.18}$$

where the “random force” $K^+(t)$ is defined by

$$\begin{aligned}K^+(t) &= e^{i(\mathbf{1-P})\mathbf{L}t}(\mathbf{1-P})i\mathbf{L}\,A(0)\\ &= e^{i(\mathbf{1-P})\mathbf{L}t}\,\underline{F}_B(0)\cdot\nabla_R\,A(0)\end{aligned} \tag{5.19}$$

However, for an *isotropic fluid*,

$$\langle K^+(t)\rangle_0 = 0 \tag{5.20}$$

and thus,

$$\nabla_R\langle K^+(t)\rangle_0 = \langle\nabla_R\,K^+(t)\rangle_0 + (1/k_BT)\langle F\,K^+(t)\rangle_0 = 0 \tag{5.21}$$

and

$$\langle i\mathbf{L}_B K^+(t)\rangle_0 = [\underline{\nabla}_P - \underline{P}_B/(k_BTM_B)]\cdot\langle\underline{F}_B\,K^+(t)\rangle_0$$

Hence, eq. (5.18) becomes

$$\begin{aligned}dA(t)/dt = e^{i\mathbf{L}t}(\underline{P}_B/M_B)\cdot\nabla_R\,A(0) + \int_0^t e^{i\mathbf{L}(t-\tau)}[\nabla_P - \underline{P}/(kTM_B)]\\ \cdot\langle\underline{F}_B\,e^{i(\mathbf{1}-P)\mathbf{L}t}\,\underline{F}_B\rangle_0\cdot\nabla_P\,A(0)\,d\tau + K^+(t)\end{aligned} \tag{5.22}$$

which is another, exact equation of the GLE type.

It should be emphasised that both equations (5.18) and (5.22) are exact and can, in principle, be used to treat systems other than those for which the Brownian limit is appropriate.

Thus, as mentioned in the Introduction to this chapter the equations may be used for other purposes, e.g. to separate a system into system and surroundings or to treat two or more component systems which may not be approximated as being in "The Brownian Limit" (see Chapter 6). Furthermore, the projection operator may be generalised to deal with other problems.

5.1.2 Langevin equation for *A*(*t*)

To obtain an equation appropriate to the Brownian limit we once again follow Albers, Deutch and Oppenheim[2] and firstly write,

$$\begin{aligned} e^{i(1-P)Lt} &= e^{iL_0 t} + [e^{i(1-P)Lt} - e^{iL_0 t}] \\ e^{iL(t-\tau)} &= e^{iLt} + [e^{iL(t-\tau)} - e^{iLt}] \end{aligned}$$

Thus,

$$\begin{aligned} &\int_0^t e^{iL(t-\tau)}[\underline{\nabla}_P - \underline{V}_B/k_BT]\cdot\langle \underline{F}_B e^{i(1-P)Lt}\, \underline{F}_B(0)\rangle_0\cdot\nabla_P A(0)\,d\tau \\ &= e^{iLt}\int_0^t [\underline{\nabla}_P - \underline{V}_B/k_BT]\cdot\langle \underline{F}_B e^{iL_0 t}\, \underline{F}_B(0)\rangle_0\cdot\nabla_P A(0)d\tau \\ &\quad+\int_0^t e^{iL(t-\tau)}[\underline{\nabla}_P - \underline{V}_B/k_BT]\cdot[\langle \underline{F}_B e^{i(1-P)Lt}\, \underline{F}_B(0)\rangle_0\cdot\nabla_P A(0) - \langle \underline{F}_B e^{iL_0 t}\, \underline{F}_B(0)\rangle_0\cdot\nabla_P A(0)]d\tau \qquad (5.23) \\ &\quad+\int_0^t [e^{iL(t-\tau)} - e^{iLt}][\underline{\nabla}_P - \underline{V}_B/k_BT]\cdot\langle \underline{F}_B e^{iL_0 t}\, \underline{F}_B(0)\rangle_0\cdot\nabla_P A(0)\,d\tau \\ &= M_B I_0(t) + M_B I_1(t) + M_B I_2(t) \end{aligned}$$

Albers, Deutch and Oppenheim[2] show that $I_1(t)$ and $I_2(t)$ may be neglected relative to $I_0(t)$ by investigating the orders of magnitude of these integrals with respect to the parameter $\lambda = (m/M_B)^{1/2}$ which is assumed to be small. The basic assumption is as follows:

$$\langle A e^{iL_0 t}\cdot B\rangle_0 = \langle A\rangle_0 \langle B\rangle_0 \text{ for } t > \tau_b$$

where A and B are arbitrary dynamical variables which depend on the bath co-ordinates and/or momenta and τ_b is the largest relaxation time for the bath in the presence of the fixed Brownian particle. More precisely, the following three conditions must be met:

1. $O(P_B(t)) = \lambda^{-\alpha} O(p)$,
   where $0 < \alpha < 2$
2. $O[e^{iLt}\, p[\underline{\nabla}_P\, \underline{\nabla}_R{}^n\, \underline{\nabla}_P{}^{n'} A] = \lambda\, O(e^{iLt}\, \underline{\nabla}_R{}^n\, \underline{\nabla}_P{}^{n'} A)$
   where $\varepsilon > 0$, $n = 0,1,2,\ldots$, $n' = 0,1,2,\ldots$
3. $O[e^{iLt}\, R[\underline{\nabla}_R\, \underline{\nabla}_R{}^n\, \underline{\nabla}_P{}^{n'} A] = \lambda^{\delta}\, O(e^{iLt}\, \underline{\nabla}_R{}^n\, \underline{\nabla}_P{}^{n'} A)$
   where $\delta \geq 0$, $n = 0,1,2,\ldots$, $n' = 0,1,2,\ldots$.

The first condition means that the magnitude of $\underline{P}_{\mathrm{B}}$ must be large compared with the magnitude of $\underline{p}$ which is the momentum of a typical bath particle but the magnitude of $\underline{V}_{\mathrm{B}}$ must be small compared to that of a typical bath particle, $\underline{v}$.

The second condition means that the property $A(\underline{P}_{\mathrm{B}},\underline{R}_{\mathrm{B}})$ and its derivatives vary infinitesimally when $\underline{P}_{\mathrm{B}}$ changes to $\underline{P}_{\mathrm{B}} \pm \underline{p}$.

Finally, the third condition means that spatial variations of $A(\underline{P}_{\mathrm{B}},\underline{R}_{\mathrm{B}})$ and its derivatives are at most of the order of unity.

Now in the limit $\lambda \rightarrow 0$ then

$$I_0(t) = \mathrm{e}^{\mathrm{i}\mathbf{L}t} \int_0^t [\underline{\nabla}_{\mathrm{P}} - \underline{P}_{\mathrm{B}}/(M_{\mathrm{B}}k_{\mathrm{B}}T)] \cdot \langle \underline{F}_{\mathrm{B}} \mathrm{e}^{\mathrm{i}\mathbf{L}_0 t} \, \underline{F}_{\mathrm{B}}(0) \rangle_0 \cdot \nabla_{\mathrm{P}} \, A(0) \mathrm{d}\tau \tag{5.24}$$

but for an isotropic system,

$$\begin{aligned} \langle \underline{F}_{\mathrm{B}} \mathrm{e}^{\mathrm{i}\mathbf{L}_0 t} \, \underline{F}_{\mathrm{B}}(0) \rangle_0 &= 1/3 \langle \underline{F}_{\mathrm{B}} \cdot \mathrm{e}^{\mathrm{i}\mathbf{L}_0 t} \, \underline{F}_{\mathrm{B}}(0) \rangle_0 \, \Pi \\ &= 1/3 \langle \underline{F}_{\mathrm{B}} \cdot \underline{F}_{\mathrm{B}}^0(t) \rangle_0 \, \Pi \end{aligned} \tag{5.25}$$

where we define

$$\underline{F}_{\mathrm{B}}^{0}(t) = \mathrm{e}^{\mathrm{i}\mathbf{L}_0 t} \, \underline{F}_{\mathrm{B}}(0)$$

and

$$\Pi = \begin{bmatrix} 1 & 0 & 0 \\ 0 & 1 & 0 \\ 0 & 0 & 1 \end{bmatrix} \quad \text{is the unit tensor.}$$

So,

$$\begin{aligned} I_0(t) &= \mathrm{e}^{\mathrm{i}\mathbf{L}t} \int_0^t [\underline{\nabla}_{\mathrm{P}} - \underline{P}_{\mathrm{B}}/(M_{\mathrm{B}}k_{\mathrm{B}}T)] \cdot \langle \underline{F}_{\mathrm{B}} \mathrm{e}^{\mathrm{i}\mathbf{L}_0 t} \, \underline{F}_{\mathrm{B}}(0) \rangle_0 \cdot \nabla_{\mathrm{P}} \, A(0) \mathrm{d}\tau \\ &= \gamma(t) \mathrm{e}^{\mathrm{i}\mathbf{L}t} [\underline{\nabla}_{\mathrm{P}} - \underline{P}_{\mathrm{B}}/(M_{\mathrm{B}}k_{\mathrm{B}}T)] \cdot \nabla_{\mathrm{P}} \, A(0) \end{aligned} \tag{5.26}$$

where

$$\gamma(t) = 1/3 \int_0^t \langle \underline{F}_{\mathrm{B}} \cdot \underline{F}_{\mathrm{B}}^0(\tau) \rangle_0 \mathrm{d}\tau \tag{5.27}$$

Now,

$$\begin{aligned}\langle K^+(t)K^+(0)\rangle_0 &= \langle K^+(t)K(0)\rangle_0\\ &= \langle \mathrm{e}^{\mathrm{i}(\mathbf{1}-\mathbf{P})\mathbf{L}t}\, \underline{F}_\mathrm{B}(0)\cdot\nabla_\mathrm{P} A(0)\,\mathrm{e}^{\mathrm{i}(\mathbf{1}-\mathbf{P})\mathbf{L_0 x_0}}\, \underline{F}_\mathrm{B}(0)\rangle_0\cdot\nabla_\mathrm{P} A(0)\\ &= \langle \mathrm{e}^{\mathrm{i}(\mathbf{1}-\mathbf{P})\mathbf{L}t}\, \underline{F}_\mathrm{B}(0)\cdot\nabla_\mathrm{P} A(0)\,\underline{F}_\mathrm{B}(0)\rangle_0\cdot\nabla_\mathrm{P} A(0)\\ &\approx 1/3\langle \underline{F}_\mathrm{B}(t)\cdot\underline{F}_\mathrm{B}(0)\rangle_0\,\nabla_\mathrm{P} A(0)\cdot\nabla_\mathrm{P} A(0)\end{aligned} \tag{5.28}$$

where the operator identity below has been used,

$$\mathrm{e}^{\mathrm{i}(\mathbf{1}-\mathbf{P})\mathbf{L}t} = \mathrm{e}^{\mathrm{i}\mathbf{L_0}t} + \int_0^t \mathrm{e}^{\mathrm{i}\mathbf{L_0}(t-\tau)}\,\mathrm{i}(\mathbf{1}-\mathbf{P})\mathbf{L_B}\,\mathrm{e}^{\mathrm{i}(\mathbf{1}-\mathbf{P})\mathbf{L}\tau}\,\mathrm{d}\tau \tag{5.29}$$

then,

$$\begin{aligned}&\langle \mathrm{e}^{\mathrm{i}(\mathbf{1}-\mathbf{P})\mathbf{L}t}\, \underline{F}_\mathrm{B}(0)\,\underline{F}_\mathrm{B}(0)\cdot\nabla_\mathrm{P} A(0)\rangle_0\cdot\nabla_\mathrm{P} A(0)\\ &= \langle \mathrm{e}^{\mathrm{i}\mathbf{L_0}t}\, \underline{F}_\mathrm{B}(0)\cdot\nabla_\mathrm{P} A(0)\,\underline{F}_\mathrm{B}(0)\rangle_0\cdot\nabla_\mathrm{P} A(0)\\ &+\left(\left\langle\int_0^t \mathrm{e}^{\mathrm{i}\mathbf{L_0}(t-\tau)}\mathrm{i}(\mathbf{1}-\mathbf{P})\mathbf{L_B}\,\mathrm{e}^{\mathrm{i}(\mathbf{1}-\mathbf{P})\mathbf{L}\tau}\, \underline{F}_\mathrm{B}(0)\cdot\nabla_\mathrm{P} A(0)\underline{F}_\mathrm{B}(0)\right\rangle_0 \mathrm{d}\tau\right)\cdot\nabla_\mathrm{P} A(0)\\ &\approx 1/3\langle \underline{F}_\mathrm{B}(t)\cdot \underline{F}_\mathrm{B}(0)\rangle_0\,\nabla_\mathrm{P} A(0)\cdot\nabla_\mathrm{P} A(0)\end{aligned} \tag{5.30}$$

So the exact equation (5.22) becomes as $\lambda = (m/M_\mathrm{B})^{1/2} \to \infty$,

$$\mathrm{d}A(t)/\mathrm{d}t = \mathrm{e}^{\mathrm{i}\mathbf{L}t}\{\underline{P}_\mathrm{B}/(M_\mathrm{B}k_\mathrm{B}T)\}\cdot\nabla_\mathrm{R} A(0) + \gamma(t)\,\mathrm{e}^{\mathrm{i}\mathbf{L}t}\,[\nabla_\mathrm{P} - \underline{P}_\mathrm{B}/(kTM_\mathrm{B})]\cdot\nabla_\mathrm{P} A(0) + K^+(t) \tag{5.31}$$

where the friction coefficient is given by

$$\gamma(t) = 1/3\int_0^t \langle \underline{F}_\mathrm{B}\cdot\underline{F}^0_\mathrm{B}(\tau)\rangle_0\,\mathrm{d}\tau \tag{5.32}$$

and the random force by

$$\begin{aligned}\mathrm{K}^+(t) &= \mathrm{e}^{\mathrm{i}(\mathbf{1}-\mathbf{P})\mathbf{L}t}(\mathbf{1}-\mathbf{P})\mathrm{i}\mathbf{L}A(0)\\ &= \mathrm{e}^{\mathrm{i}(\mathbf{1}-\mathbf{P})\mathbf{L}t}\,\underline{F}_\mathrm{B}(0)\cdot\nabla_\mathrm{R} A(0)\end{aligned} \tag{5.33}$$

The properties which determine $K^+(t)$ are now

$$\langle K^+(t)\rangle = 0$$

and

$$\langle K^{+}(t)K^{+}(0)\rangle \approx 1/3\langle \underline{F}_{B}^{0}(t)\cdot \underline{F}_{B}\rangle_{0}\, \nabla_{P} A(0)\cdot \nabla_{P} A(0)$$

So in the limit of long times we may write the Langevin equation (5.31) with

$$\gamma = \gamma(t) = \gamma(\infty) = 1/3\int_{0}^{\infty} \langle \underline{F}_{B}\bullet \underline{F}_{B}^{0}(\tau)\rangle_{0}\, d\tau \tag{5.34}$$

and $K^{+}(t)$ will be described by a Gaussian random variable with mean and mean-square value given by

$$\langle K^{+}(t)\rangle = 0 \tag{5.35}$$

and

$$\langle K^{+}(t)K^{+}(0)\rangle \approx 2\gamma A(0)\cdot \nabla_{P} A(0)\,\delta(t)$$

5.1.3 Langevin equation for velocity

If we choose as our dynamical variable the velocity of the Brownian particle, i.e. $A(t) = \underline{V}_{B}(t) = \underline{P}_{B}(t)/M_{B}$, then we will obtain the original Langevin equation and as this is such an important case we will summarise the main points here.

Firstly,

$$\begin{aligned}
&\mathrm{i}\mathbf{L}_{0}\, \underline{V}_{B} = 0\\
&\mathrm{i}\mathbf{L}_{B}\, \underline{V}_{B} = \underline{F}_{B}/M_{B} = (1/M_{B})(d/dt)\underline{V}_{B}\\
&\mathrm{i}\mathbf{P}\mathbf{L}_{0}\, \underline{V}_{B} = 0\\
&\mathrm{i}\mathbf{P}\mathbf{L}_{B}\, \underline{V}_{B} = 0\\
&\mathrm{i}\mathbf{P}(\mathbf{L}_{0} + \mathrm{L}_{B})\underline{V}_{B} = 0
\end{aligned}$$

$$\begin{aligned}
K^{+}(t) &= \mathbf{e}^{\mathrm{i}(\mathbf{1}-\mathbf{P})\mathbf{L}t}[(\mathbf{1}-\mathbf{P})\mathrm{i}\mathbf{L}\, \underline{V}_{B}(0)] = \mathrm{e}^{\mathrm{i}(\mathbf{1}-\mathbf{P})\mathbf{L}t}\, \underline{F}_{B}(0)/M_{B}\\
\langle \underline{F}_{B}\, K^{+}(t)\rangle_{0} &= \langle \underline{F}_{B}\{\mathrm{e}^{\mathrm{i}(\mathbf{1}-\mathbf{P})\mathbf{L}t}[(\mathbf{1}-\mathbf{P})\mathrm{i}\mathbf{L}\, \underline{V}_{B}(0)]\}\rangle_{0}\\
&= \langle \underline{F}_{B}\, \mathrm{e}^{\mathrm{i}(\mathbf{1}-\mathbf{P})\mathbf{L}t}\, \underline{F}_{B}(0)\rangle_{0}/M_{B}
\end{aligned} \tag{5.36}$$

So the exact equation of motion in this case is

$$d\underline{V}_{B}(t)/dt = K^{+}(t) + \int_{0}^{t} \mathrm{e}^{\mathrm{i}\mathbf{L}(t-\tau)}\langle \mathrm{i}\mathbf{L}_{B}\, K^{+}(\tau)\rangle_{0}\, d\tau \tag{5.37}$$

or

$$= K^{+}(t) + \frac{1}{M_B}\int_0^t e^{i\mathbf{L}(t-\tau)}[\underline{\nabla}_P - \underline{V}_B/kT]\cdot\langle \underline{F}_B e^{i(\mathbf{1}-\mathbf{P})\mathbf{L}t}\, \underline{F}_B(0)\rangle_0 \, d\tau$$

where $K^{+}(t)$ is given by eq. (5.36) above.

The conditions which must be met for the integral I_0 to be the only significant one is now that $\langle A\, e^{i\mathbf{L}_0 t}B\rangle_0 = \langle A\rangle_0\langle B\rangle_0$ for $t > \tau_b$ where τ_b is the longest relaxation time for the bath in the presence of the n Brownian particles fixed in position, $|\underline{P}(t)| \gg |p|$ and $|\underline{V}_B| \ll |\underline{v}|$.

So in this case,

$$\begin{aligned} I_0(t) &= (1/M_B)e^{i\mathbf{L}t}\int_0^t [\underline{\nabla}_P - \underline{V}_B/k_BT]\cdot\langle \underline{F}_B e^{i\mathbf{L}_0 t}\, \underline{F}_B(0)\rangle_0 \, d\tau \\ &= -\gamma(t)e^{i\mathbf{L}t}\, \underline{V}_B(0)/(k_BTM_B) \\ &= -\gamma(t)\underline{V}_B(t)/(k_BTM_B) \end{aligned}$$

where $\gamma(t) = 1/3\int_0^t \langle \underline{F}_B\cdot \underline{F}_B^0(\tau)\rangle_0 d\tau$
and

$$\begin{aligned} &\langle K^{+}(t)K^{+}(0)\rangle_0 \\ &= \langle K^{+}(t)K\rangle_0 \\ &= \langle e^{i(\mathbf{1}-\mathbf{P})\mathbf{L}t}\, \underline{F}_B(0)/M_B\, e^{i(\mathbf{1}-\mathbf{P})\mathbf{L}_0}\, \underline{F}_B(0)/M_B\rangle_0 \\ &= \langle e^{i(\mathbf{1}-\mathbf{P})\mathbf{L}t}\, \underline{F}_B(0)\,\underline{F}_B(0)\rangle_0/M_B^2 \end{aligned}$$

Using the operator identity (eq. (5.29)) then,

$$\begin{aligned} &\langle e^{i(\mathbf{1}-\mathbf{P})\mathbf{L}t}\, \underline{F}_B(0)\,\underline{F}_B(0)\rangle_0/M_B^2 \\ &= \langle e^{i\mathbf{L}_0 t}\, \underline{F}_B(0)\,\underline{F}_B(0)\rangle_0/M_B^2 + \left(\left\langle\int_0^t e^{i\mathbf{L}_0(t-\tau)}\, i(\mathbf{1}-\mathbf{P})\mathbf{L}_B\, e^{i(\mathbf{1}-\mathbf{P})\mathbf{L}\tau}\, \underline{F}_B(0)\,\underline{F}_B(0)\right\rangle_0 d\tau\right)\Big/M_B^2 \\ &\approx 1/3\langle \underline{F}_B(t)\cdot\underline{F}_B(0)\rangle_0/M_B^2 \end{aligned}$$

So,

$$\langle K^{+}(t)K^{+}(0)\rangle_0 \approx 1/3\langle \underline{F}_B^0(t)\cdot \underline{F}_B\rangle_0/M_B^2$$

The exact equation (3.37) becomes as $\lambda = (m/M_B)^{1/2} \to \infty$,

$$d\underline{V}_B(t)/dt = -\gamma(t)/(k_BTM_B)\underline{V}_B(t) + K^{+}(t) \tag{5.38}$$

where

$$\gamma(t) = 1/3\int_0^t \langle \underline{F}_B \cdot \underline{F}_B(\tau)\rangle_0 \, d\tau$$

and in the limit of long times we may write

$$d\underline{V}_B(t)/dt = -\gamma/(k_B T M_B)\underline{V}_B(t) + K^+(t) \tag{5.39}$$

where

$$\gamma = \gamma(t) = \gamma(\infty) = 1/3\int_0^\infty \langle \underline{F}_B \cdot \underline{F}_B(\tau)\rangle_0 \, d\tau$$

$$\langle K^+(t)\rangle = 0$$

and

$$\langle K^+(t)K^+(0)\rangle \approx 2\gamma\delta(t)$$

where in arriving at eq. (5.39) we have assumed that $\lambda = (m/M_B)^{1/2} \rightarrow \infty$ and $t \rightarrow \infty$.

This is the same equation as first arrived at by Langevin using heuristic arguments and derived in Section 2.5 from the Mori–Zwanzig approach. This equation and generalisations of it have been extensively used to describe the dynamics of many systems, e.g. as the basis of the so-called Brownian dynamics (BD) numerical scheme (see Chapter 7).

From eq. (5.39) we may take the dot product with $\underline{V}_B(0)$, ensemble average and use the fact that $K^+(t)$ and $\underline{V}_B(t)$ are uncorrelated in the Brownian limit (cf. the derivation of the GLE for $\langle \underline{V}_B(0)\cdot\underline{V}_B(t)\rangle$ in Section 2.4) and we obtain

$$d\langle \underline{V}_B(0)\cdot\underline{V}_B(t)\rangle/dt = -\gamma/(k_B T M_B)\langle \underline{V}_B(0)\cdot\underline{V}_B(t)\rangle \tag{5.40}$$

The solution of eq. (5.40) is obviously,

$$\langle \underline{V}_B(0)\cdot\underline{V}_B(t)\rangle = \langle \underline{V}_B(0)\cdot\underline{V}_B(0)\rangle e^{-\gamma/(k_B T M_B)t} \tag{5.41}$$

as was obtained from the Mori continued-fraction scheme.

Note that we still have to find a method for calculating the parameter γ which requires some other assumptions to be made. Often it is assumed that it may be obtained by use of the Stokes–Einstein relationship, i.e.

$$D = k_B T/\zeta \tag{5.42}$$

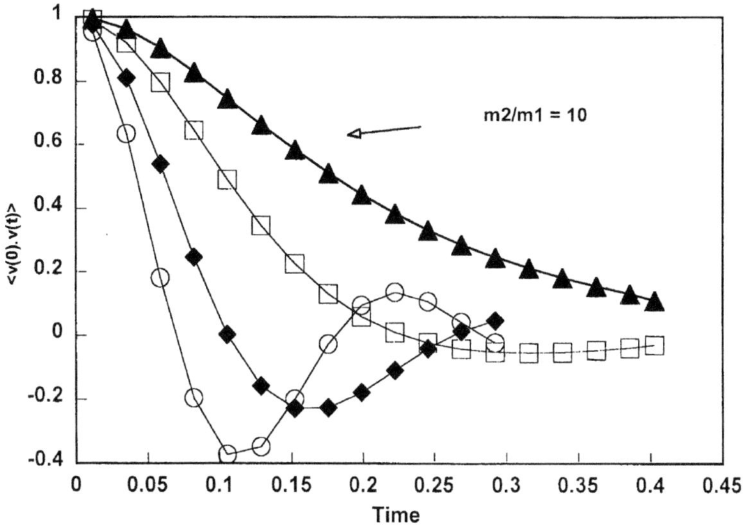

Figure 5.1 Normalized velocity autocorrelation function, $\langle v(0) \cdot v(t)\rangle / \langle v(0) \cdot v(0)\rangle$ versus delay time for one large particle in a bath of smaller particles calculated by the MD method (see text for further explanation).

It should be noted that the Brownian limit arises when the conditions discussed above are valid which essentially means that all dynamical processes have ceased in the bath system in the presence of the Brownian particle, i.e. the bath system is at equilibrium before the Brownian particle has changed its dynamical state significantly. This does not, however, imply that the dynamical processes involving the velocity of the Brownian particle have become uncorrelated and in particular it is not assumed that $\langle \underline{V}_B(0) \cdot \underline{V}_B(t)\rangle = 0$ before the Brownian particle has significantly changed its position. If on the contrary this were also assumed then we would have also made the assumption that we were working in the Diffusive (or 'Overdamped') limit. This latter case will be discussed further in Chapters 6–9.

In order to illustrate the approach to the Brownian limit Figure 5.1 shows the normalised velocity autocorrelation function $\langle v(0) \cdot v(t)\rangle / \langle v(0) \cdot v(0)\rangle$ versus delay time for one large particle in a bath of smaller particles interacting *via* a CWA potential with a core calculated by the MD method. The large particles core radius was three times that of the bath particles and the results are for mass ratios m_2/m_1 of 0.5 (○), 1.0 (◆), 2.0 (□) and 10 (▲). As can be seen from Figure 5.1 as the mass ratio of heavy to light particles increases, the velocity autocorrelation function losses the negative backscatter region at intermediate times and it decays much more slowly in time.

5.2 MANY-BODY LANGEVIN EQUATION

In order to generalise the results of the preceding section to many (n) heavy particles of mass M in a bath of N light particles of mass m we will have to firstly modify some of the basic definitions used in Section 5.1.

To re-iterate we represent the collection of positions and momenta of the bath particles by

$$\Gamma_0 = (\underline{r}^{(N)}, \underline{p}^{(N)})$$

where

$$(\underline{r}^{(N)}) = (\underline{r}_1, \underline{r}_2, \cdots, \underline{r}_N)$$

$$(\underline{p}^{(N)}) = (\underline{p}_1, \underline{p}_2, \cdots, \underline{p}_N)$$

and of the Brownian particles by,

$$\Gamma_1 = (\underline{R}^{(n)}, \underline{P}^{(n)})$$

where

$$(\underline{R}^{(n)}) = (\underline{R}_1, \underline{R}_2, \cdots, \underline{R}_n)$$

$$(\underline{P}^{(n)}) = (\underline{P}_1, \underline{P}_2, \cdots, \underline{P}_N)$$

The Hamiltonian of the system is now given by

$$H = H_0 + H_B \tag{5.43}$$

where H_0 is the Hamiltonian for the N bath particles in the potential field of the n Brownian particles held fixed at $(\underline{R}^{(n)})$,

$$H_0 = \sum_{k=1}^{N} \underline{p}_k \cdot \underline{p}_k / 2m_k + U(\underline{r}^{(N)}) + \sum_{\mu=1}^{n} \Phi_\mu(\underline{r}^{(N)}, \underline{R}_\mu) \tag{5.44a}$$

or if all the m_k's are the same

$$= \underline{p}^{(N)} \cdot \underline{p}^{(N)} / 2m + U(\underline{r}^{(N)}) + \sum \Phi_\mu(\underline{r}^{(N)}, \underline{R}_B) \tag{5.44b}$$

and H_B is the Hamiltonian for the n Brownian particles in the absence of the bath particles,

$$H_B = \sum \underline{P}_\mu \cdot \underline{P}_\mu / 2M_B + V(\underline{R}^{(n)}) \tag{5.45}$$

where $\underline{P}_\mu = M_\mu \underline{V}_\mu$.

The potential energies U and V are as follows:

1. $U(\underline{r}^{(N)})$ the short-ranged intermolecular potential energy of interaction of the bath particles, and
2. $V(\underline{R}_B)$ is the interaction potential between the Brownian particles which will be assumed to be given by

$$V(\underline{R}^{(n)}) = \sum_{\mu,\nu=1}^{n} v(|\,\underline{R}_\mu - \underline{R}_\nu\,|) \tag{5.46}$$

and

3. $\Phi_\mu(\underline{r}^{(N)},\underline{R}_B)$ is the short-range potential energy of interaction between the Brownian particle μ and all the N bath particles which we will assume to be the sum of pair interactions,

$$\Phi_\mu(\underline{r}^{(N)},\underline{R}_B) = \sum_{i=1}^{N} \phi(|\,\underline{R}_\mu - \underline{r}_l\,|)$$

Once again H_0 is a function of $(\underline{r}^{(N)}, \underline{p}^{(N)}, \underline{R}^{(n)})$ not of $(\underline{P}_B{}^{(n)})$, i.e. $H_0 = H_0(\underline{r}^{(N)}, \underline{p}^{(N)}, \underline{R}^{(n)})$. The Louville operator can thus be written as

$$\mathbf{iL} = \mathbf{iL}_0 + \mathbf{iL}_B \tag{5.47}$$

where

$$\begin{aligned} \mathbf{iL}_0 &= \sum_{k=1}^{N} [\underline{p}_k/m_k \cdot \underline{\nabla}_{rk} + \underline{F}' \cdot \underline{\nabla}_{pk}] \\ &= \sum_{k=1}^{N} \left[\underline{p}_k/m_k \cdot \underline{\nabla}_{rk} - \underline{\nabla}_{rk}\left(U + \sum \Phi_\mu\right) \cdot \underline{\nabla}_{pk}\right] \\ \underline{F}' &= -\underline{\nabla}_{rk}\,(U(\underline{r}^{(N)}) + \Sigma\Phi_\mu) \end{aligned} \tag{5.48}$$

and

$$\mathbf{iL}_B = \sum [P_\mu/M_\mu \cdot \nabla_{R\mu} + F_\mu \cdot \nabla_{P\mu}]$$

and if all the Brownian particles are of the same mass,

$$\begin{aligned} &= [\underline{P}/M_B \cdot \underline{\nabla}_R + \underline{F}_\mu \cdot \underline{\nabla}_P] \\ &= [\underline{P}/M_B \cdot \underline{\nabla}_R - \underline{\nabla}_R V \cdot \underline{\nabla}_P] \end{aligned} \tag{5.49}$$

and $\underline{F}_\mu = -\underline{\nabla}_{R\mu}[V(\underline{r}^{(N)}, \underline{R}^{(n)}) + \Phi_\mu]$

Now, once again unlike Deutch and Oppenheim[3] I have used a subscript 0 on the averaging for reasons given in Section 5.1,

$$\begin{aligned}\langle A(\boldsymbol{\Gamma})\rangle_0 &= \int f_0^{\mathrm{eq}}(\underline{p}^{(N)},\underline{r}^{(N)},\underline{R}^{(n)})A(\underline{p}^{(N)},\underline{r}^{(N)},\underline{R}^{(n)},\underline{P}^{(n)})\mathrm{d}\underline{p}^{(N)}\,\mathrm{d}\underline{r}^{(N)} \\ &= \int f_0^{\mathrm{eq}}(\boldsymbol{\Gamma}_0,\underline{R}^{(n)})A(\boldsymbol{\Gamma})\,\mathrm{d}\boldsymbol{\Gamma}_0 \\ &= B(\underline{R}^{(n)},\underline{P}^{(n)})\end{aligned} \tag{5.50}$$

where

$$\begin{aligned}&f_0^{\mathrm{eq}}(\underline{p}^{(N)},\underline{r}^{(N)},\underline{R}^{(n)}) \\ &= f_0^{\mathrm{eq}}(\boldsymbol{\Gamma}_0,\underline{R}^{(n)}) \\ &= \mathrm{e}^{-H0/kT}\Big/\left(\int \mathrm{e}^{-H0/kT}\,\mathrm{d}\boldsymbol{\Gamma}_0\right)\end{aligned}$$

Introducing the projection operator defined by

$$\begin{aligned}\mathbf{P}A(\boldsymbol{\Gamma}) &= \mathbf{P}\,A(\boldsymbol{\Gamma}_0\boldsymbol{\Gamma}_1) \\ &= \int f_0^{\mathrm{eq}}(\boldsymbol{\Gamma}_1)A(\boldsymbol{\Gamma})\,\mathrm{d}\boldsymbol{\Gamma}_0 \\ &= B(\boldsymbol{\Gamma}_1) \\ &= \langle A(\boldsymbol{\Gamma}_0\boldsymbol{\Gamma}_1)\rangle_0\end{aligned} \tag{5.51}$$

this operator will average over the phase space relevant to the N bath particles for fixed positions of the n B particles only, and the discussion in Section 5.1 on this operator is again relevant in the n-body case.

Returning to the projection operator of Deutch and Oppenheim for the particular case $A = \underline{V}_\mu$ we have

$$\begin{aligned}\mathbf{P}\underline{V}_\mu &= \mathbf{P}\,\underline{V}_\mu(\boldsymbol{\Gamma}_1) \\ &= \int f_0^{\mathrm{eq}}\,\underline{V}_\mu\,\mathrm{d}\boldsymbol{\Gamma}_0 \\ &= \underline{V}_\mu \int f_0^{\mathrm{eq}}\,\mathrm{d}\boldsymbol{\Gamma}_0\end{aligned} \tag{5.52}$$

as $\underline{V}_\mu$ is a function of $\underline{P}_\mu$ only which is not in phase space Γ_0.

Also we have

$$\begin{aligned}\mathbf{P}\mathrm{i}\mathbf{L}_{\mathrm{B}}\,\underline{V}_\mu &= \int f_0^{\mathrm{eq}}\,\mathrm{i}\mathbf{L}_{\mathrm{B}}\underline{V}_\mu\,\mathrm{d}\boldsymbol{\Gamma}_0 \\ &= \int f_0^{\mathrm{eq}}(\underline{F}_\mu/M_\mu)\,\mathrm{d}\boldsymbol{\Gamma}_0 \\ &= \langle \underline{F}_\mu\rangle_0/M_\mu \\ &= 0 \text{ for an isotropic system}\end{aligned} \tag{5.53}$$

As in this case we have n-Brownian particles in a bath of N light particles it is necessary to introduce the potential of mean force χ defined below which represents the bath or 'solvent' mediated static interaction between the Brownian particles,

$$Z(\underline{R}^{(n)}) = C(\beta, N, V)\exp[-\beta\chi(\underline{R}^{(n)})] \tag{5.54}$$

so that using

$$\begin{aligned}\langle \underline{\nabla}_{R\mu}\Phi_{\mu}\rangle_0 &= (-1/k_B T)^{-1}\,\underline{\nabla}_{R\mu}\ln Z(\underline{R}^{(n)})\\ &= \underline{\nabla}_{R\mu}\chi(\underline{R}^{(n)})\end{aligned} \tag{5.55}$$

from which we may define the bath averaged force on Brownian particle μ by

$$\langle \underline{F}_{\mu}\rangle_0 = -\langle \underline{\nabla}_{R\mu}[V(\underline{r}^{(N)}, \underline{R}^{(n)}) + \Phi_{\mu}]\rangle_0 = -\underline{\nabla}_{R\mu}[V + \chi(\underline{R}^{(n)})] \tag{5.56}$$

5.2.1 Exact equations of motion for $A(t)$

Starting with the operator identity,

$$\mathbf{G}(t) = \mathrm{e}^{\mathrm{i}\mathbf{L}t} = \mathrm{e}^{\mathrm{i}(\mathbf{1}-\mathbf{P})Lt} + \int_0^t \mathrm{e}^{\mathrm{i}\mathbf{L}(t-\tau)}\,\mathrm{i}\mathbf{PL}\,\mathrm{e}^{\mathrm{i}(\mathbf{1}-\mathbf{P})\mathbf{L}\tau}\,\mathrm{d}\tau \tag{5.57}$$

and let this operate on the variable $A(t)$ where $A(t)$ an arbitrary function of the phase space variables for the large particles and $K(t) = (\mathbf{1} - \mathbf{P})\mathrm{i}\mathbf{L}\,A(0)$ which gives an exact GLE-type equation of motion for $A(t)$,

$$\mathrm{d}A(t)/\mathrm{d}t = B(t) + \mathrm{K}^{+}(t) + \int_0^t \mathrm{e}^{\mathrm{i}\mathbf{L}(t-\tau)}\langle \mathrm{i}\mathbf{L}_{\mathbf{B}}\,K^{+}(\tau)\rangle_0\,\mathrm{d}\tau \tag{5.58}$$

where

$$\mathrm{d}A(t)/\mathrm{d}t = \mathrm{e}^{\mathrm{i}\mathbf{L}t}\,\mathrm{i}\mathbf{L}\,A(0) \tag{5.59}$$

$$B(t) = \mathrm{e}^{\mathrm{i}\mathbf{L}t}\langle \mathrm{i}\mathbf{L}_{\mathbf{B}}\rangle A(0) \tag{5.60}$$

and

$$K^{+}(t) = \mathrm{e}^{\mathrm{i}(\mathbf{1}-\mathbf{P})Lt}[\mathrm{i}\mathbf{L}\,A(0) - \langle \mathrm{i}\mathbf{L}_{\mathbf{B}}\rangle_0\,A(0)] \tag{5.61}$$

These equations are obtained by using

$$\mathbf{P}\mathrm{i}\mathbf{L}_0(\cdots) = \langle \mathrm{i}\mathbf{L}_0(\cdots)\rangle_0 = 0$$

It may be noted that the solvent averaged force is introduced by the definition of $B(t)$ and,

$$D(t) = \mathrm{e}^{\mathrm{i}\mathbf{L}t} \sum_{\mu=1}^{n} [(\underline{P}_{\mu}/M_{\mu})\cdot\underline{\nabla}_{\underline{R}\mu} + \langle \underline{F}_{\mu}\rangle_0 \cdot \underline{\nabla}_{\underline{P}\mu}] A(0)$$

Now,

$$\langle \mathrm{i}\mathbf{L}_{\mathbf{B}}\, K^{+}(\tau)\rangle_0 = \sum_{\mu=1}^{n} [(\underline{P}_{\mu}/M_{\mu})\cdot\langle \underline{\nabla}_{\underline{R}\mu}\, K^{+}(\tau)\rangle_0 + \underline{\nabla}_{\underline{P}\mu}\cdot\langle \underline{F}_{\mu} K^{+}(\tau)\rangle_0] \tag{5.62}$$

and using,

$$\langle K^{+}(\tau)\rangle_0 = \langle K^{+}(0)\rangle_0 = 0$$

we have,

$$\langle \underline{\nabla}_{\underline{R}\mu}\, K^{+}(\tau)\rangle_0 = -(1/k_{\mathrm{B}}T)\langle \underline{E}_{\mu}\, K^{+}(\tau)\rangle_0 \tag{5.63}$$

where $\underline{E}_{\mu} = \underline{F}_{\mu} - \langle \underline{E}_{\mu}\rangle_0$.

Thus,

$$\langle \mathrm{i}\mathbf{L}_{\mathbf{B}}\, K^{+}(\tau)\rangle_0 = \sum_{\mu=1}^{n} [\underline{\nabla}_{\underline{P}\mu} - (1/k_{\mathrm{B}}T\, M_{\mu})\underline{P}_{\mu}]\cdot\langle\, \underline{E}_{\mu}\, K^{+}(\tau)\rangle_0] \tag{5.64}$$

Now $K^{+}(\tau)$ may be re-written by the use of

$$(\mathbf{1}-\mathbf{P})\mathrm{i}\mathbf{L}\, A(0) = (\mathbf{1}-\mathbf{P})\mathrm{i}\mathbf{L}_{\mathbf{B}}\, A(0) = \sum \underline{E}_{\nu}\cdot\underline{\nabla}_{\underline{P}\nu}\cdot A(0)$$

as,

$$K^{+}(\tau) = \mathrm{e}^{\mathrm{i}(\mathbf{1}-\mathbf{P})\mathbf{L}t} \sum_{\nu=1} \underline{E}_{\nu}(0)\cdot\underline{\nabla}_{\underline{P}\nu}\, A(0) \tag{5.65}$$

Thus, eq. (5.58) may now be written as

$$\begin{aligned} \mathrm{d}A(t)/\mathrm{d}t &= \sum_{\mu=1}^{n} \{(\underline{P}_{\mu}/M_{\mu})\cdot\underline{\nabla}_{\underline{R}\mu(t)} + \underline{S}_{\mu}(t)\cdot\underline{\nabla}_{\underline{P}\nu(t)}\}A(t) + K^{+}(t) \\ &+ \sum_{\mu=1}^{n}\sum_{\nu=1}^{n}\int_0^t \mathrm{e}^{\mathrm{i}\mathbf{L}(t-\tau)}\cdot\{[\underline{\nabla}_{\underline{P}\mu} - (1/(k_{\mathrm{B}}TM_{\mu})\underline{P}_{\mu}]\cdot\langle \underline{E}_{\mu}\, \mathrm{e}^{\mathrm{i}(\mathbf{1}-\mathbf{P})\mathbf{L}\tau}\underline{E}_{\nu}\rangle_0\cdot\underline{\nabla}_{\underline{P}\nu}\, A(0)\}\,\mathrm{d}\tau \end{aligned} \tag{5.66}$$

where

$$\underline{S}_\mu(t) = \mathrm{e}^{\mathrm{i}\mathbf{L}t} \langle \underline{F}_\mu \rangle_0 \tag{5.67}$$

Eq. (5.66) is another exact GLE-type equation of motion for $A(t)$ and may be re-written using the same techniques as in Section 5.1.2.

5.2.2 Many-body Langevin equation for $A(t)$

Thus, as in Section 5.1.2 we may write,

$$\mathrm{d}A(t)/\mathrm{d}t = B(t) + K^+(t) + \sum_{\mu=1}^{n}\sum_{\nu=1}^{n} \mathrm{e}^{\mathrm{i}\mathbf{L}t}[\underline{\nabla}_{P\mu} - (1/k_\mathrm{B}T\,M_\mu)\underline{\mathrm{P}}_\mu] \cdot \int \langle \underline{E}_\mu\, \underline{E}_\nu^0(\tau)\rangle_0\, \underline{\nabla}_{P\nu}\, A(0)\,\mathrm{d}\tau + I_1 + I_2 \tag{5.68}$$

where

$$\underline{E}_\nu^0(\tau) = \mathrm{e}^{\mathrm{i}\mathbf{L}\tau}\, \underline{E}_\nu \tag{5.69}$$

$$I_1 = \sum_{\mu=1}^{n}\sum_{\nu=1}^{n}\int_0^t \mathrm{e}^{\mathrm{i}\mathbf{L}(t-\tau)}[\underline{\nabla}_{P\mu} - (1/k_\mathrm{B}T\,M_\mu)\underline{P}_\mu]\cdot\{\langle \underline{E}_\mu\, \mathrm{e}^{\mathrm{i}(\mathbf{1}-\mathbf{P})\mathbf{L}t}\underline{E}_\nu\rangle_0 - \langle \underline{E}_\mu \underline{E}_\nu^0(\tau)\rangle_0 \cdot \underline{\nabla}_{P\nu} A(0)\}\,\mathrm{d}\tau$$

and

$$I_2 = \sum_{\mu=1}^{n}\sum_{\nu=1}^{n}\int_0^t (\mathrm{e}^{\mathrm{i}\mathbf{L}(t-\tau)} - \mathrm{e}^{\mathrm{i}\mathbf{L}t})[\underline{\nabla}_{P\mu} - (1/k_\mathrm{B}T\,M_\mu)\underline{P}_\mu]\cdot\langle \underline{E}_\mu \underline{E}_\nu^0(\tau)\rangle_0 \cdot \underline{\nabla}_{P\nu} A(0)\}\,\mathrm{d}\tau$$

As in the case of a single Brownian particle we assume that $\langle A\ \mathrm{e}^{\mathrm{i}\mathbf{L}_0 t}\mathrm{B}\rangle_0 = \langle A\rangle_0 \langle B\rangle_0$ for $t > \tau_\mathrm{b}$, where τ_b is the longest relaxation time for the bath in the presence of the n Brownian particles fixed in position and this time will be roughly $|R_{ij}|/c$ where c is the speed of sound in the bath. If also conditions 1–3 in Section 5.1.2 are met, then to order $\lambda = (m/M)^{1/2} \ll 1$ I_1 and I_2 may be neglected so eq. (5.68) may be written as

$$A(t)/\mathrm{d}t = B(t) + \mathrm{K}^+(t) + \sum_{\mu=1}^{n}\sum_{\nu=1}^{\mathrm{n}} \mathrm{e}^{\mathrm{i}\mathbf{L}t}[\underline{\nabla}_{P\mu} - (1/k_\mathrm{B}T\,M_\mu)\underline{P}_\mu]\cdot\gamma_{\mu\nu}(t)\cdot\underline{\nabla}_{P\nu}\, A(0)\,\mathrm{d}\tau \tag{5.68a}$$

where

$$\gamma_{\mu\nu}(t, \underline{R}^{(n)}(t)) = \int_0^t \langle \underline{E}_\mu \underline{E}_\nu^0(\tau)\rangle_0\, \mathrm{d}\tau \tag{5.68b}$$

This is the many-body Langevin equation for the evolution of any phase variable A which is a function of $\Gamma_1 = (\underline{R}^{(n)}, \underline{P}^{(n)})$.

It should be remarked that general forms of the Langevin equation for any function of the dynamical variables characterising the state of the large particles, i.e. eqs. (5.58) and (5.66) are exact. Thus, we could use one of these equations which would not involve making the assumptions and taking the limits involved in arriving at eq. (5.68). Even the assumption that the bath particles are at equilibrium could be eliminated by use of the Kawasaki formalism as mentioned in Chapter 4 and refining the projection operator (eq. (5.51).

5.2.3 Many-body Langevin equation for velocity

If we choose A as the velocity of a large particle $\underline{V}_v$ then we have an exact equation for the time evolution of $\underline{V}_v$,

$$M_v\,\mathrm{d}\underline{V}_v/\mathrm{d}t = B(t) + K^+(t) + \int_0^t e^{\mathrm{i}\mathbf{L}(t-\tau)}\langle \mathrm{i}\mathbf{L}_\mathbf{B}\,K^+(\tau)\rangle_0\,\mathrm{d}\tau \tag{5.70}$$

where

$$\begin{aligned} M_v\,\mathrm{d}\underline{V}_v/\mathrm{d}t &= e^{\mathrm{i}\mathbf{L}t}\,\mathrm{i}\mathbf{L}\,\underline{V}_v(0) \\ B(t) &= e^{\mathrm{i}\mathbf{L}t}\langle \mathrm{i}\mathbf{L}_\mathbf{B}\rangle M_v\,\underline{V}_v(0) \end{aligned} \tag{5.71}$$

and

$$\begin{aligned} K^+(t) &= e^{\mathrm{i}\mathbf{L}(t-\tau)}[\mathrm{i}\mathbf{L}\,M_v\underline{V}_v(0) - \langle \mathrm{i}\mathbf{L}_\mathbf{B}\rangle_0\,M_v\underline{V}_v(0)] \\ &= e^{\mathrm{i}(1-\mathbf{P})\mathbf{L}t}\,\underline{E}_v(0)\cdot\underline{\nabla}_{Pv}\cdot M_v\underline{V}_v(0) \end{aligned}$$

Thus, eq. (5.70) may be written as another exact equation,

$$\begin{aligned} \mathrm{d}\underline{V}_v/\mathrm{d}t = {} & \underline{S}_v(t)\cdot\underline{\nabla}_{Pv(t)}\,\underline{V}_{v(t)} + K^+(t) \\ & + \sum_{\mu=1}^{n}\int_0^t e^{\mathrm{i}\mathbf{L}(t-\tau)}\cdot\{[\underline{\nabla}_{P\mu} - (1/k_\mathrm{B}TM_\mu)\cdot\underline{P}_\mu]\cdot\langle \underline{E}_\mu e^{\mathrm{i}(1-\mathbf{P})\mathbf{L}t}\,\underline{E}_v\rangle_0\cdot\underline{\nabla}_{Pv}\,\underline{V}_v(0)\}\,\mathrm{d}\tau \end{aligned} \tag{5.72}$$

and this reduces to

$$M_v\,\mathrm{d}\underline{V}_v/\mathrm{d}t = A(t) + K^+(t) + \sum_{\mu=1}^{n} e^{\mathrm{i}\mathbf{L}t}\cdot\{[\underline{\nabla}_{P\mu} - (1/k_\mathrm{B}TM_\mu)\cdot\underline{P}_\mu]\cdot\int_0^t \langle \underline{E}_\mu\underline{E}_v^0(\tau)\rangle_0\cdot\underline{\nabla}_{Pv}\,\underline{V}_v(0)\,\mathrm{d}\tau \tag{5.73}$$

where $\underline{E}_v^{\,0}(\underline{\tau}) = e^{\mathrm{i}\mathbf{L}\tau}\underline{E}_v$ if, as before, we assume that $\langle A\,e^{\mathrm{i}\mathbf{L}_0 t}B\rangle_0 = \langle A\rangle_0\langle B\rangle_0$ for $t > \tau_b$, where τ_b is the longest relaxation time for the bath in the presence of the n Brownian particles fixed in position, $|\underline{P}(t)| \gg |\underline{p}|$ and $|\underline{V}_B| \ll |\underline{v}|$.

If these conditions are met then to order $\lambda=(m/M)^{1/2} \ll 1$ I_1 and I_2 may be neglected so eq. (5.72) may be written as follows:

$$
\begin{aligned}
&M_\nu \, \mathrm{d}\underline{V}_\nu/\mathrm{d}t \\
&= \underline{A}(t) + K^+(t) + \sum_{\mu=1}^{n} \mathrm{e}^{\mathrm{i}\mathbf{L}t} \cdot \{[\underline{\nabla}_{\underline{P}\mu} - (1/k_\mathrm{B}TM_\mu)\underline{P}_\mu] \cdot \gamma_{\mu\nu}(t) \cdot \underline{\nabla}_{\underline{P}\nu} \underline{V}_\nu(0)\,\mathrm{d}\tau \\
&= \underline{S}_\nu(t) + \underline{E}_\nu^+(t) - \sum \underline{P}_\mu \cdot \zeta_{\mu\nu}(t, \underline{R}^{(n)}(t))
\end{aligned}
\tag{5.74}
$$

where

$$
\begin{aligned}
&\underline{E}_\nu^+(t) = \mathrm{e}^{\mathrm{i}(\mathbf{1}-\mathbf{P})Lt} \, \underline{E}_\nu(0) \\
&\zeta_{\mu\nu}(t, \underline{R}^{(n)}(t)) = (1/k_\mathrm{B}TM_\mu)\mathrm{e}^{\mathrm{i}\mathbf{L}t} \gamma_{\mu\nu}(t, \underline{R}^{(n)}) = (1/k_\mathrm{B}TM_\mu)\int_0^t \langle \underline{E}_\mu \underline{\mathrm{E}}_\nu^0(\tau)\rangle_0 \,\mathrm{d}\tau
\end{aligned}
\tag{5.75}
$$

and for $t \gg \tau_\mathrm{b}$, eq. (5.75) may be written as

$$
\zeta_{\mu\nu}(t, \underline{R}^{(n)}(t)) = \zeta_{\mu\nu}(\infty, \underline{R}^{(n)}(t))
$$

Thus, we have a Langevin-type equation for $\underline{\mathrm{V}}_\nu$ which says that the rate of change of momentum is given by the sum of three forces on the Brownian particle ν,

1. $\underline{S}_\nu(t) = \mathrm{e}^{\mathrm{i}\mathbf{L}t} \langle \underline{F}_\nu\rangle_0$ which is the time-dependent solvent averaged force due to the other Brownian particles.
2. $\underline{E}_\nu{}^+(t) = \mathrm{e}^{\mathrm{i}(\mathbf{1}-\mathbf{P})\mathrm{L}t}\, \underline{E}_\nu(0)$ which is the fluctuating or "Random" force,where $\underline{E}_\mu = \underline{F}_\mu - \langle \underline{E}_\mu\rangle_0$ is the difference between the direct and solvent averaged forces on the Brownian particle ν.
3. $-\Sigma\, \underline{P}_\mu \cdot \zeta_{\mu\nu}(t, \underline{R}^{(n)}(t))$ which is related to the time-correlation function of $\underline{E}_\mu$ which involves the $\underline{R}^{(n)}(t)$ dependent friction coefficients $\zeta_{\mu\nu}(t, \underline{R}^{(n)}(t))$.

In order to understand the Langevin equation (eq. (5.74)) we must find the properties of the random force $\underline{E}_\nu{}^+(t)$ and secondly the friction coefficients $\zeta_{\mu\nu}(t, \underline{R}^{(n)}(t))$.

Now we have

$$
\langle \underline{E}_\nu^+(t)\rangle_0 = \langle \underline{E}_\nu^+(0)\rangle_0 = 0 \tag{5.76}
$$

and furthermore Mazur and Oppenheim show that[1]

$$
\begin{aligned}
\langle \underline{E}_\mu \, \underline{E}_\nu^+(t)\rangle_0 &\approx \langle \underline{E}_\mu \, \underline{E}_\nu^0(\tau)\rangle_0 \\
&= 2(k_\mathrm{B}TM_\mu)\zeta_{\mu\nu}(\underline{R}^{(n)})\delta(t)
\end{aligned}
\tag{5.77}
$$

which is valid for $t \gg \tau_\mathrm{b}$ and they also show that in the weak-coupling limit, i.e. $\lambda = (m/M)^{1/2} \to 0$, $t \to \infty$ and $(\lambda^2 t) = $ constant that $\underline{E}_\mu$ can be described by a Gaussian stochastic process with mean zero (i.e. eq. (5.76)) and second moment given by eq. (5.77).

The remarks about the distinction between the Brownian limit derived above and the Diffusive limit discussed at the end of Section 5.1 should be emphasised again.

5.2.4 Langevin equation for the velocity and the form of the friction coefficients

The friction coefficients $\zeta_{\mu\nu}(t, \underline{R}^{(n)}(t))$ (or friction tensor $\boldsymbol{\zeta}_{\mu\nu}(t, \underline{R}^{(n)}(t))$ in general) may, in principle, be evaluated by either a microscopic or a macroscopic method. The microscopic approach is discussed at length by many authors, e.g. see Deutch and Oppenheim[3] (see Section 5.5). In practice, the macroscopic approach is probably the more viable one for most circumstances and we will devote some time to discussion of this approach in Chapter 7. However, it is useful to consider the relationship between the many-body Langevin equation derived from a microscopic starting point, i.e. eq. (5.74) and the macroscopic approach to the derivation. This relationship may be discussed by summarising the excellent discussion of the macroscopic approach given by Mazur[6] which he, in essence,

1. Develops methods to calculate the hydrodynamic drag force $\underline{F}_\nu^H$ on n spheres suspended in a Newtonian fluid whose hydrodynamics may be described by the linear Navier–Stokes ("creeping flow") equations for incompressible fluid flow at low velocity.
2. Combines this with the Landau–Lifshitz equation governing fluctuating hydrodynamics to give a Langevin equation.

The former he does by generalising the well-known Faxen theorem for the force on one sphere in a fluid to that of n-spheres in a fluid using the method of induced forces. This method gives this force as

$$\underline{F}_\nu^{\mathrm{H}} = \sum_{\mu=1}^{\mathrm{n}} \boldsymbol{\zeta}_{\nu\mu}^{\mathbf{H}} \cdot (\underline{V}_\mu - \langle \underline{v}_{0S\mu} \rangle) \tag{5.78}$$

where $\underline{V}_\mu$ is the velocity of the μ sphere, $\langle \underline{v}_{0S\mu} \rangle$ is the value of the unperturbed fluid velocity field averaged over the surface of the μth sphere and $\zeta_{\nu\mu}{}^{\mathbf{H}}$ is the hydrodynamic friction tensor which is related to the inverse of the mobility tensor $\boldsymbol{\mu}_{\mu\nu}^{\mathbf{H}}$ which relates $\underline{V}_\mu$ to the $\langle \underline{v}_{0S\nu} \rangle$ and the sum of the $\underline{F}_\mu^H$ by

$$\underline{V}_\nu = \langle \underline{v}_{0S\nu} \rangle - \sum_{\mu=1}^{\mathrm{n}} \boldsymbol{\mu}_{\nu\mu}^{\mathbf{H}} \cdot \underline{F}_\nu^{\mathrm{H}} \tag{5.79}$$

The latter requires the definition of a random stress (which is added to the normal linearised Navier–Stokes equation) by

$$\begin{aligned} &\langle \sigma_{ij}(\underline{R},t) \rangle = 0 \\ &\langle \sigma_{kl}(\underline{R},t)\sigma_{mn}(\underline{R}',t) \rangle = 2k_{\mathrm{B}}T\eta\{\delta_{km}\delta_{ln} + \delta_{kn}\delta_{lm} - 2/3\,\delta_{kl}\delta_{mn}\}\delta(\underline{R}-\underline{R}')\delta(t-t') \end{aligned} \tag{5.80}$$

where ⟨...⟩ denotes an average over an equilibrium ensemble average of fluid systems. This leads to the following macroscopic many-body Langevin equation:

$$M_\nu \, \mathrm{d}\underline{V}_\nu/\mathrm{d}t = \underline{F}_\nu^{\mathrm{Ext}} + \underline{F}_\nu^{\mathrm{R}} - \sum_{\mu=1}^{n} \zeta_{\nu\mu} \cdot \underline{V}_\mu \tag{5.81}$$

where $\underline{F}_\nu^{\mathrm{Ext}}$ is the total external force due to solvent averaged interactions with other particles and external sources, $\underline{F}_\nu^{\mathrm{R}}$ is the random force arising from the random stress $\sigma_{ij}(\underline{r},t)$ and $\underline{F}_\nu^{\mathrm{HD}} = -\sum\zeta_{\nu\mu} \cdot \underline{V}_\mu$ is the hydrodynamic drag force due to the solvent in the presence of the other particles. This is exactly the same form as the microscopically derived equation (5.74) and, thus, we may make the identifications (ignoring the contribution of external systems to $\underline{F}_\nu^{\mathrm{Ext}}$),

$$\begin{aligned} \underline{F}_\nu^{\mathrm{Ext}} &= \underline{S}(t) = \mathrm{e}^{\mathrm{i}\mathbf{L}t} \langle \underline{F}_\nu \rangle_0 \\ \underline{F}_\nu^{R} &= \underline{F}_\nu^{+}(t) = \mathrm{e}^{\mathrm{i}(\mathbf{1}-\mathbf{P})Lt}\, \underline{F}_\nu(0) \\ \underline{F}_\nu^{H} &= -\sum_{\mu=1}^{n} \zeta_{\nu\mu}(t, \underline{R}^{(n)}(t)) \cdot \underline{P}_\mu = -\sum \zeta_{\nu\mu}^{H} \cdot \underline{V}_\mu \end{aligned} \tag{5.82}$$

Mazur[6] shows that the random force $\underline{F}_\nu^{\mathrm{R}}$ is given by

$$\underline{F}_\nu^{\mathrm{R}} = \sum_{\mu=1}^{n} \zeta_{\nu\mu}^{\mathbf{H}} \cdot \langle \underline{v}_{0S\mu} \rangle \tag{5.83}$$

and has the following statistical properties:

$$\langle \underline{F}_\nu^{\mathrm{R}} \rangle = 0 \tag{5.84}$$

and

$$\langle \underline{F}_\nu^{\mathrm{R}}(t)\, \underline{F}_\nu^{\mathrm{R}}(t') \rangle = 2 k_{\mathrm{B}} T\, \zeta_{\nu\mu}^{\mathbf{H}} \langle \delta(t-t') \rangle$$

Thus, in the Brownian limit, i.e. when $\lambda = (m/M)^{1/2} \to 0$ we may use macroscopic hydrodynamics to calculate the hydrodynamic drag and random, fluctuating forces on suspended particles. The other force $\underline{F}_\nu^{\mathrm{Ext}}$ may be calculated by equilibrium statistical mechanical methods.

If one puts the fluctuating part of the pressure tensor **P** equal to zero, then this treatment leads to a purely macroscopic method of obtaining the friction tensors $\boldsymbol{\zeta}_{ij}$ (or $\boldsymbol{\mu}_{ij}$). In essence, this treatment is then just the hydrodynamic description of n macroscopic spheres of mass M_i and radius a_i immersed in an otherwise unbounded, incompressible, viscous fluid with coefficient of viscosity η_f.

5.3 GENERALISATION TO NON-EQUILIBRIUM

Shea and Oppenheim[7] have derived a Langevin-type equation via the non-equilibrium distribution function approach of Oppenheim and Levine.[8] This approach is based on the assumption of local thermodynamic equilibrium and has been used to derive a Langevin and a Fokker–Planck (F-P) equation for one Brownian particle in a non-equilibrium bath. This approach has not been generalised yet to treat many Brownian particles.

From a microscopic starting point this has been done in the case of the GLE as detailed in Chapter 4 and the limiting case of Brownian motion when hydrodynamic interactions may be neglected has been given in Section 4.2.

Dickinson[9] in deriving a generalised the BD algorithm of the Ermak and McCammon type[10] started by adding a term to the LE of Mazur, i.e. eq. (5.81) to represent the effect of a shear field. However, the most complete extension of the BD method of Ermak and McCammon[10] has been given by Brady and Bossis[11] starting from a macroscopic point of view which essentially generalises Mazur's result based on the Landau–Lifshitz fluctuating hydrodynamics equations.[6] Thus, Brady and Bossis[11] start with the Newton equation of motion for the combined translational and rotational degrees of freedom,

$$\mathbf{M}\cdot \mathrm{d}\mathbf{U}/\mathrm{d}t = \mathbf{F}^{\mathrm{P}} + \mathbf{F}^{\mathrm{B}} + \mathbf{F}^{\mathrm{H}} = \mathbf{F} \tag{5.85}$$

Here, $\mathbf{M}$ is a generalised mass/moment-of-inertia $6N \times 6N$ dimension matrix, $\mathbf{U}$ is the particle translational/rotational $6N$ dimensional vector and $\mathbf{F}$ is the $6N$ dimensional vector representing the total force/torque on the particles. This latter term is made up of the following:

(a) $\mathbf{F}^{\mathrm{P}}$ the sum of the interparticle and external, non-hydrodynamic forces,
(b) $\mathbf{F}^{\mathrm{B}}$ the stochastic, Brownian force, and
(c) $\mathbf{F}^{\mathrm{H}}$ the hydrodynamic force which is now generalised to include the shear field.

This latter force is then written as

$$\mathbf{F}^{\mathrm{H}} = -\mathbf{R}_{\mathrm{FU}}\cdot(\mathbf{U} - \mathbf{U}^{\infty}) + \mathbf{R}_{\mathrm{FE}} : \mathbf{E}^{\infty} \tag{5.86}$$

This is a generalisation of the hydrodynamic interactions present in the previous Langevin equation to include the effects of the bulk shear flow. The term $\mathbf{R}_{\mathrm{FU}}(\mathbf{x})$ is the configuration-dependent resistance matrix generalised to include both translational and rotational terms, i.e. it depends on the generalised configuration vector $\mathbf{x}$ which specifies the position and orientation of all the particles. Thus, this term merely collects together all the previously defined terms. The shear field enters *via* the terms $\mathbf{U}^{\infty}$, $\mathbf{R}_{\mathrm{FE}}$ and $\mathbf{E}^{\infty}$ which are as follows:

1. The term $\mathbf{U}^{\infty}$ is the velocity of the bulk shear flow evaluated at the particle centre where $\mathbf{U}^{\infty} = \Omega^{\infty}$ for rotation, $\mathbf{U}^{\infty} = \mathbf{E}^{\infty}\cdot\mathbf{x}_{\alpha}$ for translation where $\mathbf{x}_{\alpha}$ is the position vector of the αth particle and $|\mathbf{E}^{\infty}| = \dot{\gamma}$ is the magnitude of the shear rate.

2. The tensors $\mathbf{E}^\infty$ and Ω^∞ are, respectively, the position independent but time-dependent, anti-symmetric parts of the velocity-gradient tensor.
3. The tensor $\mathbf{R}_{FE}$ is the configuration-dependent resistance matrix representing the hydrodynamic forces and torques due to the imposed shear field.

It is assumed as before that the Brownian force has the following properties:

$$\langle \mathbf{F}^{\mathrm{B}} \rangle = 0 \tag{5.87}$$

and

$$\langle \mathbf{F}^{\mathrm{B}}(0)\mathbf{F}^{\mathrm{B}}(t)\rangle = 2k_{\mathrm{B}}T\mathbf{R}_{\mathrm{FU}}\delta(\mathrm{t})$$

This equation has been used to derive Brownian dynamics algorithms in the diffusive limit as will be discussed in Section 7.5.

5.4 THE FOKKER–PLANCK EQUATION AND THE DIFFUSIVE LIMIT

If we choose as our dynamical variable $A(t)$ the variable $D(t)$ defined by[3]

$$\begin{aligned} A(t) = D(t) &= D(\underline{R}^{(n)'}, \underline{P}^{(n)'}, t \mid \underline{R}^{(n)}, \underline{P}^{(n)}) \\ &= \mathrm{e}^{\mathrm{i}\mathbf{L}t}\,\delta(\underline{R}^{(n)} - \underline{R}^{(n)'})\delta(\underline{P}^{(n)} - \underline{P}^{(n)'}) \end{aligned} \tag{5.88}$$

where $\underline{R}^{(n)'}$ and $\underline{P}^{(n)'}$ denote arbitrary values of the positions and momenta of the large particles then we may obtain the Fokker–Planck equation for the reduced distribution function $f^{(n)}(\underline{R}^{(n)},\underline{P}^{(n)})$. This distribution function gives the probability of finding a particular dynamical state of the system but only for the large particles unlike the full distribution function which obeying the Liouville equation for the whole system of small and large particles, i.e. eq. (1.17). The Fokker–Planck governing the time evolution of $f^{(n)}$ and the details of this procedure may be found in Section IV of Ref. 3.

If we use eq. (5.88) then eq. (5.68) yields the following equation:

$$\begin{aligned} \mathrm{d}D(t)/\mathrm{d}t = {} & \mathrm{e}^{\mathrm{i}\mathbf{L}t}\sum_{\mu=1}^{n}[(\underline{P}_\mu/M_\mu)\cdot\underline{\nabla}_{R\mu} + \langle \underline{F}_\mu\rangle_0\cdot\underline{\nabla}_{P\mu}]D(0) \\ & + \sum_{\mu=1}^{n}\sum_{\nu=1}^{n}\mathrm{e}^{\mathrm{i}\mathbf{L}t}[\underline{\nabla}_{P\mu} - (1/k_{\mathrm{B}}T\,M_\mu)\underline{P}_\mu]\cdot\gamma_{\mu\nu}(t)\cdot\underline{\nabla}_{P\mu}D(0) + \mathrm{G}^{+}(t) \end{aligned} \tag{5.89}$$

where the random force $G^{+}(t)$ is given by

$$G^{+}(t) = \mathrm{e}^{\mathrm{i}(\mathbf{1}-\mathbf{P})\mathrm{i}\mathbf{L}t}\sum_{\nu=1}^{n}\underline{E}_\nu\cdot\underline{\nabla}_{P\nu}\,D(0) \tag{5.90}$$

Using the properties of the delta function eq. (5.89) may be re-written as

$$\partial D(t)/\partial t+\sum_{\mu=1}^{n}[(\underline{P}_{\mu}/M_{\mu})\cdot\underline{\nabla}(\underline{R}_{\mu})+\langle\underline{F}_{\mu}\rangle_0\cdot\underline{\nabla}(\underline{P}_{\mu})]D(t)$$
$$=\sum_{\mu=1}^{n}\sum_{\nu=1}^{n}\underline{\nabla}(\underline{P}_{\mu})\cdot\gamma_{\mu\nu}(t,\underline{R}^{(n)'})\cdot[\underline{\nabla}(\underline{P}_{\mu})+(1/k_{\mathrm{B}}T\,M_{\mu})\underline{P}_{\mu}]D(t)+G^{+}(t) \tag{5.91}$$

Then for times $t \gg \tau_{\mathrm{b}}$ we have

$$\partial D(t)/\partial t+\sum_{\mu=1}^{n}[(\underline{P}_{\mu}/M_{\mu})\cdot\underline{\nabla}(\underline{R}_{\mu})+\langle\underline{F}_{\mu}\rangle_0\cdot\underline{\nabla}(\underline{P}_{\mu})]D(t)$$
$$=\sum_{\mu=1}^{n}\sum_{\nu=1}^{n}\underline{\nabla}(\underline{P}\nu)\cdot\zeta_{\mu\nu}(\underline{R}^{(n)'})\cdot[(k_{\mathrm{B}}T\,M_{\mu})\underline{\nabla}(\underline{P}_{\mu})+\underline{P}_{\mu}]D(t)+G^{+}(t) \tag{5.92}$$

where

$$\zeta_{\mu\nu}(\underline{R}^{(n)'})=(1/k_{\mathrm{B}}T\,M_{\mu})\gamma_{\mu\nu}(\infty,\underline{R}^{(n)'}) \tag{5.93}$$

Now,

$$f^{(n)}(\underline{R}^{(n)},\underline{P}^{(n)})$$
$$=\int f^{(N+n)}(\underline{r}^{(N)},\underline{p}^{(N)},\underline{R}^{(n)},\underline{P}^{(n)};t=0)D(t)\,\mathrm{d}\underline{r}^{(N)}\mathrm{d}\underline{p}^{(N)}\mathrm{d}\underline{R}^{(n)}\mathrm{d}\underline{P}^{(\mathrm{n})}$$
$$=\int g(\underline{R}^{(N)},\underline{P}^{(N)})\langle D(t)\rangle_0\,\mathrm{d}\underline{R}^{(n)}\,\mathrm{d}\underline{P}^{(n)} \tag{5.94}$$

where the last step is assuming that the bath particles are in equilibrium in the field of the fixed large particles and so,

$$\partial\langle D(t)\rangle/\partial t=-\sum_{\mu=1}^{n}[(\underline{P}_{\mu}/M_{\mu})\cdot\underline{\nabla}(\underline{R}_{\mu})+\langle\underline{F}_{\mu}\rangle_0\cdot\underline{\nabla}(\underline{P}_{\mu})]\langle D(t)\rangle$$
$$+\sum_{\mu=1}^{n}\sum_{\nu=1}^{n}\underline{\nabla}(\underline{P}_{\mu})\cdot\zeta_{\mu\nu}(t,\underline{R}^{(n)'})\cdot[(k_{\mathrm{B}}T\,M_{\mu})\underline{\nabla}(\underline{P}_{\mu})+\underline{P}_{\mu}]\langle D(t)\rangle \tag{5.95}$$

where the random force term vanishes as $\langle G^{+}(t)\rangle = 0$ and now

$$\zeta_{\mu\nu}(t,\underline{R}^{(n)'})=(1/k_{\mathrm{B}}T\,M_{\mu})\gamma_{\mu\nu}(t,\underline{R}^{(n)'}) \tag{5.96}$$

and once again if $t \gg \tau_{\mathrm{b}}$ then $\zeta_{\mu\nu}(\underline{R}^{(n)'}) = \zeta_{\mu\upsilon}(t,\underline{R}^{(n)'}) = (1/k_{\mathrm{B}}T\,M_{\mu})\gamma_{\mu\upsilon}(\infty,\underline{R}^{(n)'})$.

From eqs. (5.94) and (5.96) we have

$$
\begin{aligned}
&\partial f^{(n)}(\underline{R}^{(n)}, \underline{P}^{(n)})/\partial t \\
&= -\sum_{\mu=1}^{n} [(\underline{P}_{\mu}/M_{\mu})\cdot\underline{\nabla}(\underline{R}_{\mu}) + \langle \underline{F}_{\mu} \rangle_0 \cdot \underline{\nabla}(\underline{P}_{\mu})] f^{(n)}(\underline{R}^{(n)}, \underline{P}^{(n)}) \\
&+ \sum_{\mu=1}^{n}\sum_{\nu=1}^{n} \underline{\nabla}(\underline{P}_{\nu})\cdot\zeta_{\mu\nu}(t, \underline{R}^{(n)'})\cdot[(k_{\mathrm{B}}T\, M_{\mu})\underline{\nabla}(\underline{P}_{\mu}) + \underline{P}_{\mu}] f^{(n)}(\underline{R}^{(n)}, \underline{P}^{(n)})
\end{aligned}
\tag{5.97}
$$

and eq. (5.97) is the many-body Fokker–Planck equation which shows the direct relationship and equivalence of the Langevin and Fokker–Planck approaches.

Just as with the general forms of the Langevin equation for any function of the dynamical variables characterising the state of the large particles, i.e. eq. (5.58) or (5.66) we could start with one of these equations and obtain an exact form of the Fokker–Planck equation. This form would not involve making the assumptions and taking the limits involved in arriving at eq. (5.97).

At the end of Section 5.1 we stated that the Brownian limit is valid when all dynamical processes have ceased in the bath system in the presence of the Brownian particle before the Brownian particle has changed its dynamical state significantly. This does not, however, imply that the dynamical processes involving the velocity of the Brownian particle have ceased before the Brownian particle has significantly changed its position. If on the contrary this were also assumed then we would have also made the assumption that we were working in the Diffusive (or "Overdamped") limit. Often it is valid to make this second assumption as will be discussed in relationship to obtaining algorithms to describe the motion of colloidal particles suspended in a simple fluid (see Chapters 7 and 9) and in this case we may assume that $\langle \underline{V}_{\mathrm{B}}(0) \cdot \underline{V}_{\mathrm{B}}(t) \rangle = 0$.

5.5 APPROACH TO THE BROWNIAN LIMIT AND LIMITATIONS

Having established the form of the LE both for a single Brownian particle and for many Brownian particles the question must be asked, "When is the LE applicable?" At present there seems to be no definitive answer to this question. However, a large number of studies have been made of issues arising from this question using microscopic theories, kinetic theories, MD simulations and by the analysis of Mori coefficients, and the topic continues to be of great interest. A majority of these studies have considered a single Brownian particle but the case of two-particle systems has been studied by microscopic theory, kinetic theory and by MD whilst many-particle systems have been treated by microscopic theory and by MD. Since there are so many of these studies we will not attempt to review them all but shall only try to provide some general results which have been obtained. The selection of these particular results may be personal but the references given at the end of this chapter will hopefully act to somewhat balance this view, see Refs. 1–5,8,10,12–74.

5.5.1 A basic limitation of the LE and FP equations

The first question to investigate is that most microscopic theories use an expansion in terms of $(m/M)^{1/2}$, however, the LE and FP equation can only be strictly valid if $\rho/\rho_B \ll 1$.[29] This is often not met in practice as in most experiments one wishes to have $\rho/\rho_B \approx 1$ in order to avoid sedimentation of the large particles. This leads to incorrect behaviour of the velocity autocorrelation function for the heavy B particles $C_V = \langle \underline{V}_B(0) \cdot V_B(t) \rangle$ as can be seen from eq. (5.41) as the LE gives an exponential decay at large times for $C_V(t) \propto \exp(-\zeta t)$ but from simulations, theory and experiment we know that $C_V(t) \to -At^{-3/2}$ as $t \to \infty$ (see Section 3.7). However, it should be pointed out that the contribution of this long time tail to $C_V(t)$ is very small compared to the exponential component of its decay. Now a correct description will result in a GLE with a non-Markovian memory function.[29,36,75–79] Thus, the use of the LE or FP equation for large particles in the diffusive limit assumes that the velocity autocorrelation function for the heavy B particles $\langle \underline{V}_B(0) \cdot V_B(t) \rangle$ has relaxed to zero before the spatial positions of the large particles have changed significantly (see Sections 5.1.3 and 6.2.5). However, experiments have detected a long time tail for both dilute suspensions[66](essentially single-particle relaxation) as well as for concentrated suspensions of colloidal particles.[67] Once again the contribution to $C_V(t)$ is small but it does lead to some fascinating new physics which does not arise from the LE.

5.5.2 The friction coefficient

One of the interesting results obtained from kinetic theory based on the very useful time scale separation technique for hard particles[31,32] is that the friction coefficient may be expressed as

$$\gamma = M\zeta = \gamma_1 + \gamma_2 \tag{5.98}$$

where

$$\gamma_1 = ((\sigma+\Sigma)/2)^2\, 8/3(2\pi m k_B T)^{1/2}\, \rho^{eq}((\sigma+\Sigma)/2)$$

and

$$\gamma_2 = 1/(3k_B T)\int_0^\infty \langle \underline{F}_+(0)\cdot \underline{F}_-(-\tau)\rangle_0\, d\tau$$

where σ and Σ are the bath and Brownian particle diameters and $\rho^{eq}((\sigma+\Sigma)/2)$ is the contact value of the equilibrium density profile of the bath particles in the field of the Brownian particle fixed at $\underline{R}$. The "forces" $\underline{F}_-$ and $\underline{F}_+$ are given by

$$\underline{F}_\pm = \Sigma((\sigma+\Sigma)/2)^2 \int 2m(\underline{v}_i(t)\cdot\underline{\sigma})^2\, \theta(\pm \underline{v}_i(t)\cdot\underline{\sigma})\underline{\sigma}\, \delta\{\underline{R}-(\sigma+\Sigma)/2)\underline{\sigma}-\underline{r}_i(t)\}\, d\underline{\sigma} \tag{5.99}$$

where the summation is over the bath particles, θ is the Heaviside step function and the integration is over the unit vector $\underline{\sigma}$ along the direction joining the centres of the colliding spheres. Hansen et al.[32] evaluated γ_2 by the use of MD simulations which showed that

1. The Enskog approximation breaks down as the packing fraction and/or the size ratio increases Σ/σ
2 For a fixed packing fraction, the friction coefficient agrees with the Stoke's estimate $d_H \approx \Sigma$ independent of Σ/σ, provided a stick boundary condition is used. This is somewhat surprising as one would expect slip boundary conditions to be more appropriate as most other studies find.

A study by Bhattacharyya and Bagchi[34] using re-normalised Kinetic Theory (giving results essentially equivalent to mode-coupling theory) for a system of particles with continuous L–J interactions rather than hard interactions obtained exact expression for the friction on a tagged particle which was decomposed into a short-time and a long-time part by using the separation of time scales between binary collisions and the repeated re-collisions. Thus, $\zeta(z) = \zeta_D(z) + \zeta_R(z)$, where $\zeta_D(z)$ is the short-time part which arises from the direct collisions between the solute and the solvent particles and $\zeta_R(z)$ is the long-time part which arises from the correlated re-collisions of the solute particle with the solvent particles. It is predicted that the crossover from the microscopic to hydrodynamic regime should occur when the solute is about 2–3 times larger than the solvent molecules.

Essentially the same problem was studied by Brey and J. Gomez-Ordonez[46] who made an MD study of 863 bath particles and one heavy particle of diameter σ_{BB} greater than that of the solvent fluid particles σ_{ff} and varied $\lambda = (m/M)^{1/2}$. The friction constant ζ was calculated from

$$\zeta = 1/(3k_B T) \int_0^{\tau_{max}} \langle \underline{F}(0)\cdot\underline{F}(s)\rangle_0 \, ds \tag{5.100}$$

where the upper limit on the integral τ_{max} was either taken as finite or infinite and they found that

1. For a given value of σ_{BB}/σ_{ff} the value of ζ becomes independent of mass beyond a given value of $\lambda = (m/M)^{1/2}$.
2. The value of $\lambda = (m/M)^{1/2}$ for which ζ reaches a stationary value is not always the same.
3. The above-mentioned behaviour can be explained if we take into account the parameter $\lambda = (\rho_f/\rho_B)^{12}$ and anomalous values of ζ occur when $\gamma' > 0.2$, i.e. the Langevin description is valid only if γ' is small.
4. There is linear dependence on σ_{BB} as is required by Stokes' law and if a hydrodynamic radius of $E\sigma_{BB}/2$ is used, then $\zeta = 4\pi E\sigma_{BB}/2$ with $E = 1.7 \pm 0.4.0$ and using the Stokes–Einstein relationship as $D_0 = k_B T/(c\pi\eta R)$ they found that with $c = 4$, i.e. a slip boundary.

So to summarize if $\lambda \ll 1$ and $\lambda' \ll 11$, then ζ is mass independent.

5.5.3 Self-diffusion coefficient (D_s)

Heyes and co-workers have made extensive studies of a variety of particles in various background fluids by means of MD.[51–55] Many types of particles, e.g. particles differing only in mass from those of the fluid, particles with cores ("smooth particles") and composite particles ("clusters") which have rough surfaces were studied for various combinations of interaction parameters and masses under a variety of physical conditions. However, the following two results of rather general interest were as follows;

1. One system[51]consisted of identically sized particles of which one is a heavy particle of mass M_B and the other N were light particles of mass m but all of which interact *via* purely repulsive interactions. Here it was found that D_s does not depend on mass for $M_B/m \geq 10$ provided N is large enough.
2. Cluster calculations[55] in which the ratio of cluster atom mass, m_c to solvent atom mass, m varies in the range $0.2 < m_c/m < 4.0$ so as to vary the relative densities of cluster to solvent, i.e. $\rho_c/\rho_s = m_s/m$ were made. The effective diameters of the clusters was $\sigma_c/\sigma = 3.3$, 4.5, 6.1, 6.4 and 9.1. The results indicate that

   (i) At liquid-like densities D_s and the rotational diffusion coefficient D_R show virtually no dependence on the density of the cluster (both smaller and larger than the solvent).
   (ii) $\langle \underline{V}(0) \cdot \underline{V}(t) \rangle$ decays monotonically and more rapidly with increasing cluster density.
   (iii) The density of the nano-colloidal particle has virtually no effect on D_s at liquid-like solvent densities.
   (iv) D_s is sensitive to the solvent bulk density and to the size of the solute particle and therefore it would appear to be the bulk characteristics of the solvent that largely determines the value of D_s.
   (v) Density has no influence on D_R.

Another MD study of diffusion using a modified LJ126 potential with $N = 5324$ by Ould–Kaddour and Levesque[62] found that the size and mass values for a solute diffusion coefficient to obey the Stokes–Einstein formula are as follows:

1. The crossover to the hydrodynamic regime for equal masses occurs when the particle to background particle size ratio $D/d > 4$.
2. For fixed size ratio of 0.5 it occurs at a mass ratio of more than 40.
3. For equal masses the slip boundary condition is found to be most appropriate.
4. As the size ratio increases, the hydrodynamic limit is reached at a lower mass ratio, e.g. $D/d = 4$, the appropriate value of $M/m = 5$.
5. Mass and size effects are strongly coupled.
6. The diffusion constant was also evaluated from the Stokes–Einstein relationship using eq. (5.100) for ζ but was only reasonably accurate (error about 30%) for large mass and size ratios. The best result was when these ratios were greatest (8.6%).

A conclusion is that behaviour of the Brownian type only occurs when the size ratio of particles is >3 for practically any mass ratio. For size ratios of the order of 1, the Brownian result is questionable for mass ratios <10. The theoretical results of Bhattacharyya and Bagchi[34] agree with this work.

Some results which illustrate several of the points made above are shown below in Figures 5.2–5.4 and are from an MD study of a system of heavy particles, of varying percent by number, in a bath of lighter but otherwise identical particles for various mass ratios α. All particles interacted *via* a CWA potential.

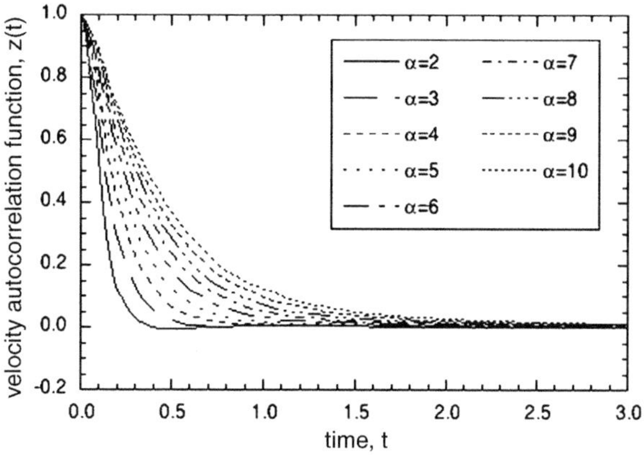

Figure 5.2 Normalised velocity autocorrelation function, $\langle v(0)\cdot v(t)\rangle/\langle v(0)\cdot v(0)\rangle$ versus delay time for the heavy particles in a system of 4% by number of heavy particles in a bath of lighter but otherwise identical particles for various mass ratios α (see text for further explanation).

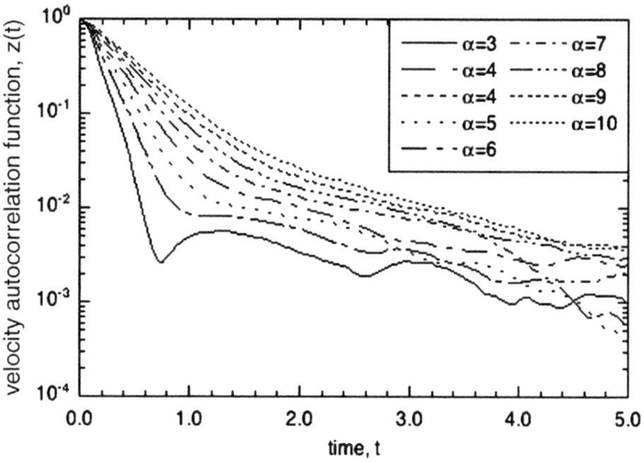

Figure 5.3 Normalised velocity autocorrelation function, $\langle v(0)\cdot v(\mathrm{t})\rangle/\langle v(0)\cdot v(0)\rangle$ versus delay time for the heavy particles in a system of 4% by number of heavy particles in a bath of lighter but otherwise identical particles for various mass ratios α (see text for further explanation).

From Figure 5.2 it may be seen that $\langle v(0)\cdot v(t)\rangle/\langle v(0)\cdot v(0)\rangle$ decays increasingly slowly as the mass ratio increases as predicted by the Langevin equation.

However, upon plotting the results shown in Figure 5.3 on a log-linear scale one can see the following:[69]

1. The velocity autocorrelation function $c(t)$ is never exponential over its whole range, in particular, showing a long-time tail consistent with $t^{-3/2}$ behaviour.
2. The self-diffusion coefficient D_2 for the heavy particle does appear to be approaching a mass-independent value for $M/m > 10$ at low volume fraction but not at high volume fractions.

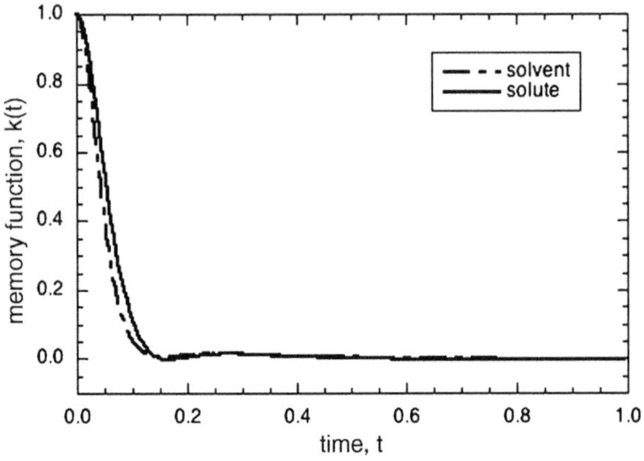

Figure 5.4 Memory functions versus delay time for the heavy (solute) and light (solvent) particles in a system of 4% heavy particles in a bath of lighter but otherwise identical particles for mass ratios $\alpha = 10$ (see text for further explanation).

3. The self-diffusion coefficient D_1 of the solvent is dependent on the volume fraction of the heavy particles.
4. The memory function of the solute is of much shorter range than that of $c(t)$ and the discrepancy grows as M/m increases. However, the memory function corresponding to D_1 and D_2 are rather similar (see Figure 5.4)

5.5.4 The intermediate scattering function F(q,t)

Hoheisel[47] showed by means of MD simulations that the intermediate scattering function $F(k,t)$ and the longitudinal current autocorrelation function show the time-scale separation required for the Fokker–Planck, Smoluchowski or Langevin equations to be applicable. The trends found undoubtedly indicate that large particle-volume ratios together with large mass ratios lead to timescale separation even for the relatively small differences in the solvent and solute particle's size and mass. A large mass ratio alone suffices as well to produce time-scale separation. These results contradict the theoretical work by Masters.[77]

5.5.5 Systems in a shear field

A study of viscous flow by NEMD for a system of N_1 particles of mass m and N_2 particles of mass M of the same size interacting according to the CWA potential showed the following results:[69]

1. The value of the suspension viscosity at low shear rates, η_0 is independent of shear rate indicating Newtonian flow.
2. The suspension viscosity decreases at high shear rates indicating shear thinning. However, when this first occurs the solvent is still Newtonian and the shear rate at which the shear thinning occurs it decreases with increasing value of M/m.

3. The low shear rate viscosity, η_0 at low volume fraction tends to a constant value as (M/m) increases which is slightly smaller than that given by the Einstein theory, as can be seen from Figure 5.5.

Figure 5.6 shows the results of an NEMD calculation[80] of the low shear rate coefficient of viscosity η_0 for a mixture of particles which differ both in mass and size. The interaction potential energy function used was again the CWA potential but with a core and the relative core radii and particle masses were chosen so that the large particles were neutrally buoyant. As can be seen from Figure 5.6 the calculated results show quite reasonable agreement with experimental estimates of η_0 despite the fact that the large particles are only about twice the size of the background particles.[80]

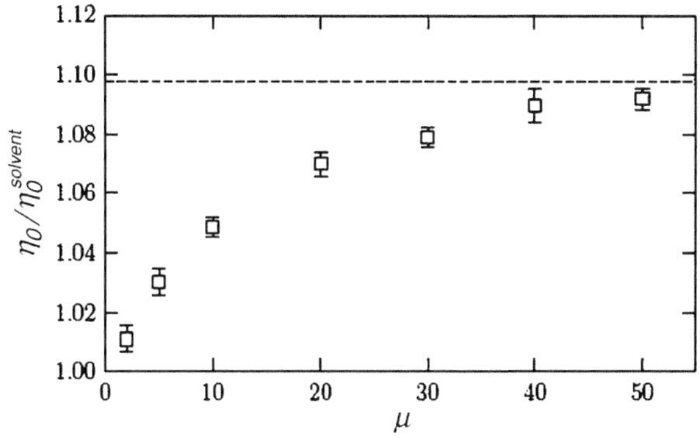

Figure 5.5 The low shear rate coefficient of viscosity versus mass ratio μ for a system of heavy (solute) and light (solvent) particles in a system of which 4% by weight were heavy particles (see text for further explanation).[80]

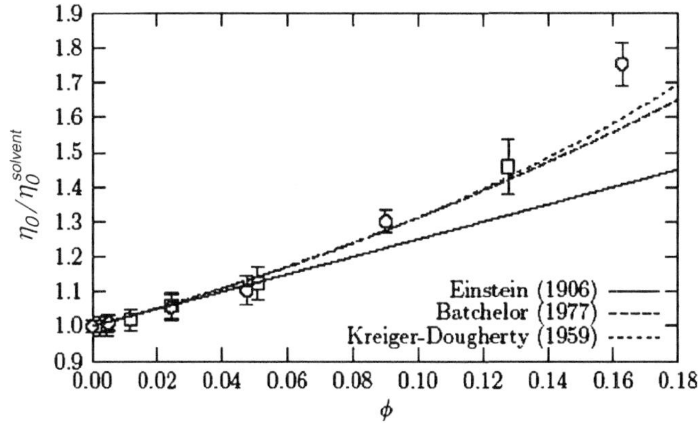

Figure 5.6 The results of an NEMD calculation of the low shear rate coefficient of viscosity η_0 for a mixture of neutrally buoyant particles which differ both in mass and size[80] (see text for further details).

5.6 SUMMARY

Since the many-body Langevin equation is central to many theoretical and simulation studies of colloids and polymers, we will now summarise the extensive work on its derivation.

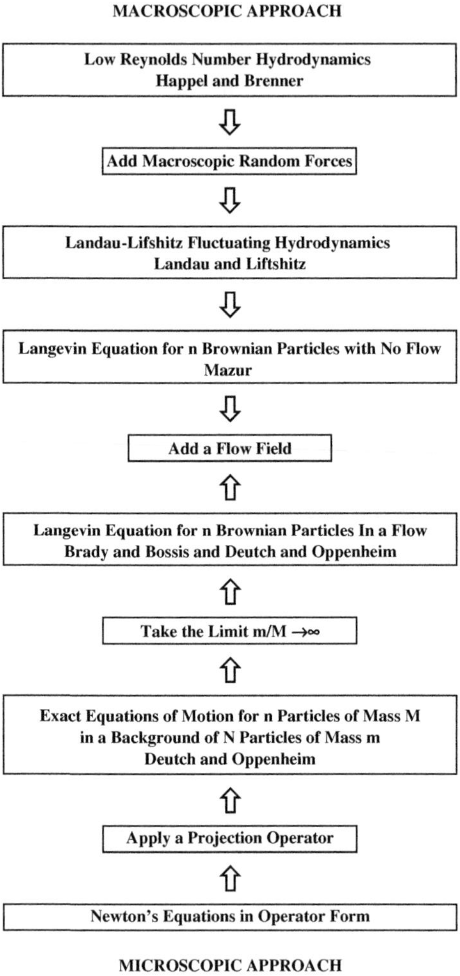

5.7 CONCLUSIONS

In this chapter we derived a generalised Langevin equation for any arbitrary function $A(\Gamma_1,t)$ of the dynamical variable Γ_1 characterising the state of a selected group of n particles of mass M in a state where the other N particles of mass m are at equilibrium in the presence of the selected particles held at fixed positions. This was done by averaging over an equilibrium distribution function $f_{eq}^{(N)}$ for the N particles in the presence of the other n particles held in fixed positions. This approach given by Mazur, Deutch and Oppenheim[1–3] allowed the GLEs to be reduced to many-body Langevin equations by taking the limit $(m/M) \to 0$. A particular choice of $A(t)$ given by eq. (5.88) was shown to lead to the n-body

Fokker–Planck equation which describes the probability of finding the n-particles in state Γ_1. This demonstrates the equivalence of the Langevin and Fokker–Planck approaches.

Since the n-body Langevin equations and Fokker–Planck equations play such a central role in describing and calculating the properties of colloidal suspensions and polymer solutions, some time was spent discussing what has been learned about the limits of validity of the Langevin equation.

REFERENCES

1. P. Mazur and I. Oppenheim, *Physica*, **50**, 241 (1970).
2. J. Albers, J.M. Deutch and I. Oppenheim, *J. Chem. Phys.*, **54**, 3541 (1971).
3. J.M. Deutch and I. Oppenheim, *J. Chem. Phys.*, **54**, 3547 (1971).
4. R. Zwanzig, *Adv. Chem. Phys.*, **15**, 325 (1969).
5. T.J. Murphy and J.L. Aguirre, *J. Chem. Phys.*, **57**, 2098 (1972).
6. P. Mazur, *Physica*, **110A**, 128 (1982).
7. J.-E. Shea and I. Oppenheim, *J. Phys. Chem.*, **100**, 19035 (1996).
8. I. Oppenheim and R.D. Levine, *Physica A*, **99**, 383 (1979).
9. G.C. Ansell, E. Dickinson and M. Ludvigsen, *J. Chem. Soc., Faraday Trans.* 2, **81**, 1269 (1985).
10. D.E. Ermak and J.A. McCammon, *J. Chem. Phys.*, **69**, 1352 (1978).
11. L. Durlofsky, J.F. Brady and G. Bossis, *J. Fluid Mech.*, **180**, 21 (1987); J. Brady and G. Bossis, *Ann. Rev. Fluid Mech.*, **20**, 111 (1988); *J. Chem. Phys.* **87**, 5437 (1987); R.J. Phillips, J.F. Brady and G. Bossis, *Phys. Fluids*, **31**, 3462 (1988); *ibid.* **31**, 3473 (1973); L.J. Durlofsky and J.F. Brady, *J. Fluid Mech.*, **200**, 39 (1989); J.F. Brady, R.J. Phillips, J.C. Lester and G. Bossis, *J. Fluid Mech.*, **195**, 257 (1988).
12. J.L. Lebowitz and E. Rubin, *Phys. Rev.*, **131**, 2381 (1963).
13. J.L. Lebowitz and P. Resibois, *Phys. Rev.*, **139**, 1101 (1965).
14. P. Resibois and H.T. Daivis, *Physica*, **30**, 1077 (1964).
15. R.I. Cukier and J.M. Deutch, *Phys. Rev.*, **177**, 240 (1969)
16. R.M. Mazo, *J. Stat. Phys.*, **1**, 559 (1969).
17. J.J. Brey, *Physica*, **90A**, 574 (1978).
18. J. Mercer and T. Keyes, *J. Stat. Phys.*, **32**, 35 (1983).
19. J. R. Dorfman, H. van Beijeren and C.F. McClure, *Archiw. Mech. Stos.*, **28** 333 (1976); *J. Stat. Phys.*, **23**, 35 (1980).
20. A. Masters and T. Keyes, *J. Stat. Phys.*, **33**, 149 (1983).
21. M.H. Ernst and J.R. Dorfman, *Physica*, **61**, 156 (1972).
22. A.J. Masters and P.A. Madden, *J. Chem. Phys.*, **74**, 2450 (1981); **75**, 127 (1981).
23. A.J. Masters and T. Keyes, *Phys. Rev. A*, **27**, 2603 (1983).
24. A.J. Masters and T. Keyes, *J. Stat. Phys.*, **36**, 401 (1984).
25. R.G. Cole and T. Keyes, *J. Stat. Phys.*, **51**, 249 (1988).
27. M. Pagitsas, J.T. Hynes and R. Kapral, *J. Chem. Phys.*, **71**, 4492 (1979).
28. R.I. Cukier, R. Kapral, J.R. Lebenhaft and J.R. Mehaffey, *J. Chem. Phys.*, **73**, 5244 (1980).
29. M. Tokuyma and I. Oppenheim, *Physica A*, **94**, 501 (1978).
30. M.J. Lindenfield, *J. Chem. Phys.*, **73**, 5817 (1980).
31. L. Bocquet, J. Piasecki and J.-P. Hansen, *J. Stat. Phys.*, **76**, 505 (1994).
32. L. Bocquet, J.-P. Hansen and J. Piasecki, *J. Stat. Phys.*, **76**, 527 (1994).
33. L. Bocquet, J-P. Hansen and J. Piasecki, *Il Nuovo Cimento*, **16**D, 981 (1994).
34. S. Bhattacharyya and B. Bagchi, *J. Chem. Phys.*, **106**, 1757 (1997).
35. J. Schofield and I. Oppenheim, *Physica A*, **187**, 210 (1992).
36. L. Bocquet and J. Piasecki, *J. Stat. Phys.*, **87**, 1005 (1997).

37. R.I. Cukier, J.R. Mehaffey and R. Kapral, *J. Chem. Phys.*, **69**, 4962 (1978).
38. R.I. Cukier, R. Kapral and J.R. Mehaffey, *J. Chem. Phys.*, **73**, 5254 (1980).
39. R.I. Cukier, R. Kapral and J.R. Mehaffey, *J. Chem. Phys.*, **74**, 2494 (1981).
40. J. Piasecki, L. Bocquet and J.-P. Hansen, *Physica A*, **218**, 125 (1995).
41. P.T. Herman and B.J. Alder, *J. Chem. Phys.*, **56**, 987 (1972).
42. W.E. Alley and B.J. Alder, *Phys. Rev. A*, **27**, 3158 (1983).
43. B.J. Alder, W.E. Alley and J.H. Dymond, *J. Chem. Phys.*, **61**, 1415 (1974).
44. J.J. Erpenbeck, *Phys. Rev. A*, **39**, 4718 (1989); **45**, 2298 (1992); **48**, 223 (1993).
45. A.J. Masters and T. Keyes, *Phys. Rev. A*, **27**, 2603 (1983).
46. J.J. Brey and J. Gomez-Ordonez, *J. Chem. Phys.*, **76**, 3260 (1982).
47. C. Hoheisel, *J. Phys.: Condens. Matter*, **2**, 5849 (1990)
48. A.J. Masters, *Mol. Phys.*, **57**, 303 (1986).
49. H.M. Schaink and C. Hoheisel, *Phys. Rev. A*, **45**, 8559 (1992).
50. L.F. Rull, E. de Miguel, J.J. Morales and M.J. Nuevo, *Phys. Rev. A*, **40**, 5856 (1989).
51. M.J. Nuevo, J.J. Morales and D.M. Heyes, *Phys. Rev. E*, **51**, 2026 (1995).
52. D.M. Heyes, M.J. Nuevo and J.J. Morales, *Mol. Phys.*, **88**, 1503 (1996).
53. M.J. Nuevo, J.J. Morales and D.M. Heyes, *Mol. Phys.*, **91**, 769 (1997).
54. D.M. Heyes, M.J. Nuevo and J.J. Morales, *Mol. Phys.*, **93**, 985 (1998).
55. D.M. Heyes, M.J. Nuevo and J.J. Morales, *J. Chem. Soc., Faraday Trans.*, **94**, 1625 (1998).
56. H.-N. Roux, *Physica A*, **188**, 526 (1992).
57. D.M. Heyes, M.J. Nuevo, J.J. Morales and A.C. Branka, *J. Phys.: Condens. Matter*, **10**, 10159 (1998).
58. M. J. Nuevo, J.J. Morales and D.M. Heyes, *Phys. Rev. E*, **58**, 5845 (1998).
59. D.M. Heyes and A.C. Branka, *Mol. Phys.*, **96**, 1757 (1999).
60. K. Kerl and M. Willeke, *Mol. Phys.*, **97**, 1255 (1999); **99**, 471.
61. T. Yamaguchi, Y. Kimura and N. Hirota, *Mol. Phys.* **94**, 527 (1998).
62. F. Ould-Kaddour and D. Levesque, *Phys. Rev. E*, **63**, 011205 (2000).
63. A. Malevanets and R. Kapral, *J. Chem. Phys.*, **112**, 7260 (2000).
64. S.H. Lee and R. Kapral, *J. Chem. Phys.*, **121**, 11163 (2004).
65. S.H. Lee and R. Kapral, *J. Chem. Phys.*, **122**, 214916 (2005).
66. G.L. Paul and P.N. Pusey, *J. Phys. A: Math. General*, **14**, 3301 (1981).
67. W. van Megen, *J. Phys: Condens. Matter*, **14**, 7699 (2002).
68. D.J. Evans and H.J.M. Hanley, *Phys. Rev. A*, **20**, 1648 (1979); H.J.M. Hanley, D.J. Evans and S. Hess, *J. Chem. Phys.*, **78**, 1440 (1983).
69. I. Snook, B. O'Malley, M. McPhie and P. Daivis, *J. Mol. Liquids*, **103–104**, 405 (2003)
70. S. Toxvaerd, *Mol. Phys.*, **56**, 1017 (1985).
71. L.F. Rull, E. de Miguel, J.J. Morales and M.J. Nuevo, *Phys. Rev. A*, **40**, 5856 (1989).
72. M.J. Nuevo and J.J. Morales, *Phys. Lett.* A, **178**, 114 (1993).
73. K. Tankeshwar, *J. Phys.: Condens. Matter*, **7**, 9715 (1995).
74. M. J. Nuevo, J.J. Morales and D.M. Heyes, *Phys. Rev. E*, **58**, 5845 (1998).
75. E.H. Hauge and A. Martin-Lof, *J. Stat. Phys.*, **7**, 259 (1973).
76. E.J. Hinch, *J. Fluid Mech.*, **72**, 303 (1975).
77. A.J. Masters, *Mol. Phys.*, **57**, 303 (1986).
78. J.N. Roux, *Physica A*, **188**, 526 (1992).
79. D. Bedeaux and P. Mazur, *Physica A*, **76**, 247 (1974).
80. M. McPhie, *Thermal and Transport Properties of Shearing Binary Fluids*, PhD Thesis, RMIT University, Melbourne, Australia, 2003.
81. A. Einstein, *Investigations on the Theory of the Brownian Motion* (edited by R. Furth and translated by A.D. Cowper), Dover, New York, 1956; I.M. Krieger and T.H. Dougherty, *Trans. Soc. Rheol.*, **3**, 137 (1959); G.K. Batchelor, *J. Fluid Mech.*, **83**, 97 (1977).

– 6 –

Langevin and Generalised Langevin Dynamics

Having established the theoretical basis for the GLE and LE, and having discussed approximate schemes to analyse time correlation functions, we will now turn our attention to the implementation of numerical methods to solve these equations. In this chapter we will cover methods based on the GLE-equations that are most appropriate to be used for atomic, molecular and small polymer simulations. In the next chapter we will turn to those methods that are needed for systems with much greater disparities in size and mass of the two classes of particles being modelled. Initially, though, we will need to discuss some limitations of the GLE and its extensions.

6.1 EXTENSIONS OF THE GLE TO COLLECTIONS OF PARTICLES

The GLE for a particle of each species i is from the Mori–Zwanzig approach is given by eq. (2.28),

$$m_i(\mathrm{d}/\mathrm{d}t)\underline{v}_i(t) = -m_i\int_0^t K_v(\tau)\underline{v}_i(t-\tau)\,\mathrm{d}\tau + \underline{F}_i^{\mathrm{R}}(t) \tag{6.1}$$

However, it should be noted that these equation do not explicitly contain the direct or bath averaged interaction between the selected particles. These interactions are implicitly contained in the memory functions and random force terms in eq. (6.1). The same comments apply to the exact GLEs for a single particle in a background from the Albers–Deutch–Oppenheim approach by using eq. (5.18) and putting $A(t) = \underline{V}_{\mathrm{B}}(\mathrm{t})$, i.e.

$$\mathrm{d}\underline{V}_{\mathrm{B}}(t)/\mathrm{d}t = K^+(t) + \int_0^t \mathrm{e}^{\mathrm{i}\mathbf{L}(t-\tau)}\langle \mathrm{i}\mathbf{L}_{\mathrm{B}}K^+(\tau)\rangle_0\,\mathrm{d}\tau \tag{6.2}$$

and equivalently by eq. (5.22),

$$d\underline{V}_B(t)/dt = e^{i(1-P)Lt}\,\underline{F}_B(0)/M_B + 1/M_B\int_0^t e^{iL(t-\tau)}$$
$$[\underline{\nabla}_P - \underline{V}_B/k_BT]\cdot\langle \underline{F}_B\, e^{i(1-P)Lt}\,\underline{F}_B(0)\rangle_0\, d\tau$$

or

$$= K^+(t) + 1/M_B\int_0^t e^{iL(t-\tau)}[\underline{\nabla}_P - \underline{V}_B/k_BT]\cdot\langle \underline{F}_B\, e^{i(1-P)Lt}\,\underline{F}_B(0)\rangle_0\, d\tau \tag{6.3}$$

By contrast the equation for many particles in a background of other particles does have explicitly terms referring to these 'direct' interactions among the selected particles. We now write these equations below and while retaining the term bath for the N_b particles to be averaged over and call the N_s particles to be explicitly considered as labelled by lower case symbols in order of emphasis that they are not necessarily Brownian particles, i.e. eqs. (5.70) and (5.72),

$$m_i\, d\underline{v}_i/dt = A(t) + K^+(t) + \int_0^t e^{iL(t-\tau)}\langle iL_B K^+(\tau)\rangle_0 d\tau \tag{6.4}$$

and

$$m_i\, d\underline{v}_i/dt = \underline{S}_i(t)\cdot\underline{\nabla}_{\underline{p}_i(t)}\underline{v}_i(t) + K^+(t)$$
$$+\sum_{j=1}^{N_s}\int_0^t e^{iL(t-\tau)}\cdot\left\{\left[\underline{\nabla}_{\underline{p}_j} - (1/k_BT/m_j)\cdot\underline{p}_j\right]\cdot\langle \underline{E}_j\, e^{i(1-P)Lt}\,\underline{E}_i\rangle_0\cdot\underline{\nabla}_{\underline{p}_i}\underline{v}_i(0)\right\}d\tau \tag{6.5}$$

where i = 1 to N_s.

The potential of mean force χ defined now represents the bath or 'solvent' mediated static interaction between the n selected particles,

$$Z\left(R^{(N_s)}\right) = C(\beta,N,V)\exp\left[-1/k_BT\chi\left(\underline{r}^{(N_s)}\right)\right] \tag{6.6}$$

so that using

$$\begin{aligned}\langle -\underline{\nabla}_{ri}\Phi_i\rangle_0 &= -k_BT^{-1}\underline{\nabla}_{ri}\ln Z\left(\underline{r}^{(N_s)}\right)\\ &= \underline{\nabla}_{ri}\chi\left(\underline{R}^{(n)}\right)\end{aligned} \tag{6.7}$$

from which we may define the bath averaged force on a selected particle μ by,

$$\langle \underline{F}_i\rangle_0 = -\left\langle \underline{\nabla}_{ri}\left[V\left(\underline{r}^{(N)},\underline{r}^{(N_s)}\right)+\Phi_i\right]\right\rangle_0 = -\underline{\nabla}_{ri}\left[V + \chi\left(\underline{r}^{(N_s)}\right)\right] \tag{6.8}$$

and define $\underline{E}_i = \underline{F}_i - \langle \underline{F}_i\rangle_0$.

However, it is quite difficult to use eq. (6.4) or eq. (6.5) explicitly to treat the general case because it is not clear how the terms involved can be accurately represented. Also even though we may use the simpler equation (5.74) this equation is a limiting case of the above equations and is strictly only valid in the limit $m/M \to \infty$ while also requiring elaborate many-body hydrodynamic calculations to obtain the terms appearing therein.

It would be desirable to find a method that combines the simplicity of eq. (6.1) but containing terms explicitly representing the interactions amongst the particles which have been chosen to describe (averaged over the bath particles or not) and having the effect of the background particles contained in a memory function. Unfortunately, this is a very difficult task and may not be possible to accomplish in a general way, which is both rigorous and practical.[1–4]

Clearly it should be possible to write an equation of the form we wish if for no other reasons than a combination of physical insight based on the Deutch–Oppenheim method and dimensional analysis. However, the problem is to find expressions for the memory function $K_v(t)$ and the random force $\underline{F}_i^{\mathrm{R}}(t)$ or at least for some of their characteristics such as the short time expansion of $K_v(t)$ and the moments of $\underline{F}_i^{\mathrm{R}}(t)$.

The simplest such method is to merely postulate that the equation of motion is[4],

$$m_i(\mathrm{d}/\mathrm{d}t)\underline{v}_i(t) = -m_i\int_0^t K_v(\tau)\underline{v}_i(t-\tau)\,\mathrm{d}\tau + \underline{F}_i(t) + \underline{F}_i^{\mathrm{R}}(t) \tag{6.9}$$

for $i = 1$ to N_s where $\underline{F}_i(t)$ is some postulated force law representing the direct interaction between the members of the chosen particles. Usually, $\underline{F}_i(t)$ is taken to be a 'solvent-averaged' potential of mean force which is obtained from some physical model, e.g. the DLVO theory for the interaction of charged colloidal particles.

An even simpler approach which is often used is to combine eq. (6.9) with the assumption that $K(t)$ is delta correlated and $\underline{F}_i^{\mathrm{R}}(t)$ is a Gaussian random variable so we have[4]

$$m_i(\mathrm{d}/\mathrm{d}t)\underline{v}_i(t) = -\zeta\underline{v}_i(t) + \underline{F}_i(t) + \underline{F}_i^{\mathrm{R}}(t) \tag{6.10}$$

where $i = 1$ to N_s

$$\langle \underline{F}_i^{\mathrm{R}}(0)\rangle = 0$$

$$\begin{aligned}\langle \underline{F}_i^{\mathrm{R}}(0)\cdot\underline{F}_i^{\mathrm{R}}(t)\rangle &= 6\zeta\, k_{\mathrm{B}}T\,\delta(t)\\ &= 6\gamma M_{\mathrm{B}}\, k_{\mathrm{B}}T\,\delta(t)\end{aligned}$$

This is, of course, the limiting form obtained for N_s particles in a background of small particles in the limit that $m/M \to \infty$ and neglecting many-body hydrodynamics. This approach has been used to describe colloidal suspensions and to maintain constant temperature in atomic systems, for example, in simulating the deposition of ions onto solid surfaces.

Bossis, Quentrec and Boon[2,3] have introduced an equation of the type given by eq. (6.9) for treating a single polymer chain, which will be discussed in Chapter 8.

6.2 NUMERICAL SOLUTION OF THE LANGEVIN EQUATION

To illustrate the methods used to solve the GLE, we will start by discussing the numerical solution of the limiting case of the LE equation (6.10). We will use the rather formal approach due to Ermak and Buckhloz[4] in order to fully illustrate all the subtleties involved. From eq. (6.10) the LE for a particle i acted is obtained by assuming that the memory function $K(t)$ is a delta function, i.e.

$$m_i(\mathrm{d}/\mathrm{d}t)\,\underline{v}_i = \underline{F}_i - m_i\beta\,\underline{v}_i + \underline{F}_i^{\mathrm{R}} \tag{6.11}$$

where $\underline{F}_i$ an external force, $-m_i\beta\underline{v}_i = -\zeta\underline{v}_i$ a drag force and $\underline{F}_i^{\mathrm{R}}$ is a Gaussian random force defined by (using β instead of ζ as used previously so as to be consistent with reference 4)

$$\langle \underline{F}_i^{\mathrm{R}}(t)\rangle = 0 \tag{6.12}$$

and

$$\langle \underline{F}_i^{\mathrm{R}}(t)\cdot\underline{F}_i^{\mathrm{R}}(t')\rangle = 6k_{\mathrm{B}}T\,m_i\beta\,\delta(t-t').$$

In order to derive a numerical scheme for solving this equation we must be able to integrate both the terms $\underline{F}_i$ and $\underline{F}_i^{\mathrm{R}}$. The former term may simply be approximated by some assumed form for $\underline{F}_i$ over the time step Δt as is done in MD simulations, e.g. it may be assumed to be constant or of quadratic form. The integration of the random force term $\underline{F}_i^{\mathrm{R}}$ requires a little more thought as to what such an integral means and this will be discussed below.

To make eq. (6.11) easier to solve we multiply it by $\mathrm{e}^{-\beta t_0}$ and integrate from t_0 to $t_0 + t$ which leads to

$$\begin{aligned}\underline{v}_i(t_0+t)-\underline{v}_i(t_0)\mathrm{e}^{-\beta t} &= 1/m_i\int_0^t \mathrm{e}^{-\beta(t-s)}\,\underline{F}_i(s+t_0)\,\mathrm{d}s\\ &\quad+(1/m_i)\int_0^t \mathrm{e}^{-\beta(t-s)}\,\underline{F}_i^{\mathrm{R}}(s+t_0)\,\mathrm{d}s\end{aligned} \tag{6.13}$$

By integrating eq. (6.13) with respect to time we obtain

$$\begin{aligned}\underline{r}_i(t_0+t)-\underline{r}_i(t_0) &= \underline{v}_i(t_0)/\beta(1-\mathrm{e}^{-\beta t})+(1/m_i)\int_0^t \mathrm{d}s\int_0^s \mathrm{e}^{-\beta(s-s')}\,\underline{F}_i(s'+t_0)\,\mathrm{d}s'\\ &\quad+(1/m_i)\int_0^t \mathrm{d}s\int_0^s \mathrm{e}^{-\beta(s-s')}\,\underline{F}_i^{\mathrm{R}}(s'+t_0)\,\mathrm{d}s\end{aligned} \tag{6.14}$$

The above two equations (6.13) and (6.14) are formally exact so in order to obtain a practical numerical scheme we must evaluate the integrals involving $\underline{F}_i(s + t_0)$ and $\underline{F}_i^{\mathrm{R}}(s + t_0)$. If we assume that the external force $\underline{F}_i$ is constant over the time interval t_0 to $t + t_0$ then eq. (6.13) becomes

$$\underline{v}_i(t_0 + t) = \underline{v}_i(t_0)\mathrm{e}^{-\beta t} + \underline{F}(t)/\beta(1 - \mathrm{e}^{-\beta t}) + (1/m_i)\int_0^t \mathrm{e}^{-\beta(t-s)}\,\underline{F}_i^{\mathrm{R}}(t_0 + s)\mathrm{d}s \tag{6.15}$$

and eq. (6.14) then yields as equation for the particle displacement over t_0 to $t_0 + t$,

$$\begin{aligned}\underline{r}_i(t_0 + t) = \underline{r}_i(t_0)\mathrm{e}^{-\beta t} + m_i\,\underline{v}_i(t_0)/\beta(1 - \mathrm{e}^{-\beta t}) + \underline{F}_i(t_0)/(m_i\beta)[t - (1/\beta)(\mathrm{e}^{-\beta t} - 1)]\\ + 1/(m_i\beta)\int_0^t \left[1 - \mathrm{e}^{-\beta(t-s)}\right]\underline{F}_i^{\mathrm{R}}(s)\mathrm{d}s\end{aligned} \tag{6.16}$$

where the last integral arises from integration by parts of the last term in eq. (6.14).

6.2.1 Gaussian random variables[5,6]

Thus, we have to evaluate two integrals involving the random force $\underline{F}_i^{\mathrm{R}}$; now these terms cannot be evaluated by standard methods of integration due to the random nature of $\underline{F}_i^{\mathrm{R}}$ which means that all we know about it are its statistical properties given by eq. (6.12).

Now if $\underline{A}(\lambda)$ is a random variable then[5,6], $\int_{t_1}^{t_2}\underline{A}(\lambda)f(\lambda)\mathrm{d}\lambda = \mathrm{B}(\lambda)$ is a random variable whose statistical properties are related to those of $\underline{A}(\lambda)$. We may use theorems due to Chandrasekhar[5] to find these statistical properties.

If $\underline{A}(\lambda)$ is a Gaussian random variable for which, $\langle\underline{A}\rangle = 0$ and $\langle\underline{A}(t)\,\underline{A}(t')\rangle = 6\zeta k_{\mathrm{B}}T\delta(t - t')$ and if $f(\lambda) = 1$ then the probability of occurrence of different values of $B(t)$ is governed by the distribution function,

$$w\{\underline{B}(t)\} = 1/(4\pi qt)^{3/2}\exp\{-|\underline{B}(t)|^2/(4qt)\} \tag{6.17}$$

where $q = \beta(k_{\mathrm{B}}T/m_i)$.

In the more general case $\int_0^t \underline{A}(\lambda)\psi(\lambda)\mathrm{d}\lambda = \underline{R}(\lambda)$ is a random variable whose probability distribution is given by

$$w\{\underline{R}(t)\} = 1/(4\pi qA)^{3/2}\exp\{-|\underline{R}(\mathrm{t})|^2/(4qA)\} \tag{6.18}$$

where $A = \int_0^t \psi^2(\lambda)\mathrm{d}\lambda$.

Thus, $\underline{R}$ may be chosen from a Gaussian distribution with mean zero and mean-square value given by $\langle|\underline{R}(t)|^2\rangle = 2q\int\psi^2(\lambda)\mathrm{d}\lambda$.

If we have two random variables $\underline{R}$ and $\underline{V}$ given by

$$\underline{R}(\lambda) = \int_0^t \underline{A}(\lambda)\psi(\lambda)\,d\lambda$$

and

$$\underline{V}(\lambda) = \int_0^t \underline{A}(\lambda)\phi(\lambda)\,d\lambda$$

then $\underline{R}$ and $\underline{V}$ are governed by the bi-variate probability distribution given by[4–6]

$$\begin{aligned} W(\underline{R},\underline{V}) &= 1/\{(8\pi^3)(EG - H^2)^{3/2}\}\exp[\{G\,|\,\underline{R}(t)|^2 - 2H\,\underline{R}\cdot\underline{V} + E\,|\,\underline{V}(t)|^2\}/\{2(EG - H^2)\}] \\ &= 1/\{(8\pi^3)(E^{3/2}G^{3/2}(1 - H^2/EG)^{3/2}\}\exp[1/\{2(1 - H^2/EG)\} \\ &\quad \times\{|\,\underline{R}(t)|^2/E - 2H\,\underline{R}\cdot\underline{S}/(EG) + |\,\underline{V}(t)|^2/G\}] \end{aligned} \tag{6.19}$$

where

$$E = 2q\int_0^t \psi^2(\lambda)\,d\lambda \; G = 2q\int_0^t \phi^2(\lambda)\,d\lambda \quad \text{and} \quad H = 2q\int_0^t \psi(\lambda)\phi(\lambda)\,d\lambda$$

with $\langle\underline{R}\rangle = \langle\underline{V}\rangle = 0$, $\langle|\underline{R}(t)|^2\rangle = 3E$, $\langle|\underline{R}(t)|^2\rangle = 3G$ and $\langle\underline{R}\cdot\underline{V}\rangle = 3H$

Thus, if one wishes one can sample $\underline{R}$ and $\underline{V}$ from the bi-variate (joint) probability function (6.19), which means that the variables are statistically interdependent. Alternatively and, in practice, somewhat more simply, one may choose to sample one of the variables from an independent Gaussian (unconditional) distribution and then sample the second from a conditional probability distribution whose properties depend on the first, unconditional probability distribution. For example, we may choose $\underline{V}$ from the Gaussian distribution,[4–6]

$$\begin{aligned} W(\underline{V}) &= \int W(\underline{R},\underline{V})\,d\underline{r} \\ &= 1/\{(2\pi G)^{3/2}\}\exp\left[|-\underline{V}(t)|^2/2G\right] \end{aligned} \tag{6.20}$$

and then choosing $\underline{R}$ from the conditional probability distribution,

$$\begin{aligned} W(\underline{R}\,|\,\underline{V} = \underline{V}) &= W(\underline{R},\underline{V})/W(\underline{V}) \\ &= (2\pi G)^{3/2}/\{(2\pi(EG(1 - H^2/EG)\}^{3/2} \\ &\quad \exp[1/\{2(1 - H^2/EG)\}\{|\,\underline{R}(t)|^2/E \\ &\quad - 2H\,\underline{R}\cdot\underline{S}/(EG) + |\,\underline{V}(t)|^2/G\} + 1/(2G)\,|\,\underline{V}(t)|^2] \\ &= 1/\{(E(1 - H^2/EG)\}^{3/2}\exp[1/\{2E(1 - H^2/EG)\}\{|\,\underline{R}(t) - H/G\,\underline{V}(t)|^2] \end{aligned} \tag{6.21}$$

Thus, $\underline{V}$ is sampled from a simple Gaussian distribution function and the variable $\{\underline{R}(t) - H/G\underline{V}(t)\}$ (but not $\underline{R}$) may then also chosen from a simple Gaussian distribution function of mean value zero and mean-square value $3E(1 - H^2/EG) = 3(E - H^2/G)$.

Similarly, if we chose to sample $\underline{R}$ from a simple Gaussian distribution distribution,

$$\begin{aligned} W(\underline{R}) &= \int W(\underline{R},\underline{V})\mathrm{d}\underline{v} \\ &= 1/\{(2\pi E)^{3/2}\}\exp[|-\underline{R}(t)|^2/2E] \end{aligned} \tag{6.22}$$

and then choosing $\underline{V}$ from the conditional probability distribution,

$$\begin{aligned} W(\underline{V} \mid \underline{R} = \underline{R}) &= W(\underline{R},\underline{V})/W(\underline{R}) \\ &= (2\pi E)^{3/2}/\{(2\pi(EG)(1-H^2/EG))\}^{3/2} \\ &\quad \exp\Big[1/\{2(1-H^2/EG)\}\{|\underline{R}(t)|^2/E - 2H\,\underline{R}\cdot\underline{S}/(EG) + |\underline{V}(t)|^2/G\} \\ &\quad + 1/(2E)|\underline{R}(t)|^2\Big] \\ &= 1/\{(G(1-H^2/EG)\}^{3/2}\exp\left[1/\{2G(1-H^2/EG)\}\{|\underline{V}(t) - 2H/E\,R(t)|^2\}\right] \end{aligned} \tag{6.23}$$

i.e. we may sample the variable $\{\underline{V}(t) - 2H/E\ \underline{R}(t)\}$ (but not $\underline{V}$) from a simple Gaussian distribution of mean zero and mean-square value $3G(1 - H^2/EG) = 3(G - H^2/E)$.

6.2.2 A BD algorithm to first-order in Δt^4

Using the above results we may now evaluate the integrals in eqs. (6.15) and (6.16) or more precisely we may find the statistical properties of these new random variables from those of the random variable $\underline{F}_i^{\mathrm{R}}(t)$. Let us consider eq. (6.15), by the above theorems the random variable $\underline{V}$ given by

$$\begin{aligned} \underline{V} &= \int_0^t \mathrm{e}^{-\beta(t-s)}\underline{F}_i^{\mathrm{R}}(s+t_0)/m_i\,\mathrm{d}s \\ &= \underline{v}_i(t_0+t) - \underline{v}_i(t_0)\mathrm{e}^{-\beta t} + 1/m_i\int_0^t \mathrm{e}^{-\beta(t-s)}\underline{F}_i(s+t_0)\mathrm{d}s \end{aligned}$$

is a Gaussian random variable with probability density,

$$W(\underline{V}) = 1/\{4\pi q\,A\}^{3/2}\exp\{-|\underline{V}|^2/(-4qA)\}$$

where $q = \beta(k_{\mathrm{B}}T/m)$ and

$$\begin{aligned} A &= \int_0^t (\mathrm{e}^{-\beta(t-s)})^2\,\mathrm{d}s \\ &= 1/(2\beta)[1-\mathrm{e}^{-2\beta t}] \end{aligned}$$

So the probability distribution for $\underline{V}$ is

$$\begin{aligned} W(\underline{V}) &= W(\underline{v}_i(t)-\underline{v}_i(t_0)\mathrm{e}^{-\beta t}-1/(m_i\beta)\,\underline{F}_i(t_0)[1-\mathrm{e}^{-\beta t}]) \\ &= 1/\{(2\pi\beta k_B T/m_i)[1-\mathrm{e}^{-2\beta t}]^{3/2}\}\exp\{-|\underline{V}|^2/\{(2k_B T/m_i)[1-\mathrm{e}^{-2\beta t}]\}\} \end{aligned} \tag{6.24}$$

which is a Gaussian of mean zero and mean-square value $\langle |V|^2 \rangle = (3k_B T/m_i)\,[1-\mathrm{e}^{-2\beta t}]$.

Hence, the random variable $\underline{v}_i(t) - \underline{v}_i(t_0)\,\mathrm{e}^{-\beta t} - 1/(m_i\beta)\,\underline{F}_i(t_0)[1-\mathrm{e}^{-\beta t}]$ may be chosen from a 3-D Gaussian distribution of mean value zero and mean-square value $(3k_B T/m_i)$ $[1-\mathrm{e}^{-2\beta t}]$, i.e.

$$\underline{v}_i(t) = \underline{v}_i(t_0)\mathrm{e}^{-\beta t} - 1/(m_i\beta)\underline{F}_i(t_0)[1-\mathrm{e}^{-\beta t}] + \underline{V}(t) \tag{6.25}$$

where $\underline{V}(t)$ is a random number chosen from a 3-D Gaussian distribution of mean zero and mean-square value $(3k_B T/m_i)$ $[1-\mathrm{e}^{-2\beta t}]$.

Now eq. (6.16) may be re-written as

$$\begin{aligned} \underline{R}(t) &= \int_0^t [1-\mathrm{e}^{-\beta(t-s)}]/\beta\,\underline{F}_i^{\mathrm{R}}(s)/m_i\,\mathrm{d}s \\ &= \underline{r}_i(t_0+t)-\underline{r}_i(t_0)+\underline{v}_i(t_0)/\beta(1-\mathrm{e}^{-\beta t})+\underline{F}_i(t_0)/(m_i\beta)[t-1/\beta(\mathrm{e}^{-\beta t}-1)] \end{aligned}$$

Thus, $\underline{R}(t)$ is a Gaussian random variable whose distribution function is

$$W(\underline{R}) = 1/[4\pi q A]\exp\{-|\underline{R}|^2/(4qA)\}$$

where $q = \beta(k_B T/m_i)$
and

$$\begin{aligned} A &= \int_0^t \left[1-\mathrm{e}^{-\beta(t-s)}\right]^2 \mathrm{d}s \\ &= 1/(2\beta^3)\left[2\beta t-3+4\mathrm{e}^{-\beta t}-\mathrm{e}^{-2\beta t}\right] \end{aligned}$$

Thus,

$$\begin{aligned} W(\underline{R}) = {} & 1/\{2\pi k_B T/(m_i\beta^2)[2\beta t-3+4\mathrm{e}^{-\beta t}-\mathrm{e}^{-2\beta t}]\}^{3/2} \\ & \exp\{-|\underline{r}_i(t_0+t)-\underline{r}_i(t_0)+\underline{v}_i(t_0)/\beta(1-\mathrm{e}^{-\beta t})+\underline{F}_i(t_0)/(m_i\beta)[t-1/\beta(\mathrm{e}^{-\beta t}-1)]|^2 \\ & /[2k_B T/(m_i\beta^2)[2\beta t-3+4\mathrm{e}^{-\beta t}-\mathrm{e}^{-2\beta t}]\} \end{aligned} \tag{6.26}$$

The Gaussian random variables $\underline{V}$ and $\underline{R}$ (and thus, $\underline{r}$, $\underline{v}$) are correlated and hence obey a Bivariate Gaussian distribution[4–6]and from the theorems above the appropriate distribution is

$$\begin{aligned} W(\underline{R},\underline{V}) &= 1/\{(8\pi^3)(EG-H^2)^{3/2}\}\exp[\{G\,|\,\underline{R}(t)\,|^2 - 2H\,\underline{R}\cdot\underline{V} + E\,|\,\underline{V}(t)\,|^2\}/\{2(EG-H^2)\}] \\ &= 1/\{(8\pi^3)(E^{3/2}\,G^{3/2}\,(1-H^2/EG)^{3/2}\}\exp[1/\{2(1-H^2/EG)\} \\ &\quad \times\{|\,\underline{R}(t)\,|^2/E - 2H\,\underline{R}\cdot\underline{V}/(EG) + |\,\underline{V}(t)\,|^2/G\}] \end{aligned} \tag{6.27}$$

with

$$\begin{aligned} E &= 2q\int_0^t \{[1-\mathrm{e}^{-\beta(t-\lambda)}]/\beta\}^2\,\mathrm{d}\lambda \\ &= k_\mathrm{B}T/(m_i\beta^2)(2\beta t - 3 + 4\mathrm{e}^{-\beta t} - \mathrm{e}^{-2\beta t}) \end{aligned}$$

$$\begin{aligned} G &= 2q\int_0^t (\mathrm{e}^{-\beta(t-\lambda)})^2\,\mathrm{d}\lambda \\ &= k_\mathrm{B}T/m_i\left(1-\mathrm{e}^{-2\beta t}\right) \end{aligned}$$

and

$$\begin{aligned} H &= 2q\int_0^t \{[1-\mathrm{e}^{-\beta(t-\lambda)}]/\beta\}\left(\mathrm{e}^{-\beta(t-\lambda)}\right)\mathrm{d}\lambda \\ &= k_\mathrm{B}T/(m_i\beta)(1-\mathrm{e}^{-\beta t})^2 \end{aligned}$$

So,

$$\langle\underline{R}\rangle = \langle\underline{V}\rangle = 0$$

$$\left\langle|\underline{R}(\mathrm{t})|^2\right\rangle = 3E = 3k_\mathrm{B}T/(m_i\beta^2)\left(2\beta t - 3 + 4\mathrm{e}^{-\beta t} - \mathrm{e}^{-2\beta t}\right)$$

$$\left\langle|\underline{V}(t)|^2\right\rangle = 3G = 3k_\mathrm{B}T/m_i\left(1-\mathrm{e}^{-2\beta t}\right)$$

and

$$\langle\underline{R}\cdot\underline{V}\rangle = 3H = 3\,k_\mathrm{B}T/(m_i\beta)(1-\mathrm{e}^{-\beta t})^2$$

Thus,

$$\underline{v}_i(t_0+t) = \underline{v}_i(t_0)\mathrm{e}^{-\beta t} + 1/(m_i\beta)\,\underline{F}_i(t_0)[1-\mathrm{e}^{-\beta t}] + \underline{A}_1(t) \tag{6.28}$$

and

$$\underline{r}_i(t_0+t)=\underline{r}_i(t_0)+\underline{v}_i(t_0)/\beta(1-e^{-\beta t})+\underline{F}_i(t_0)/(m_i\beta)[t-1/\beta(e^{-\beta t}-1)]+\underline{A}_2(t) \tag{6.29}$$

where $\underline{A}_1$ and $\underline{A}_2$ are 3-D Gaussian random variables for which

$$\langle\underline{A}_1\rangle=\langle\underline{A}_2\rangle=0 \tag{6.30}$$

$$\left\langle|\underline{A}_1|^2\right\rangle=3G=3k_BT/m_i\left(1-e^{-2\beta t}\right)$$

$$\left\langle|\underline{A}_2|^2\right\rangle=3E=3k_BT/\left(m_i\beta^2\right)\left(2\beta t-3+4e^{-\beta t}-e^{-2\beta t}\right)$$

and

$$\langle\underline{A}_1\cdot\underline{A}_2\rangle=3H=3k_BT/(m_i\beta)\left(1-e^{-\beta t}\right)$$

The algorithm represented by eq. (6.28)–(6.30) is not usually used as it is easier to work with uncorrelated Gaussian variables.

6.2.3 A second first-order BD algorithm[4]

We may now derive another algorithm for generating $\underline{r}_i$ and $\underline{v}_i$ by using the distribution

$$\begin{aligned} W(\underline{V})&=W\left(\underline{v}_i(t+t_0)-v_i(t_0)e^{-\beta t}-1/(m_i\beta)\underline{F}_i(t_0)[1-e^{-\beta t}]\right)\\ &=\int W(\underline{R},\underline{V},t)d^3\underline{r}\\ &=1/\{8\pi^3G^{3/2}\}\exp\{-|\underline{V}|^2/2G\}\end{aligned} \tag{6.31}$$

which is a Gaussian distribution of mean zero and

$$\left\langle|\underline{V}|^2\right\rangle=3G=3k_BT/m_i\left(1-e^{-2\beta t}\right)$$

Thus,

$$\underline{v}_i(t+t_0)=v_i(t_0)e^{-\beta t}-1/(m_i\beta)\underline{F}_i(t_0)[1-e^{-\beta t}])+\underline{B}_1 \tag{6.32}$$

where

$$\langle\underline{B}_1\rangle=0$$

and

$$\left\langle |\underline{B}_1|^2 \right\rangle = 3G = 3k_BT/m_i\left(1-e^{-2\beta t}\right)$$

Then the variable $\underline{R} = \underline{r}_i(t_0 + t) - \underline{r}_i(t_0) + \underline{v}_i(t_0)/\beta(1 - e^{-\beta t}) + \underline{F}_i(t_0)/(m_i\beta)\ [t - 1/\beta(e^{-\beta t} - 1)]$ must be chosen from the conditional probability,

$$\begin{aligned}W(\underline{R} \mid \underline{V} = \underline{V}) &= W(\underline{R},\underline{V},t)/W(\underline{V},t)\\ &= 1/\{8\pi^3 E(1-H^2/EG)\}^{3/2}\exp\{-1/\{2E(1-H^2/EG)\}[\mid \underline{R} - r\sigma_2/\sigma_1\underline{V}\mid^2]\}\end{aligned} \tag{6.33}$$

where

$$r\sigma_2/\sigma_1 = H/G = (1/\beta)(1-e^{-\beta t})^2/(1-e^{-2\beta t})$$

and

$$\sigma_2^2(1-r^2) = E(1-H^2/EG) = (E-H^2/G)$$

Thus, the variable $\underline{R} - r\sigma_2/\sigma_1\underline{V}$ must be chosen from a Gaussian distribution of mean zero and mean-square value $3\sigma_2^2(1 - r^2) = 3(E - H^2/G)$, hence,

$$\begin{aligned}\underline{B}_2 &= \underline{R} - r\sigma_2/\sigma_1\underline{V}\\ &= \underline{r}_i(t_0+t) - \underline{r}_i(t_0) + \underline{v}_i(t_0)/\beta(1-e^{-\beta t}) + \underline{F}_i(t_0)/(m_i\beta)[t - 1/\beta(e^{-\beta t}-1)]\\ &\quad -(1/\beta)(1-e^{-\beta t})^2/(1-e^{-2\beta t})\{\underline{v}_i(t+t_0) - \underline{v}_i(t_0)e^{-\beta t} - 1/(m_i\beta)\underline{F}_i(t_0)[1-e^{-\beta t}]\}\\ &= \underline{r}_i(t_0+t) - \underline{r}_i(t_0) + 1/\beta\{\underline{v}_i(t+t_0) - \underline{v}_i(t_0) - 2\underline{F}_i(t_0)/(m_i\beta)(1-e^{-\beta t})/(1+e^{-\beta t})\\ &\quad -\underline{F}_i(t_0)/(m_i\beta)t\end{aligned}$$

Thus, $\underline{B}_2$ is a Gaussian random variable for which $\langle \underline{B}_2 \rangle = 0$ and $\langle |\underline{B}_2|^2 \rangle = 3(E - H^2/G) = 6k_BT/(m_i\beta^2)\ \{\beta t - 2\ (1 - e^{-\beta t})/(1 + e^{-\beta t})\}$

To summarise,

$$\underline{v}_i(t+t_0) = \underline{v}_i(t_0)e^{-\beta t} - 1/(m_i\beta)\underline{F}_i(t_0)\left[1-e^{-\beta t}\right]) + \underline{B}_1 \tag{6.34}$$

where $\underline{B}_1$ is a Gaussian random variable with,

$$\langle \underline{B}_1 \rangle = 0 \text{ and } \langle |\underline{B}_1|^2 \rangle = 3k_BT/m_i(1-e^{-2\beta t}) \tag{6.35}$$

and

$$\begin{aligned}\underline{r}_i(t_0+t) = \underline{r}_i(t_0) + 1/\beta\{&\underline{v}_i(t+t_0) + \underline{v}_i(t_0)\\ &-2\underline{F}_i(t_0)/(m_i\beta)(1-e^{-\beta t})/(1+e^{-\beta t})\\ &-\underline{F}_i(t_0)/(m_i\beta)t + \underline{B}_2\end{aligned} \tag{6.36}$$

where $\underline{B}_2$ is a Gaussian random variable with

$$\langle \underline{B}_2 \rangle = 0 \text{ and } \langle |\underline{B}_2|^2 \rangle = 6k_BT/(m_i\beta^2)\{\beta t - 2(1-e^{-\beta t})/(1+e^{-\beta t})\} \tag{6.37}$$

and

$$\langle \underline{B}_1 \cdot \underline{B}_2 \rangle = 0 \tag{6.38}$$

Eqs. (6.34)–(6.38) provide an algorithm for performing BD simulations.

6.2.4 A third first-order BD algorithm[4]

Alternatively, we may choose the variable $\underline{R} = \underline{r}_i(t_0 + t) - \underline{r}_i(t_0) + \underline{v}_i(t_0)/\beta(1 - e^{-\beta t}) + \underline{F}_i(t_0)/(m_i\beta)[t - 1/\beta(e^{-\beta t} - 1)]$ from a Gaussian random distribution with

$$\begin{aligned} W(\underline{R}) &= W(\underline{r}_i(t_0+t) - \underline{r}_i(t_0) + \underline{v}_i(t_0)/\beta(1-e^{-\beta t}) + \underline{F}_i(t_0)/(m_i\beta)[t - 1/\beta(e^{-\beta t}-1)] \\ &= \int W(\underline{R}, \underline{V}, t)\,d^3\underline{v} \\ &= 1/\{8\pi^3 G^{3/2}\}\exp\{-|R|^2/2G\} \end{aligned} \tag{6.39}$$

where

$$\langle \underline{R} \rangle = 0 \text{ and } \langle |\underline{R}|^2 \rangle = 3E = 3k_BT/(m_i\beta^2)(2\beta t - 3 + 4e^{-\beta t} - e^{-2\beta t})$$

So,

$$\underline{R} = \underline{r}_i(t_0+t) - \underline{r}_i(t_0) - \underline{v}_i(t_0)/\beta(1-e^{-\beta t}) - \underline{F}_i(t_0)/(m_i\beta)[t - 1/\beta(1-e^{-\beta t})] = \underline{C}_1$$

where $\underline{C}_1$ is a Gaussian random variable of mean zero and mean-square value $\langle|\underline{C}_1|^2\rangle = 3k_BT/(m_i\beta^2)(2\beta t - 3 + 4e^{-\beta t} - e^{-2\beta t})$.

Then $\underline{V}$ must be chosen from the conditional probability,

$$\begin{aligned} W(\underline{V} \,|\, \underline{R} = \underline{R}) &= W(\underline{R}, \underline{V})/W(\underline{R}) \\ &= 1/\{8\pi^3 G(1 - H^2/EG)\}^{3/2} \exp \\ &\quad \{-1/[2G(1 - H^2/EG)]|\underline{V} - r\sigma_2/\sigma_1 \underline{R}|^2\} \end{aligned}$$

where defining $D = (2\beta t - 3 + 4e^{-\beta t} - e^{-2\beta t})$

$$r\sigma_2/\sigma_1 = H/E = \beta(1 - e^{-\beta t})^2/D$$

and

$$\sigma_2^2(1-r^2) = G(1-H^2/EG) = (G-H^2/E)$$

Hence, the variable ($\underline{V} - r\sigma_2/\sigma_1\underline{R}$) not $\underline{V}$ must be chosen from a Gaussian distribution of mean zero and mean-square value $3\sigma_2^2(1 - r^2) = 3(G - H^2/E)$ which means that

$$\begin{aligned}
\underline{C}_2 &= \underline{V} - r\sigma_2/\sigma_1\underline{R} \\
&= \underline{v}_i(t_0+t) - \underline{v}_i(t_0)\mathrm{e}^{-\beta t} - \underline{F}_i(t_0)/(m_i\beta)[1-\mathrm{e}^{-\beta t}] \\
&\quad -\beta(1-\mathrm{e}^{-\beta t})^2/D\{\underline{r}_i(t+t_0) - \underline{r}_i(t_0) - \underline{v}_i(t_0)/\beta[1-\mathrm{e}^{-\beta t}] \\
&\quad +\underline{F}_i(t_0)/(m_i\beta)(t-(1/\beta)(1-\mathrm{e}^{-\beta t})\} \\
&= \underline{v}_i(t_0+t) - \underline{v}_i(t_0)(2\beta t)\mathrm{e}^{-\beta t} - 1 + \mathrm{e}^{-2\beta t})/D - \underline{F}_i(t_0)/(m_i\beta)[\beta t(1-\mathrm{e}^{-\beta t}) \\
&\quad -(1-\mathrm{e}^{-\beta t})^2]/D - \beta)\{\underline{r}(t_0+t) - \underline{r}(t)\}(1-\mathrm{e}^{-\beta t})^2/D
\end{aligned}$$

Thus, $\underline{C}_2$ is a Gaussian random variable for which $\langle\underline{C}_2\rangle = 0$ and $\langle|\underline{C}_2|^2\rangle = 3\,(G - H^2/E) = (6k_BT/m_i)\{\beta t(1 - \mathrm{e}^{-2\beta t}) - 2(1 - \mathrm{e}^{-\beta t})^2\}/D$.

Using this result (6.15) and (6.16) may be written as

$$\underline{r}_i(t_0+t) = \underline{r}_i(t) + \underline{v}_i(t)/\beta(1-\mathrm{e}^{-\beta t}) + \underline{F}(t)/\beta\left[t-(1/\beta)(1-\mathrm{e}^{-\beta t})\right] + \underline{C}_1 \tag{6.40}$$

where $\underline{C}_1$ is a Gaussian random variable for which

$$\langle\underline{C}_1\rangle = 0 \text{ and } \langle|\underline{C}_1|^2\rangle = 3\mathrm{k_B T}/(m_i\beta^2)\ 2\beta\mathrm{t} - 3 + 4\mathrm{e}^{-\beta\mathrm{t}} - \mathrm{e}^{-\beta\mathrm{t}}) \tag{6.41}$$

and

$$\underline{v}_i(t_0+t) = \underline{v}_i(t)\mathrm{e}^{-\beta t} + \underline{F}(t)/\zeta(1-\mathrm{e}^{-\beta t})\underline{r}(t_0+t) + \underline{C}_2 \tag{6.42}$$

where $\underline{C}_2$ is a Gaussian random variable for which

$$\langle\underline{C}_2\rangle = 0 \text{ and } \langle|\underline{C}_2|^2\rangle = 6k_BT/(m_i)\{\beta t(1-\mathrm{e}^{-2\beta t}) - 2(1-\mathrm{e}^{-\beta t})^2\}/D \tag{6.43}$$

and $\langle\underline{C}_1{\cdot}\underline{C}_2\rangle = 0$. Eqs. (6.40)–(6.43) provide another BD algorithm.

All the algorithms derived in Sections 6.2.3 and 6.2.4 are thus based on being able to change the interdependent, Gaussian random variables $\underline{V}$ and $\underline{R}$ into new variables that are independent, Gaussian random variables, i.e. in Section 6.2.3 into the variables $\underline{V}$ and $\underline{R} - r\sigma_2/\sigma_1$ and in Section 6.2.4 into the variables $\underline{R}$ and $\underline{V} - r\sigma_2/\sigma_1$.

6.2.5 The BD algorithm in the diffusive limit

Often we are interested in the case where $t >> \beta^{-1}$ and β^{-1} is a time characterising the decay of the velocity autocorrelation function while t is a time step over which the spatial configuration of the particles has not changed significantly, this is usually termed the 'Diffusive Limit' or the 'Overdamped Case'. Then $e^{-\beta t} \approx 0$, $1 - e^{-\beta t} \approx 1$ and $\beta^{-1} \to 0$, thus, from eq. (6.28) we have

$$\underline{v}_i(t_0 + t) = \underline{A}_1(t) \tag{6.44}$$

and from eq. (6.29),

$$\underline{r}_i(t_0 + t) = \underline{r}_i(t_0) + \underline{F}_i(t_0)/(m_i\beta)t + \underline{A}_2(t) \tag{6.45}$$

where $\underline{A}_1$ and $\underline{A}_2$ are 3-D Gaussian random variables for which from eq. (6.30)

$$\langle \underline{A}_1 \rangle = \langle \underline{A}_2 \rangle = 0 \tag{6.46}$$

$$\langle |A_1|^2 \rangle = 3k_\mathrm{B}T/m_i$$

$$\begin{aligned} \langle |\underline{A}_2|^2 \rangle &= 3k_\mathrm{B}T/(m_i\beta^2)(2\beta t - 3) \\ &= [6k_\mathrm{B}T/(m_i\beta)]t \end{aligned}$$

and

$$\langle \underline{A}_1 \cdot \underline{A}_2 \rangle = 0$$

Hence, because of eq. (6.46) for $\underline{A}_1$ it is obvious that $\underline{v}_i$ (t_0+t) obeys a Maxwell distribution, i.e. $\underline{v}_i$ is 'at equilibrium', the Gaussian random variables $\underline{A}_1$ and $\underline{A}_2$ are independent, meaning that there is obviously no coupling between velocities and positions and $\underline{r}_i(t_0+t)$ is not correlated with $\underline{v}_i(t_0+t)$.

In fact $\underline{r}_i(t_0+t)$ is given by the very simple equation (6.45) and this same equation results if we take the diffusive limits of eqs. (6.34) and (6.36) or eqs. (6.40) and (6.42).

There have been several BD algorithms developed[7–17] where the opposite conditions to the diffusive limit applies, i.e. they are applicable when $t << \beta^{-1}$ so $e^{-\beta t} \approx 1 - \beta t$ and $1 - e^{-\beta t} \approx \beta t$. These have been applied to simulating problems such as electrolyte solutions and solids.

6.3 HIGHER-ORDER BD SCHEMES FOR THE LANGEVIN EQUATION

The same approach or an equivalent one may be used to generate algorithms of higher order in Δt, i.e. starting from eqs. (6.13) and (6.14). This has been done by van Gunsteren and Berendsen[18] who expanded the direct force $\underline{F}$ to second order, i.e.

$$\underline{F}(t) = \underline{F}(t_n) + \mathrm{d}\underline{F}(t_n)/\mathrm{d}t(t - t_n) + O[(t - t_n)^2] \tag{6.47}$$

where the derivative $d\underline{F}(t_n)/dt$ is evaluated at $t = t_n$ and like the Verlet algorithm used in many MD simulations[29] the results for $\Delta t = (t_{n+1} - t_n)$ and $-\Delta t$ were obtained and combined to give a third-order algorithm for each co-ordinate x which does not involve the velocity,

$$\begin{aligned} x(t_n+\Delta t) = {} & x(t_n)\left[1+e^{-\beta\Delta t}\right] - x(t_n-\Delta t)e^{-\beta\Delta t} + m^{-1}F(t_n)(\Delta t)^2(\beta\Delta t)^{-1}\left[1-e^{-\beta\Delta t}\right] \\ & + m^{-1}\,\mathrm{d}F(t_n)/\mathrm{d}t(\Delta t)^3(\beta\Delta t)^{-2}\left[1/2\beta\Delta t\{1+e^{-\beta\Delta t}\} - \{1+e^{-\beta\Delta t}\}\right] \\ & + X_n(\Delta t) + e^{-\beta\Delta t}X_n(-\Delta t) + O\left[(\Delta t)^4\right] \end{aligned} \tag{6.48}$$

Here $X_n(\Delta t)$ and $X_n(-\Delta t)$ are two correlated Gaussian random variables, the properties of which are given by van Gunsteren and Berendsen.[18] The generation of $X_n(\Delta t)$ and $X_n(-\Delta t)$ is made easy for small values of $\beta\Delta t$ as analytic expressions for the functions that determines them are given.

An algorithm for $v(t_n)$ was also obtained which is

$$\begin{aligned} v(t_n) = {} & \left[x(t_n+\Delta t) - x(t_n-\Delta t)\right] + m^{-1}F(t_n)(\Delta t)^2(\beta\Delta t)^{-2}G(\beta\Delta t) \\ & - m^{-1}\,\mathrm{d}F(t_n)/\mathrm{d}t(\Delta t)^3(\beta\Delta t)^{-3}G(\beta\Delta t) + \left[X_n(-\Delta t) - X_n(\Delta t)\right]H(\beta\Delta t)/\Delta t \end{aligned} \tag{6.49}$$

where

$$H(\beta\Delta t) = \beta\Delta t/\left[e^{\beta\Delta t} - e^{-\beta\Delta t}\right]$$

In the diffusive limit, as can be easily established, eq. (6.48) reduces to the following,

$$x(t_n+\Delta t) = x(t_n) + (m\beta)^{-1}\left[F(t_n)\Delta t + 1/2\,\mathrm{d}F(t_n)/\mathrm{d}t(\Delta t)^2\right] + X_n(\Delta t) \tag{6.50}$$

where now the variable $X_n(\Delta t)$ is a Gaussian random variable of mean zero and

$$\langle X_n^2(\Delta t)\rangle = 2k_BT(m\beta)^{-1}\Delta t \tag{6.51}$$

6.4 GENERALISED LANGEVIN EQUATION [3,4,13,19,20]

Here, we wish to find a numerical algorithm to solve the GLE (6.1) (see also Section 2.3) of the following form:

$$m\,\mathrm{d}\underline{v}_i = \underline{F}_i(t) - m\int_0^t K_v(t-t')\underline{v}_i(t')\mathrm{d}t' - \underline{F}_i^R(t) \tag{6.52}$$

where from eqs. (2.29) and (2.30) we have

$$\langle \underline{F}_i^{\mathrm{R}}(0) \rangle = 0 \tag{6.53}$$

$$\langle \underline{F}_i^{\mathrm{R}}(0) \cdot \underline{F}_i^{\mathrm{R}}(t) \rangle = 3mk_{\mathrm{B}}TK(t)$$

Thus, firstly we need to know the form of the memory function $K(t)$ and the 'random force' $\underline{F}_i^{\mathrm{R}}(t)$. If we assume that the random force $\underline{F}_i^{\mathrm{R}}$ is a Gaussian random variable then the two conditions expressed by eq. (6.53) are sufficient to determine $\underline{F}_i^{\mathrm{R}}$ or rather integrals over this variable. In fact, we have little choice but to assume that $\underline{F}_i^{\mathrm{R}}$ is a Gaussian random variable as eq. (6.53) is all that we know of $\underline{F}_i^{\mathrm{R}}$ and this assumed Gaussian form will be the least biased approximation to $\underline{F}_i^{\mathrm{R}}$ in an information theory sense.

The development of schemes to numerically solve eq. (6.52) is rather more complicated than solving the LE, i.e. eq. (6.10) because of the appearance of the memory function $K_v(t)$, which means that the properties at time t are functions of all previous values at earlier times weighted by $K_v(t)$. Many methods have been developed to obtain numerical solutions to eq. (6.52) both when the direct force $\underline{F}_i$ is present and also when it is zero. Some solutions are quite general and some assume a particular form of $K_v(t)$.

For example, Shugard, Tully and Nitzan[13] developed a scheme to solve eq. (6.52) when $K(t) = Ce^{-\beta t}\cos(\omega t + \alpha)$ since in this case $\underline{F}_i^{\mathrm{R}}(t)$ is a projection of a two-dimensional Gauss−Markov process $[\underline{F}_i^{\mathrm{R}}(t), \mathrm{d}\underline{F}_i^{\mathrm{R}}(t)/\mathrm{d}t]$.[19]

Wan, Wang and Shi[20] have developed an algorithm for the particular case of eq. (6.52) where $K_v(t)$ is the exponential memory $K_0\,\mathrm{e}^{-t/\tau}$ by using a scheme similar to that used by van Gunsteren and Berendsen and give expressions for $v(t_n + \Delta t/2)$ and $x(t_{n+1})$ in terms of $v(t_n - \Delta t/2)$, $F(t_n)$, $x(t_n)$, $F_1^{fr}(t_n)$, $F_1^{fr}(t_{n+1/2})$, $F^{\mathrm{R}}(t_{n+1/2})$ and the random variables $V_n(\Delta t/2)$, $V_n(-\Delta t/2)$, $X_{n+1/2}(\Delta t/2)$ and $X_{n+1/2}(-\Delta t/2)$ which are defined along with formulae for their calculation[20] The terms $F_1^{fr}(t_n)$ and $F_1^{fr}(t_{n+1/2})$ arise from expressing the integral over $K_v(t)$ in eq. (6.52) as a discrete sum and explicit formulae are given for the random variables.

6.4.1 The method of Berkowitz, Morgan and McCammon[21]

Turning to more general scheme Berkowitz, Morgan and McCammon[21]developed a general scheme for the case when $\underline{F}_i = 0$ by repeating $F_i^{\mathrm{R}}(t)$ periodically with period P and writing for the 1-D case,

$$F_i^{\mathrm{R}}(t) = \sum_{k=1}^{\infty} \{a_k \cos(\omega_k t) + b_k \sin(\omega_k t)\} \tag{6.54}$$

where $\omega_k = 2\pi k/P$. The first coefficient $a_0 = 0$ and the remaining a_k and b_k are independent Gaussian random variables and may be sampled from Gaussian distributions. Combining this with a modified version of the Verlet MD algorithm to obtain the velocity and position of the particle and, thus, to evaluate the integral over $K_v(t)$ they obtained a modified Verlet algorithm which for a 1-D system is

$$x_{m+1}\left[1+1/2\,K_v(0)(\Delta t)^2 w_0\right] = -x_{n-1}\left[1-(1/2)\,K_v(0)(\Delta t)^2\, w_0\right]+2x_n - (\Delta t)^3 \sum_{j=1}^{n} K_v(t_j)v_{n-j}w_j/m + f_n(\Delta t)^2 \tag{6.55}$$

In practice x_0, x_1 and v_0 need to be specified to commence this scheme, which was shown to work extremely well for the exponential, Gaussian and Levesque–Verlet forms of the memory function (see eqs. (3.24) and (3.50)).

6.4.2 The method of Ermak and Buckholz[4]

The method of Ermak and Buckholz[4] developed a method based on taking the Fourier-Laplace transform of eq. (6.52) and extending the lower limit of integral to $-\infty$, i.e.

$$\underline{v}(t) = (1/m)\int_{-\infty}^{t} \psi(t-t')\left\{\underline{F}(\underline{r},t')+\underline{F}^{\mathrm{R}}(\underline{r},t')\right\}\mathrm{d}t' \tag{6.56}$$

$$\underline{r}(t) = \underline{r}(-\infty)+(1/m)\int_{-\infty}^{t} \phi(t-t')\left\{\underline{F}(\underline{r},t')+\underline{F}^{\mathrm{R}}(\underline{r},t')\right\}\mathrm{d}t'$$

where $\psi(t - t')$ is the inverse Laplace transform of $K_v(t)$ and $\phi(t - t') = \int_0^t \psi(|t'|)\mathrm{d}t'$.

Putting the time step equal to t, $\underline{v}_n = \underline{v}(nt)$, $\Delta\underline{r}_n = \underline{r}_n - \underline{r}_{n-1}$, $\Delta\phi_n = \phi_n - \phi_{n-1}$ and assuming $\underline{F}$ is constant over a time step t then eq. (6.56) may be written as

$$\underline{v}_n = (t/m)\sum_{i=\infty}^{n-1} \psi_{n-1}\underline{F}_i + \underline{V}_n \tag{6.57}$$

$$\Delta\underline{r}_n = (t/m)\left\{\phi_1\underline{F}_{n-1} + \sum_{i=-\infty}^{n-2} \Delta\phi_{n-i}\underline{F}_i\right\} + \underline{R}_n$$

where $\underline{V}_n$ and $\underline{R}_n$ are Gaussian random deviates with the following characteristics:

$$\langle\underline{V}_i\rangle = \langle\underline{R}_i\rangle = 0 \tag{6.58}$$

$$\langle\underline{V}_i\cdot\underline{V}_j\rangle = (3k_{\mathrm{B}}T/m)\psi_{i-j}$$
$$\langle\underline{R}_i\cdot\underline{V}_j\rangle = (3k_{\mathrm{B}}T/m)\Delta\,\phi_{i-j}$$
$$\langle\underline{R}_i\cdot\underline{R}_j\rangle = (3k_{\mathrm{B}}T/m)\Delta^2\,\chi_{i-j}$$

where $\Delta^2\chi_n = \chi_{n+1} - 2\chi_n + \chi_{n-1}$ and $\chi(t) = \int_0^t \phi(t')\mathrm{d}t'$.

The random deviates $\underline{V}_n$ and $\underline{R}_n$ are, as before, statistically correlated and it is best to transform them into independent Gaussian random terms which is done by writing,

$$\underline{V}_n = \sum_{i=-\infty}^{n} \sigma_{n-i} \underline{Y}_i \tag{6.59}$$

$$\underline{R}_n = \sum_{i=-\infty}^{n} \{\rho_{n-i} \underline{Y}_i + \xi_{n-i} \underline{Z}_i\}$$

where

$$\langle \underline{Y}_i \rangle = \langle \underline{Z}_i \rangle = \langle \underline{Y}_i \cdot \underline{Z}_j \rangle = 0 \tag{6.60}$$

$$\langle \underline{Y}_i \cdot \underline{Y}_j \rangle = \langle \underline{Z}_i \cdot \underline{Z}_j \rangle = \delta_{ij}$$

The weighting parameters in eq. (6.59) are given by

$$(3k_B T/m)\psi_i = \sum_{j=0}^{\infty} \sigma_j \sigma_{i+j}$$

$$(3k_B T/m)\Delta \phi_i = \sum_{j=0}^{\infty} \sigma_j \rho_{i+j}$$

$$(3k_B T/m)\Delta^2 \chi_i = \sum_{j=0}^{\infty} \{\rho_j \rho_{i+j} + \xi_j \xi_{i+j}\} \tag{6.61}$$

where $i = 0$ to ∞.

In practice, of course, the infinite sums in the above equations must be truncated and this is done by assuming that the memory function $K_v(t)$ is zero beyond a certain time, i.e. for $i \geq N$ for some N which means that $\sigma_i = 0$ for $i \geq N$ and thus $\Delta\phi_i = \Delta^2 \chi_i = \rho_i = \xi_i = 0$ for $i \geq N$.

As eq. (6.57) for the displacement over the nth time step $\underline{\Delta r}_n(t)$ does not contain terms dependent on the velocity, we may use it to determine displacement without having to know the time history of the velocity of the particle before the nth time step and, thus,

$$\underline{\Delta r}_n = t/m \left\{ \phi_1 \underline{F}_{n-1} + \sum_{i=n-N}^{n-2} \Delta\phi_{n-i} \underline{F}_i \right\} + \sum_{i=n-N}^{N} \sigma_{n-i} \underline{Z}_i \tag{6.62}$$

where

$$\langle \underline{Z}_i \rangle = 0$$
$$\langle \underline{Z}_i \underline{Z}_j \rangle = \delta_{ij} \tag{6.63}$$

and

$$(3k_B T/m)\Delta^2 \chi_i = \sum_{j=0}^{N-i} \sigma_j \sigma_{i+j}$$

for $i = 0$ to N.

This method was applied to the case of an exponential memory $M(t) = \alpha\beta e^{-\alpha t}$ and the acceleration $\underline{a}$ introduced as a separate independent variable resulting in the algorithm dependent on a tri-variate distribution function, which is detailed in Ref. 4.

6.4.3 The method of Ciccotti and Ryckaert[22]

An alternative method is to use the Mori continued fraction method, which results in the memory function hierarchy discussed in Section 3.3. Such a scheme was introduced by Ciccotti and Ryckaert[22] for the velocity autocorrelation function $c_v(t) = \langle \underline{v}_i(0) \cdot \underline{v}_i(t) \rangle / \langle \underline{v}_i(0) \underline{v}_i(t) \rangle$ as was done in eq. (3.15) and starting from the memory function equation for $\underline{v}_i(t)$, i.e.

$$(\mathrm{d}/\mathrm{d}t)\underline{v}_i(t) = -\int_0^t K_1(t-s)\underline{v}_i(s)\mathrm{d}s + \underline{R}_i(t) \tag{6.64}$$

we have the continued fraction expression for $c_v(t)$ which they write as follows:

$$\tilde{c}_v(s) = 1/(s + K_1(0))/(s + K_2(0))/(s + K_3(0))/\cdots/(s + \tilde{K}_n(s)) \tag{6.65}$$

where $n = 1$ to ∞.

They introduce a set of random forces $\{R_i(t), i = 1 \text{ to } \infty\}$ by the generalisation of eq. (6.64),

$$\mathrm{d}/\mathrm{d}t\, R_i(t) = -\int_0^t K_{i+1}(t-s)R_i(s)\mathrm{d}s + R_{i+1}(t) \tag{6.66}$$

$i = 0$ to ∞ and as usual we have,

$$< R_i(t)R_i(s) > = K_i(|t-s|)\langle R_{i-1}{}^2 \rangle \tag{6.67}$$

for $i = 2$ to ∞.

The continued fraction (6.65) is truncated by assuming that for $i = n$ $R_n(t)$ is Gaussian white noise as was done in eq. (3.17), i.e. $R_n(t)$ is represented by a Gaussian random variable $\xi(t)$ such that

$$\langle \xi(t) \rangle = 0 \tag{6.68}$$

$$\langle \xi(t)\xi(\mathrm{s}) \rangle = 2\lambda_n \langle A_{n-1}{}^2 \rangle \delta(t-s) \tag{6.69}$$

Then defining a set of auxiliary variables $\{A_i(t),\ i = 1 \text{ to } n\}$ by

$$A_0(t) = v(t) \tag{6.70}$$

$A_i(t) = -\int_0^t K_i(t-s)A_{i-1}(s)\mathrm{d}s + R_i(t)$ which upon truncation leads to set of linear, first-order differential equations,

$$A_0(t) = A_1(t) \tag{6.71}$$

$$A_i(t) = -K_i(0)A_{i-1}(t) + A_{i+1}(t)$$

for $i = 1$ to $n - 2$

and

$$A_{n-1}(t) = -K_{n-1}(0)A_{n-2}(t) - \beta A_{n-1}(t) + \xi(t)$$

This set of differential equations may be solved by numerical integration once the coefficients $\{K_i(0),\ i = 1 \text{ to } n\}$ and the parameter λ_n are determined. This integration has been expressed as a first-order algorithm in Appendix B of the first reference in 22 and by second-order scheme in the appendix of second reference in 22. This procedure has led to good results for a variety of forms of the memory function $K_v(t)$ but it does need very large values of n to achieve accurate results and does not work for simple approximations to $K_v(t)$ such as the Levesque and Verlet form eq. (3.50).

6.4.4 Other methods of solving the GLE

A scheme has been introduced by Doll and Dion[23] to numerically integrate the GLE, eq. (6.9), but as it ignores the inter-dependence of the statistical properties of the velocities and positions it is an approximate scheme. However, it may be useful as a simple scheme to generate GLE trajectories and should be useful for generating results in the diffusive limit when the the statistical properties of the velocities and positions are independent.

Finally, as will be discussed in Section 8.1 Toxvaerd[3] has introduced a scheme for solving the GLE by a combination of the use of the Fourier series expansion of the random force as in eq. (6.54) and a finite difference algorithm for $\underline{r}(t)$ and $\underline{v}(t)$.

6.5 SYSTEMS IN AN EXTERNAL FIELD

To treat systems in a shear field Dotson[24] started with the Langevin equation obtained by adding an external field term to the LE given by eq. (6.10),

$$\mathrm{d}\underline{v}_i(t)/\mathrm{d}t = -\beta[\underline{v}_i(t) - \underline{u}_i(t)] + \underline{F}_i(t)/m + \underline{F}_i^{\mathrm{R}}(t)/m \tag{6.72}$$

where $\underline{u}_i(t)$ is the stream velocity at the position of particle i.

For a system in steady shear flow then $\underline{u}_i(t) = (\dot{\gamma}y, 0,0)$ where $\dot{\gamma}$ is the shear rate Dotson obtained the following algorithm for particle displacement over the time step t in the diffusive limit:

$$\begin{aligned} x_i(t) = {} & x_i(0) + (1/m\beta)\int_0^t F_{ix}(\{x_j\})[1-\mathrm{e}^{-\beta(t-\lambda)}]\mathrm{d}\lambda \\ & + R_{ix}(t) + \dot{\gamma}\, y_i(0)t + (\dot{\gamma}/m\beta)\int_0^t F_{iy}(\{y_j\})[1+\mathrm{e}^{-\beta(t-s)}](t-s) + (2/\beta)[1+\mathrm{e}^{-\beta(t-s)}]\mathrm{d}s \\ & + S_i(t) \end{aligned}$$

$$y_i(t) = y_i(0) + (1/m\beta)\int_0^t F_{iy}(\{y_j\})[1-\mathrm{e}^{-\beta(t-s)}]\mathrm{d}s + R_{iy}(t)$$

$$z_i(t) = z_i(0) + (1/m\beta)\int_0^t F_{iz}(\{z_j\})[1-\mathrm{e}^{-\beta(t-s)}]\mathrm{d}s + R_{iz}(t) \tag{6.73}$$

where $R_{ix}(t)$, $R_{iy}(t)$, $R_{iy}(t)$ and $S_i(t)$ are independent Gaussian random deviates whose mean values are zero and whose mean-square values are given by,

$$\langle R_{i\alpha}(t)R_{j\beta}(t')\rangle = \{2k_{\mathrm{B}}T/(m\beta)\}t\,\delta_{ij}\,\delta_{\alpha\beta}\,\delta(t-t') \tag{6.74}$$

$$\langle S_i(t)S_i(t)\rangle = \{(2k_{\mathrm{B}}T/(m\beta))\dot{\gamma}t^3$$

Thus, the displacement in the x-direction is coupled to the force $F_{iy}(\{y_j\})$ in the y-direction by the action of the shear field. The only other effect of the shear field is to introduce the extra random displacement $S_i(t)$ but as this is of third order in the time step t then it is usually neglected as all other effects are of lower order in t.

Although this algorithm neglects many-body hydrodynamic interactions and thus is strictly only applicable to dilute suspensions, it and related algorithms have been used to investigate dense colloidal systems in shear[25] and only some studies have made allowance for hydrodynamic interactions.[26] The motivation for this is that colloids show a wide variety of interesting properties in a shear field, e.g. non-Newtonian effects such as shear thinning and shear thickening[27] and the structures produced by flow are rather fascinating.[28]

6.6 BOUNDARY CONDITIONS IN SIMULATIONS

The number of particles that can be used in a simulation is relatively limited, which means that if we simulated such a system in free boundary conditions we would be simulating a rather small cluster of particles. This is fine if a finite cluster is to be simulated, but if we wish to model a bulk phase, then such a system would have a very large number of its particles at the surface and would be very inhomogeneous. To avoid this, a generalization of the Born-van Karman boundary conditions introduced in the theory of lattice dynamics is used except in the case here the particle positions in the basic cell may be ordered or disordered as determined by the equations governing the particle positions. Thus, we usually simulate a homogeneous system by use of a small cell containing N particles and surround this basic cell with periodic replicas in all the dimensions in which we wish the system to be homogeneous. These periodic boundary conditions (PBCs) or toriodal boundary conditions impose certain constraints on the system and must be constructed in such a way as to reflect the physical system to be simulated as we will discuss below.[29]

6.6.1 PBC in equilibrium

The simplest systems to simulate are equilibrium ones, or rather ones with no external field, which would impose constraints on the system. Then all that is needed are for the cells to be space-filling and often a cube cell is sufficient.[29] A simple 2-D representation of such a system is shown below in Figure 6.1.

Now all properties of the system are obtained by summation over properties of the particles, e.g. the potential energy, the forces and virials; thus, we need a summation method and since the system is periodic then this must be taken into account in these summations.

If the potential or forces involved are short ranged then a cut-off can be used, i.e. the summations will be truncated at some boundary. For example, for a spherically symmetric pairwise additive potential, the potential may be set to zero outside a sphere of radius r_{cut}. Then the usual way to do the summation is over nearest images using the minimum image convention (MIC)[29] where the interaction between a chosen particle i and another particle j is between particle i and either particle j <u>or</u> the image of j which is closest to i, i.e. the distance $r_{ij}^{'}$ used is given by

$$r_{ij}^{'} = \min\{r_{ij}, r_{ij'}\} \tag{6.75}$$

where r_{ij} is the distance between i and j and $r_{ij}^{'}$ is the distance between particle i and particle $j^{'}$ which is any image of j. It is illustrated in Figure 6.1 for a two-dimensional system that the red particle i in the centre of the circle interacts with the blue particle j on the left of i in the cell on the left of the basic cell and not the one in the basic cell on its right.

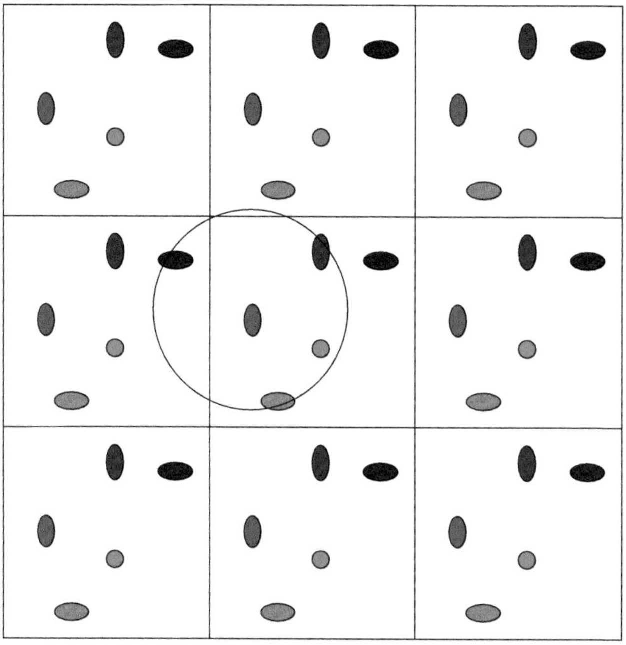

Figure 6.1 A 2-D lattice of cells used to illustrate simple PBCs and the minimum image convention (Courtesy of Dr. A. Budi, Applied Physics, RMIT).

Allowance for longer ranged interactions outside the truncation region ('long-range corrections') may be made by, for example, direct summation over other cells for a limited number of atomic configurations or by some mean field theory.

For truly long-ranged potentials and forces, e.g. Coulomb forces and dipole–dipole interactions direct summations must be used over the whole system, e.g. by the Ewald method,[29] which will be discussed in conjunction with hydrodynamic interactions in Chapter 7.

6.6.2 PBC in a shear field[29–31]

For a system in a 2-D, planar shear field as shown in Figure 6.2 we need to imposed PBCs that reflect the symmetry of this imposed field by using the Lees–Edwards[30] Boundary Conditions illustrated in Figure 6.3 in which the upper and lower layers of cell images must be shifted so that the velocity profile is continuous across the boundaries.

6.6.3 PBC in elongational flow[33–37]

To be able to simulate elongational flow a boundary condition due to Kraynik and Reinelt (KRB)[33] must be used if a continuous deformation of the system is to be made. In this method the simulation box evolves in such a manner with the flow that the lattice vectors describing

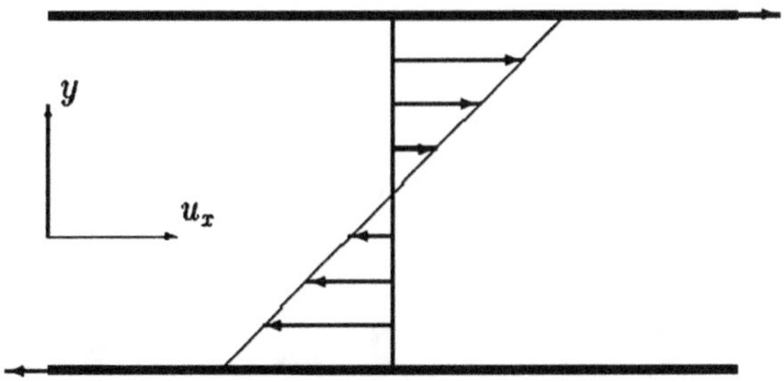

Figure 6.2 A representation of the imposed shear field on a system. This is planar shear flow in which the streaming velocity u_x with position in the *y* direction[32].

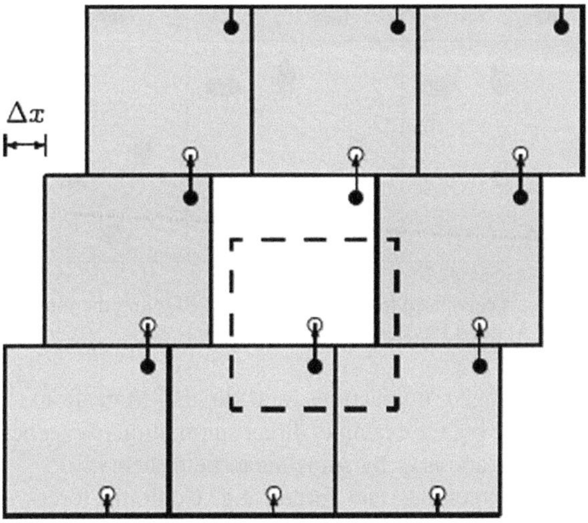

Figure 6.3 Lees-Edwards or Sliding-Brick Boundary Conditions[30,32].

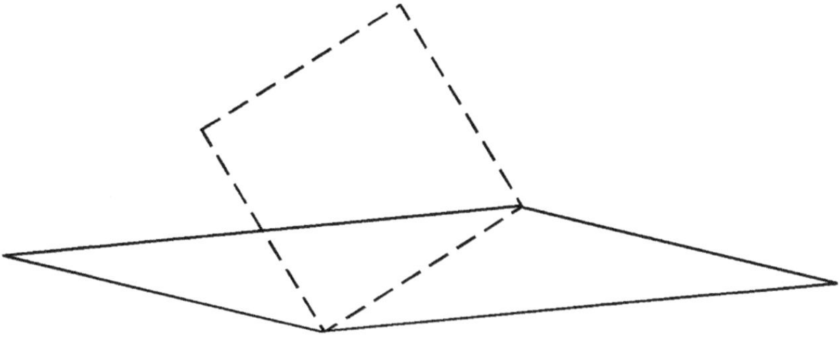

Figure 6.4 Elongational boundary conditions for a 2-D system, the original shape (dashed line) and the fully deformed system shape (solid line) for a 2-D system undergoing periodic elongational flow using Kraynik and Reinelt boundary conditions. (Figure courtesy of Dr M. Matin, RMIT University, Applied Physics[37].)

the box reach a value that is a linear combination of the original lattice vectors when the Hencky strain on the system reaches a value ε_p. After this time called the strain period by Matin, Daivis and Todd,[37] the simulation box is reset to its original shape and the simulation proceeds. The conditions on the original box vectors for this periodicity to occur are described by Todd and Daivis.[36] A pictorial representation of the cell is shown is Figure 6.4.

6.7 CONCLUSIONS

In this chapter we have explored various ways of deriving numerical algorithms to solve the LE and GLE for the velocity of a particle in a system of interacting particles.

In the case of the LE many body hydrodynamic effects were ignored and several schemes are given that enable the efficient computation of the velocity and position of the Brownian particle. A brief discussion of the generalisation of algorithms to the case of particles in a shear field but once again ignoring many-body hydrodynamic interactions was also given.

In the case of the GLE several schemes are presented for its numerical solution the most general and useful of which is probably that due to Ermak and Buckholz[4] although other schemes may also be useful especially for specific choices of the memory function $K_v(t)$. Such equations have been used to study dynamics in a variety of systems, e.g. surface processes on solids,[38] biopolymers[39] and thin film deposition.[40]

Finally, as the size of most systems simulated on a computer contain only a relatively small number of particles the system must be simulated under periodic boundary conditions. These boundary conditions will vary with the type of system being simulated and these were discussed for a system at equilibrium, for a system in shear flow and for one in elongational flow.

In the next chapter we will extend some of the methods developed in this chapter to systems with many-body hydrodynamic interactions but only in the Brownian limit.

REFERENCES

1. G. Ciccotti and J.-P. Ryckaert, *J. Stat. Phys.*, **26,** 73 (1981).
2. G. Bossis, B. Quentrec and J.P. Boon, *Mol. Phys.*, **45,** 191 (1982).
3. S. Toxvaerd, *J. Chem. Phys.* **86,** 3667 (1987).
4. D.L. Ermak and H. Buckholz, *J. Comp. Phys.*, **35,** 169 (1980).
5. S. Chandrasekhar, Stochastic problems in physics and astronomy, *Rev. Mod. Phys.*, **15,** 1–89 (1943).
6. A. Papoulis, *Probability, Random Variables and Stochastic Processes*, McGraw-Hill, New York, 1965.
7. P. Turque, F. Lantelme and H.L. Friedman, *J. Chem. Phys.*, **66,** 3039 (1977).
8. P. Turque, F. Lantelme and H.L. Friedman, *Mol. Phys.*, **37,** 223 (1979).
9. H. Posch, F. Vesely and W.A. Steele, *Mol. Phys.*, **44,** 241 (1981).
10. J.T. Hynes, R. Kapral and G.M. Torrie, *J. Chem. Phys.*, **72,** 177 (1980).
11. J.H. Weiner and R.E. Forman, *Phys. Rev. B*, **69,** 785 (1978).
12. M.R. Pear and J.H. Weiner, *J. Chem. Phys.*, **69,** 785 (1978); *Macromolecules*, **10,** 317 (1977).
13. M. Shugard, J.C. Tully and A. Nitzan, *J. Chem. Phys.*, **66,** 2534 (1977).
14. J.C. Tully, G.H. Gilmer and M. Shugard, *J. Chem. Phys.*, **71,** 1630 (1979).

15. T. Schneider and E. Stoll, *Phys. Rev. B*, **17,** 1302 (1978).
16. E. Heland, *J. Chem. Phys.*, **69,** 1010 (1978).
17. E. Helfand, Z.R. Wasserman and T.A. Weber, *Macromolecules*, **13,** 526 (1980).
18. W.F. van Gunsteren and H.J.C. Berendsen, *Mol. Phys.*, **45,** 637 (1982).
19. M.C. Wang and G.E. Uhlenbeck, *Rev. Mod. Phys.*, **17,** 323 (1945).
20. S.Z. Wan, C.X, Wang and Y.S. Shi, *Mol. Phys.*, **93,** 901 (1998).
21. M. Berkowitz, J.D. Morgan and J.A. McCammon, *J. Chem. Phys.*, **78,** 3256 (1983).
22. G. Ciccotti and J.-P. Ryckaert, *Mol. Phys.*, **40,** 141 (1980) and G. Ciccotti, M.Ferrario and J.-P. Ryckaert, *Mol. Phys.*, **46,** 875 (1982).
23. J.D. Doll and D.R. Dion, *Chem. Phys. Lett.*, **37,** 386 (1976).
24. P.J. Dotson, *J. Chem. Phys.*, **79,** 5730 (1983).
25. D.M. Heyes, *Phys. Lett. A*, **132,** 399 (1988); W. Xue and G.S. Grest, *Phys. Rev. Lett.*, **64,** 419 (1990); A.A. Rigos and G. Wilemski, *J. Phys. Chem.*, **96,** 3981(1992); S. Butler and P. Harrowell, *Phys. Rev. E*, **54,** 457 (1996).
26. J.R. Melrose, *Faraday Disc.*, **123**, 355 (2003).
27. J. Vermant and M.J. Solomon, Flow induced structure in colloidal suspensions, *J. Phys.: Condens. Matter*, **17,** R187−R216 (2005).
28. R.L. Hoffman, *J. Colloid Interface Sci.*, **46,** 491 (1974); B.J. Ackerson and N.A. Clark, *Phys. Rev. Lett.*, **46,** 123 (1981); *Physica*, **118A**, 221 (1983); J. Vermant and M.J. Solomon, Flow induced structure in colloidal suspensions, *J. Phys.: Condens. Matter*, **17,** R187−R216 (2005); H. Versmold, *Phys. Rev. Lett.*, **75,** 763 (1995); C. Dux, H. Versmold, V. Reus, T. Zemb and P. Lindner, *J. Chem. Phys.*, **104,** 6369 (1996); C. Dux and H. Versmold, *Phys. Rev. Lett.*, **78,** 1811 (1997); H. Versmold, S. Musa and A. Bierbaum, *J. Chem. Phys.*, **116,** 2658 (2002).
29. M.P. Allen and D.J. Tildesley, *Computer Simulation of Liquids,* Oxford Science Publications, Oxford, UK, 1989.
30. A.W. Lees and S.F. Edwards, *J. Phys. C: Solid State Phys.*, **5,** 1921(1972).
31. D.J. Evans, *Mol. Phys.*, **37,** 1745 (1979); D.P. Hansen and D.J. Evans, *Mol. Simul.*, **13,** 375 (1994); D.J. Evans and G.P. Morriss, *Non-Newtonian Molecular Dynamics*, *Comp. Phys. Rep.*, **1,** 297−343 (1984).
32. M. McPhie, *Thermal and Transport Properties of Shearing Binary Fluids*, PhD Thesis, RMIT University, Melbourne, Australia, 2003.
33. A.M. Kraynik and D.A. Reinelt, *Int. J. Multiphase Flow*, **18,** 1045 (1992).
34. B.D. Todd and P.J. Daivis, *Phys. Rev. Lett.*, **81,** 1118 (1998).
35. A. Baranyai and P.T. Cummings, *J. Chem. Phys.*, **110,** 42 (1999).
36. B.D. Todd and P.J. Daivis, *Comput. Phys. Commun.*, **117,** 191 (1999).
37. B.D. Todd and P.J. Daivis, *J. Chem. Phys.*, **112**, 40 (2000); M.L. Matin, P.J. Daivis and B.D. Todd, *Compt. Phys. Commun.*, **151,** 35 (2003); M.L. Matin, P.J. Daivis and B.D. Todd, *J. Chem. Phys.*, **113,** 9122 (2000); M.L. Matin, *Molecular Simulation of Polymer Rheology*, PhD Thesis, RMIT University, Melbourne, Australia, 2001; P.J. Daivis, M.L. Matin and B.D. Todd, *J. Non-Newtonian Fluid Mech.*, **111,** 1 (2003).
38. S.A. Adelman and J.D. Doll, Brownian motion and chemical dynamics on solid surfaces, *Accounts Chem. Res.*, **10,** 378 (1977); J.C. Tully, *Adv. Chem. Phys.*, **42,** 63 (1980); J.C. Tully, G.H. Gilmer and M. Shugard, *J. Chem. Phys.*, **71,** 1630 (1979).
39. J.A. McCammon and M. Karplus, *Accounts Chem. Res.*, **16,** 187 (1983).
40. C.-C. Fang, V. Prasad, R.V. Joshi, F. Jones and J.J. Hsieh, A process model for sputter deposition of thin films using molecular dynamics. In: *Thin Films. Modeling of Film Deposition for Microelectronics Applications* (Ed. S. Rossnagel), Academic Press, New York, 1996, Vol. 22, pp. 117−173.

– 7 –

Brownian Dynamics

One of the most productive techniques in the atomic theory of condensed matter have been molecular dynamics (MD) methods which use numerical techniques to solve Newton's and related equations of motion to study the equilibrium, transport and scattering behaviour of many, many systems. Although not yet as widely used as MD, methods based on the numerical solution of the many-body Langevin equation, Brownian dynamics (BD) have been very useful and their use continues to expand. In this chapter we will outline the derivation of algorithms to carry out BD simulations, discuss the calculation of hydrodynamic interactions that are needed and also mention some approximations that have been found to be useful.

7.1 FUNDAMENTALS

As shown in Chapters 2 and 5 when the ratio of the masses of the particles (the solute or suspended particles) of interest to those of the other particles (the solvent or dispersion medium) is very large, an equation of motion for the large particles may be derived which involves the direct interactions and indirect, solvent-mediated interactions between the particles. This may be conveniently re-written in the form,[1]

$$M_i \, \mathrm{d}V_i/\mathrm{d}t = -\sum_{j=1}^{3n} \xi_{ij} V_j + F_i + \sum_{j=1}^{3n} +\alpha_{ij} f_j \tag{7.1}$$

where i and j now label components ($1 \leq i, j \leq 3n$) not vectors and we have,

$$\langle f_i \rangle = 0 \tag{7.2}$$

$$\langle f_i(t) f_j(t') \rangle = 2\delta_{ij}\, \delta(t-t')$$

and

$$\zeta_{ij} = 1/k_B T \sum_{l=1}^{3n} \alpha_{il}\alpha_{jl} \tag{7.3}$$

Just as Newton's equations of motion may be used as a basis to develop numerical schemes, MD methods, to simulate atomic systems we can use this equation as the basis for a numerical scheme, called BD methods, for simulating the dynamics of heavy particles suspended in a "bath" of lighter particles.

Now F_i is the total "direct" force acting on component i, i.e. the sum of external fields and the solvent-averaged force of interaction between the suspended particles, and as this is defined as an equilibrium ensemble average given by eq. (5.56), it may be calculated by the methods of equilibrium statistical mechanics.[2–8] We will thus take this term as given and concentrate on the other interaction terms in eq. (7.1).

Physically the force $F_i^{fl} = -\Sigma_{j=1}^{3n} \zeta_{ij} V_j$ is the Hydrodynamic drag force on the heavy particles due to the suspension medium and the friction coefficients $\boldsymbol{\zeta}_{ij}$ (or friction tensor $\boldsymbol{\zeta}_{ij}$) may be thought of as hydrodynamic coupling constants.

It is also useful to define the diffusion tensor $\mathbf{D}_{ij}$ and mobility tensor $\boldsymbol{\mu}_{ij}$ by,

$$\sum_{j}^{3n} \zeta_{ij} D_{jl} = \sum_{j}^{3n} D_{ij}\zeta_{jl} = k_{\mathrm{B}}T\,\delta_{il} \tag{7.4}$$

and

$$\mathbf{D}_{ij} = k_{\mathrm{B}}T\boldsymbol{\mu}_{ij} \tag{7.5}$$

These tensors may be obtained by solving the Navier–Stokes equation and because we usually only deal with extremely low Reynolds numbers we only need to solve the quasi-static, Linearised Navier–Stokes or creeping-flow equations,[9–45] i.e.

$$\underline{\nabla}\cdot\mathbf{P_{kl}}(\underline{r},t) = 0 \text{ for } |\,\underline{r} - \underline{R}_i(t)\,| > a_i$$

$$P_{kl} = p\delta_{kl} - \eta(\partial v_1/\partial r + \partial v_k/\partial r) \tag{7.6}$$

$$\underline{\nabla}\cdot\underline{v}(\underline{r},t) = 0$$

where $\underline{v}(\underline{r},t)$ is the velocity field of the fluid, $\mathbf{P_{kl}}(\underline{r},t)$ is the pressure tensor with components P_{kl} $(k,l = 1,2,3)$, p is the hydrostatic pressure and here, $i = 1,\ldots\ldots, n$.

As these equations are linear we may write,

$$\underline{F}_i = \sum_{j=1}^{n} \zeta_{ij}^{\,\mathrm{TT}}\cdot\underline{V}_j + \sum_{j=1}^{n} \zeta_{ij}^{\,\mathrm{TR}}\cdot\underline{\Omega}_j \tag{7.7}$$

$$\underline{T}_i = \sum_{j=1}^{n} \zeta_{ij}^{\mathrm{TR}}\cdot\underline{V}_j + \sum_{j=1}^{n} \zeta_{ij}^{\mathrm{RR}}\cdot\underline{\Omega}_j$$

or

$$\mathrm{d}\underline{R}_i/\mathrm{d}t = \underline{V}_i = \sum_{j=1}^{n} \boldsymbol{\mu}_{ij}^{\mathrm{TT}} \cdot \underline{F}_j + \sum_{j=1}^{n} \boldsymbol{\mu}_{ij}^{\mathrm{TR}} \cdot \underline{T}_j \tag{7.8}$$

$$\mathrm{d}\underline{\theta}_i/\mathrm{d}t = \underline{\Omega}_i = \sum_{j=1}^{n} \boldsymbol{\mu}_{ij}^{\mathrm{TR}} \cdot \underline{F}_j + \sum_{j=1}^{n} \boldsymbol{\mu}_{ij}^{\mathrm{RR}} \cdot \underline{T}_j$$

or if we assume free rotation,

$$\mathrm{d}\underline{R}_i/\mathrm{d}t = \underline{V}_i = \sum_{j=1}^{n} \boldsymbol{\mu}_{ij}^{\mathrm{TT}} \cdot \underline{F}_j \tag{7.9}$$

$$\mathrm{d}\underline{\theta}_i/\mathrm{d}t = \underline{\Omega}_i = \sum_{j=1}^{n} \boldsymbol{\mu}_{ij}^{\mathrm{TR}} \cdot \underline{F}_j$$

where $\underline{\theta}_i$, $\underline{\Omega}_i$, $\underline{F}_j$and $\underline{T}_j$ are the angular position, the angular velocity, the force and the torque on particle i, respectively. The superscript T refers to translation and R to rotation.

Thus, the fundamental problem that arises in this description of particle motion is the calculation of the hydrodynamic interactions between the particles, which amounts to the calculation of the tensors $\boldsymbol{\zeta}_{ij}$ or $\boldsymbol{\mu}_{ij}$. Unfortunately, these quantities are functions not only of the positions of particles i and j but also of all the particles in the suspension, i.e. they are not pair-wise additive. They are also of extremely long range and diverge when particles approach one another if certain approximations are used. Thus, as discussed below they must be treated with great care in order to obtain meaningful results.

7.2 CALCULATION OF HYDRODYNAMIC INTERACTIONS

This problem reduces to finding methods to calculate $\boldsymbol{\mu}_{ij}$ or $\boldsymbol{\zeta}_{ij}$ which is done by solving eq. (7.7) or (7.8) subject to boundary conditions, e.g.

$$\underline{v}(\underline{r}) = \underline{V}_i + \underline{\Omega} \times (\underline{R} - \underline{R}_i) \tag{7.10}$$

when $|\underline{R} - \underline{R}_i| = a_i$.

This problem has been treated by many techniques, e.g.,

1. The method of reflections.[9–12]
2. The collocation technique of Ganatos and co-workes.[13,14]
3. The Legendre function scheme of Caflisch and co-workers.[15]
4. The method of induced forces (MIF) of Mazur and co-workers.[9,16–21,28–32]
5. The point source method used by Cichockci and Felderhof.[33]
6. The combined MIF and lubrication theory scheme originated by Bossis and Brady.[22–25]

The MIF has been extensively investigated by Mazur and co-workers,[9,16–21] Felderhoff, Jones and co-workers for free boundary conditions (FBC)[11] and by Brady and Bossis[22–27], Ladd[28–31] and Felderhoff, Cichocki and Hinsen for both FBC and Periodic Boundary Conditions (PBC)[33–36]. The combined results in this area have led to some very significant advances in our ability to calculate the $\boldsymbol{\mu}_{ij}$ for many-body systems.

One of the most successful and widely used methods, the MIF, accounts for the boundary condition (7.10) by placing a distribution of point forces on each sphere surface. This force distribution and $\underline{v}(\underline{r})$ are then expanded about the centre of each sphere in a series of spherical harmonics, which leads to a system of linear simultaneous equations relating the moments of the force on each sphere to the moments of the velocity on another sphere. This results in a set of equations from which $\underline{v}(\underline{r})$ may be determined which satisfies eq. (7.10) and, thence, $\boldsymbol{\zeta}_{ij}$ or $\boldsymbol{\mu}_{ij}$ may be obtained. In practice the series expansion obtained must be truncated and, thus, eq. (7.10) can only be approximately satisfied but results for $\boldsymbol{\mu}_{ij}$ may be obtained to any desired accuracy.

There are two well-developed methods for carrying this out:

1. Expansion of this truncated set as a cluster series in the co-ordinates of 1,2,…,n particles and then expanding these resultant terms in powers of the reciprocal of the particle separation, R.

or

2. Solving the truncated set of simultaneous equations obtained by numerical matrix inversion to give a multipole expansion of $\boldsymbol{\zeta}_{ij}$ or $\boldsymbol{\mu}_{ij}$.

The expansion method 1 has been used extensively, in particular by Mazur and co-workers[9,16–21] and Felderhoff and Jones and co-workers[11] for few particle systems and is a very useful method for treating small numbers of particles under free boundary conditions.

A low-order expansion of $\boldsymbol{\mu}_{ij}$ is often used (the Rotne and Prager tensor, see Appendix M,) which gives a mobility tensor that has the correct asymptotic long-ranged behaviour and is positive definite, which is essential in BD simulations (see Section 7.4.1 and Appendix N).

However, when the truncated expansion is used in PBCs it cannot guarantee that the resultant $\boldsymbol{\mu}_{ij}$'s are positive definite, which leads to problems in some applications.[37–39] This occurs with the Rotne–Prager tensor if the minimum-image convention is used in PBCs in conjunction with the Minimum Image Convention.[38] However, the Rotne–Prager tensor can be adapted to be consistent with PBCs by use of the Ewald summation technique.[26,37–39]

The method 2, which was pioneered by Ladd,[28–31] has been recently extensively refined by others.[23–27,32–36] Although Ladd's intuitive derivation[28] was criticised in the literature,[33] it was subsequently given a rigorous derivation for both free and periodic boundary conditions, which showed the essential correctness of the original derivation[32] (see Appendix N). This derivation[32] also proved that the method gives a diffusion matrix that is positive definite at all levels of truncation in either free or periodic boundary conditions and also the incompressibility condition[32] (7.6). We give an outline of the derivation of the Ladd method of calculating the mobility matrix in Appendix N, as this method is widely used if accurate results are needed.

There has been and continue to be significant development in this area, both theoretically and computationally, e.g. by Felderhoff, Cichocki and Hinsen,[33–36] Brady,[40] Sangani[41] and Higdon[42] who all have developed methods that enable very efficient computation of $\boldsymbol{\mu}_{ij}$.

However, a major problem arising from the use of this multipole method (often called the far-field many-body contribution) is that the series may converge very slowly if at all (especially at higher volume fraction, ϕ). A particular problem arises as a pair of particles approaches close contact in that the hydrodynamic forces diverge as s^{-1} and ln s^{-1} where $s = (R_{12} - 2a)/a$ is the separation between the particles relative to their radius, a. Brady, Bossis and co-workers overcame this problem[22–27] by showing that convergence can be accelerated by adding in the contributions to $\boldsymbol{\zeta}_{ij}$ given by near-field lubrication theory (see Appendix M). This theory describes the forces arising from the thin layer of viscous fluid separating the particle surfaces and, e.g. results in the relative motion of two particles approaching zero as their surfaces approach contact. This theory is the appropriate one to use when surfaces come very close together and being pairwise additive it is computationally very efficient. One must avoid counting interactions twice, so those two-body effects already included by the far-field calculations must be subtracted off.

An extremely important problem is the effect of walls on hydrodynamic interactions and this is very important in simulating the flow of colloidal suspensions and polymers through narrow tubes, gaps and between closely spaced surfaces, e.g. in simulating nanofluidic devices and flow through pores. These effects have been approached by use of method 1,[10,43,44] but a practical solution of the general problem has still to be found although less direct methods have been developed to treat this problem.[45]

7.3 ALTERNATIVE APPROACHES TO TREAT HYDRODYNAMIC INTERACTIONS

It should emphasised that the calculation of the hydrodynamic tensors describing the indirect, solvent-mediated, many-body interactions between large particles in solution is a complicated problem and requires much computer time to achieve in many-body systems except under very simple circumstances, e.g. for dilute solutions. Thus, other alternative, simpler approaches have been developed to solve this problem at some level[46] and these methods are, in fact, still being explored and developed. There are many of these alternative techniques, which is not surprising, since the general problem of fluid dynamics is an enormously important one and is central to many areas of science and engineering.

Some of the methods which have been developed are the Lattice Boltzmann (LB) approach,[47,48] Smooth Particle Hydrodynamics (SPH) or Smooth Particle Applied Mechanics (SPAM),[49] the Direct Simulation Monte Carlo (DSMC) method[50] and Dissipative Particle Dynamics (DPD).[51] We will not discuss them all here but only give an outline of the two methods that have been most used to date in treating solutions containing large particles, namely the LB approach and DPD.

7.3.1 The lattice Boltzmann approach[47,48]

The Lattice Gas and LB methods have their geneses in the "Game of Life" and Cellular Automata.[47] Such methods use a regular lattice and a "particle" on each lattice site can have a small number of different states; the particles move from site to site and their states are determined by simple rules. These rules are designed to give an average behaviour of the collection of particles, which describe the same physical behaviour as do the Navier–Stokes equations. Thus, a collection of discrete states on a regular lattice is used to mimic the behaviour of a physical system described by continuous variables. Only a few quite simple rules are required for the LB method to give essentially exact agreement with the results of continuum methods while being much easier to program and much, much less costly in computer time.[47,48] In fact, the saving in time using an LB method over traditional methods of solving continuum equations is usually enormous.[48]

In addition to providing a practical method to solve the Navier–Stokes and other problems in fluid mechanics these methods have been adapted to describe the properties of colloidal suspensions, and we will return to this topic in Section 7.7.1.

7.3.2 Dissipative particle dynamics

Dissipative particle dynamics (DPD) is a stochastic method that was introduced as an off-lattice version of the Lattice-Gas methods[51,52] described in Section 7.3.1 above, and although it may be used to simulate fluid dynamics, it has been extensively used to develop equations of motion to describe colloidal particles and polymers solutions, and we will discuss this in Section 7.7.2. Suffice to say that this method is extremely fast relative to the methods outlined above in Section 7.2 and enables the simulation of many more particles in a fluid background to be made than these methods. However, more work is still needed on the theoretical background to this method.

7.4 BROWNIAN DYNAMICS ALGORITHMS

The description in terms of eq. (7.1) is an exact in the limit $m/M \rightarrow 0$ and $t > \tau_b$ but these equations are, in general, insoluble analytically. One method of overcoming the similar problem for atomic and molecular systems is to resort to numerical techniques which lead to the method of MD. We will now outline an analogous method based on the Langevin equation, which is known as the BD method.

7.4.1 The algorithm of Ermak and McCammon[1]

Here we will follow the treatment given in the pioneering and seminal work of Ermak and McCammon.[1] In order to obtain a finite difference algorithm starting from eq. (7.1) we multiply by D_{li}, sum over i and change the dummy index we have,

$$\sum_j \tau_{ij}\, \mathrm{d}V_j/\mathrm{d}t = -V_i + (1/k_\mathrm{B}T)\sum_j D_{ij}\, F_j + \sum_j \sigma_{ij}\, f_j \tag{7.11}$$

where $\tau_{ij} = D_{ij}(M_j/k_\mathrm{B}T)$ and $\sigma_{ij} = (1/k_\mathrm{B}T)\sum_1 D_{i1}\,\alpha_{1j}$

Now from eqs. (7.4) and (7.5) we have,

$$D_{ij} = \sum_{l} \sigma_{il}\, \sigma_{jl}$$

Next use a Taylor series expansion to first order,

$$\tau_{ij} = \tau_{ij}{}^{0} + \sum_{1} \partial \tau_{ij}{}^{0} / \partial R_{1}\, \Delta R_{1} + \cdots \tag{7.12}$$

where $\Delta R_l = R_l(t) - R_l(0)$.

Similarly D_{ij}, σ_{ij} and F_j may also be expanded in Taylor series and in these cases the coefficients of ΔR_l are zero if component l is associated with different particles than components i and j. To calculate means and mean square values we thus need first-order terms τ_{ij} but only the zero-order terms in the other variables are needed for an algorithm which is first order in t. Note that the superscript 0 means that the term is to be evaluated at $t = 0$. This gives,

$$\begin{aligned} \tau_{ij}^{0}\mathrm{d}V_i/\mathrm{d}t + V_i &= \tau_{ii}^{0}\exp(-t/\tau_{ij}^{0})\, \mathrm{d}/\mathrm{d}t[\exp(-t/\tau_{ii}^{0})V_j] \\ &= -\sum_{j} \tau_{ij}^{0}\mathrm{d}V_j/\mathrm{d}t - \sum_{j}\sum_{1} \partial\tau_{ij}^{0}/\partial R_l\, \Delta R_l\, \mathrm{d}V_j/\mathrm{d}t \\ &+(1/k_{\mathrm{B}}T)\sum_{j} D_{ij}^{0}\, F_j^{0} + \sum_{j} \sigma_{ij}^{0}\, f_j \end{aligned} \tag{7.13}$$

Eq. (7.13) could be used to obtain an equation for $V_j(t)$ by multiplying eq. (7.13) by $(\tau_{ii}{}^{0})^{-1}\exp(t/\tau_{ij}{}^{0})$ and integrating over time as with the simpler LEs treated in Chapter 6. However, this is usually not of any particular interest as we are almost always interested in the diffusive limit, which were discussed in Sections 5.1.3 and 5.5.1 and all that is required is an equation for $R_i(t)$. This can be done by performing a second-time integration and eliminating the resultant double integral by integration by parts which give,

$$\begin{aligned} &\Delta R_i(t) = \tau_{ii}^{0}\, V_i(0)[1-\exp(-t/\tau_{ij}^{0})] \\ &-\sum_{j} \tau_{ij}^{0} \int_{0}^{t} [1-\exp(-(t-s)/\tau_{ii}^{0})](\mathrm{d}V_j/\mathrm{d}t)\,\mathrm{d}s \\ &+\sum_{j} \int_{0}^{t} \{[1-\exp(-(t-s)/\tau_{ii}^{0})]\} \\ &\left\{-\sum_{l} \partial\tau_{ij}^{0}/\partial R_l\, \Delta R_l \mathrm{d}V_j/\mathrm{d}t + \sigma_{ij}^{0}\, f_j + (1/k_{\mathrm{B}}T)\, D_{ij}^{0}\, F_j^{0}\right\}\mathrm{d}s \end{aligned} \tag{7.14}$$

As in Sections 5.5 and 6.2.5 we assume that $t \gg t_{ii}^{0} = D_{ii}^{0} / (M_i / k_{\mathrm{B}}T)$ which means that $\exp(-t/\tau_{ii}{}^{0}) \approx 0$ so $[1 - \exp(-(t - s)/\tau_{ii}{}^{0})] \approx 1$ for $0 \le s < t$ and $[1 - \exp(-(t - s)/\tau_{ii}{}^{0})] \approx 0$ if $s = t$.

Now, $V = O((k_BT/M))$ and $\Delta R = O((Dt)^{1/2})$ so $\tau^0 V = O((M/k_BT)^{1/2}D) \ll (Dt)^{1/2} \approx \Delta R$ which means that,

$$\begin{aligned}\Delta R_i(t) = &+\sum_j\left(-\sum_l \partial\tau_{ij}^0/R_l\int_0^t(\Delta R_i\,\mathrm{d}V_j/\mathrm{d}t)\mathrm{d}s+(1/k_BT)D_{ij}^0F_j^0t\right)\\ &+\sum_j\sigma_{ij}^0\int_0^t\{[1-\exp(-(t-s)/\mathrm{t}_{ii}^0)]\}f_j(s)\mathrm{d}s\end{aligned} \tag{7.15}$$

Note that in the random (last) term in eq. (7.15) we cannot drop the term involving the exponential as it is essential to the correct description of this random term.

It should be noted that $\Delta R_i\,\mathrm{d}V_j/\mathrm{d}t = \mathrm{d}/\mathrm{d}t(\Delta R_j\,V_j) - V_iV_j$ and so,

$$\int_0^t V_iV_j\mathrm{d}s \approx k_BT/M_i\,\delta_{ij} \tag{7.16}$$

and since we have $t \gg \tau_{ii}^0$,

$$\int_0^t(\Delta R_j\,\mathrm{d}V_j/\mathrm{d}t)\mathrm{d}s \approx (\Delta R_iV_j) - k_BT/M_it\,\delta_{ij} \approx -k_BT/M_i\,t\delta_{ij}$$

The last approximation arises because $(\Delta R_iV_j) \approx [(k_BTD/M)t]^{12} \ll (k_BT/M)t$ so eq. (7.15) may be written as

$$\begin{aligned}\Delta R_i(t) &= \sum_j D_{ij}^0/\partial R_j\,t+\sum_j(D_{ij}^0F_j^0/k_BT)t\\ &\quad+\sum_j\sigma_{ij}^0\int_0^t\{[1-\exp(-(t-s)/\tau_{ii}^0)]\}f_j(s)\mathrm{d}s\\ &=\sum_j D_{ij}^0/\partial R_j\,t+\sum_j(D_{ij}^0F_j^0/k_BT)t+X_i(t)\end{aligned} \tag{7.17}$$

Thus, we have a first-order algorithm for $\Delta R_i(t)$ which is of a form similar to the simple case (6.45) and to complete the algorithm as is usual we need to find the statistical properties of the random term, $X_i(t)$, now,

$$\begin{aligned}\langle\Delta R_i(t)f_j(t)\rangle &= \sum_l\int_0^t\{[1-\exp(-(t-s)/\tau_{ii}^0)]\}\sigma_{ij}^0\langle f_l(s)f_j(t)\rangle\mathrm{d}s\\ &=2\int_0^t\{[1-\exp(-(t-s)/\tau_{ii}^0)]\}\sigma_{ij}^0\delta(s-t)\mathrm{d}s\\ &=0\end{aligned} \tag{7.18}$$

It may be used to show that,

$$\langle \Delta R_i(t) \rangle = \sum_j \{ D_{ij}^0 / \partial R_j + D_{ij}^0 F_j^0 / (k_B T) \} t \tag{7.19}$$

and

$$\begin{aligned} \langle \Delta R_i(t) \Delta R_j(t) \rangle = 2 \sum_l \int_0^t & \{[1 - \exp(-(t-s)/\tau_{ii}^0)]\} \\ & \{[1 - \exp(-(t-s)/\tau_{jj}^0)]\} \sigma_{il}^0 \sigma_{jl}^0 \mathrm{d}s \\ & \approx 2 D_{ij}^0 t \end{aligned} \tag{7.20}$$

Thus, we have

$$\langle X_i(t) \rangle = 0 \tag{7.21}$$

and

$$\langle X_i(t) X_j(t) \rangle = 2 D_{ij}^0 t$$

Hence, the Ermak and McCammon algorithm[1] for the displacement of a particle (or rather the component R_i in time step Δt in the diffusive limit) is,

$$R_i(t + \Delta t) = R_i(t) + \sum_{j=1}^{3n} (\partial D_{ij}(t) / \partial r_j) \Delta t + \sum_{j=1}^{3n} (D_{ij}(t) / k_B T) F_j(t) \Delta t + X_i(\Delta t) \tag{7.22}$$

where

$$\langle X_i(\Delta t) \rangle = 0 \tag{7.23}$$

$$\langle X_i(\Delta t) X_j(\Delta t) \rangle = 2 k_B T \mu_{ij}(t) \Delta t = 2 D_{ij}(t) \Delta t \tag{7.24}$$

The Ermak–McCammon algorithm (7.22) is first order in the time step Δt, i.e. it is an Euler-type algorithm.

The algorithm (7.22) can also be obtained from the Smoluchowski equation,[1] and generalisation of this algorithm to freely and non-freely rotating non-spherical particles has been made[53] as we shall see in Chapter 8. Strictly speaking eq. (7.22) does not apply to systems with impulsive forces such as hard sphere systems and the above scheme has been generalised to these cases by Cichocki and Hinsen[54–56] but only in the limit of a hydrodynamically dilute system.

The algorithm says that the displacement ΔR_i in time step Δt is the sum of three terms, the later two representing the effect of the direct interactions between the particles and the random term $X_i(t)$ as in the simple case (6.45) except now these terms are summed over all the other co-ordinates making all the equations coupled. The first term involving $\partial D_{ij}^0 / \partial R_j$ is new but is zero if the diffusion tensor $\mathbf{D_{ij}}$ is divergenceless.

In practice if an infinite, bulk system is being simulated then the above algorithm must be implemented using n particles in a basic cell in Periodic Boundary Conditions (PBC); see Section 6.6.

A point which should be emphasised is that eq. (7.22) is an algorithm for generating positions for a system in the Diffusive Limit only as we have used this in arriving at eq. (7.14) from eq. (7.13) or more specifically we have assumed that Δt in eq. (7.22) satisfies $\Delta t \gg t_{ii}^{0} = M_i D_{ii}^{0}/(k_B T) > 0$. Thus, we cannot take the limit of $\Delta R_i(\Delta t) = R_i(t + \Delta t) - R_i(t)$ as $\Delta t \to 0$ and the most we can say is that $\Delta V_i(\Delta t) = (R_i(t + \Delta t) - R_i(t))/\Delta t$ is a mean velocity of some type over the time interval Δt. So, tempting as it may be, we cannot find $\langle V_i(0)V_i(t)\rangle$ from eq. (7.22).

7.4.2 Approximate BD schemes

A complete description of the properties of a suspension using eq. (7.22) is, in practice, very time-consuming; thus, approximations have been devised to treat some limiting cases or to generate approximate BD schemes.

7.4.2.1 Algorithms neglecting Brownian motion

If we assume that the random forces are negligible, then we have

$$R_i(t+\Delta t) = R_i(t) + \sum_{j=1}^{3n} (\partial D_{ij}(t)/\partial r_j)\Delta t + \sum_{j=1}^{3n} (D_{ij}(t)/kT)F_j(t)\Delta t \qquad (7.25)$$

Alternatively we can directly numerically solve the differential eqs. (7.8) (or eq. (7.9) for freely rotating particles[59]), this latter method being preferable as it leads to much greater accuracy.

This scheme is useful for simulating systems in the limit of high Peclet numbers, i.e. where the shear effects dominate over those due to the Brownian terms and for very large particles in a gravitational field.

For example, there has been a large number of studies, both by experiment and by simulation, of cluster dynamics of large sedimenting particles and a surprisingly wide variety of fascinating behaviour have been observed,[10,13–15,22–24,57–59] which may be classified be into several broad distinct classes,

1. Unstable, separating structures.[4,10,13,14,57,58]
2. Structures that represent unstable equilibria.[13,14]
3. Stable equilibrium structures.[10,15,56,57]
4. Periodic structures.[15,22,57,58]

Such a wide range of fascinating structures which form in such simple systems, the particles of which are only interacting by hydrodynamic forces, is extremely interesting and illustrates the importance of these interactions in systems of particles suspended in a fluid.

7.4.2.2 *Use of effective two-body tensors*

A simple approximation is to assume that the diffusion tensors are pair-wise additive but merely using a pair-wise additive tensor based on the truncation of the cluster MIF scheme (the cluster method 1 described in Section 7.2) may not be sufficient as this may lead to non-positive definite diffusion tensors in systems with PBC. To overcome this and to make approximate allowance for many-body effects Snook, van Megen and Tough[60–64] devised a simple volume-fraction-dependent screened two-body tensor based on the ideas of porous media[30] where unlike for freely diffusing particles screening can be rigorously shown to occur. This scheme was parameterised using experimental values of the diffusion coefficients D_L and D_c (see Appendix J) for hard-sphere-like dispersions,[60] and although this approach has been criticised on purely theoretical grounds,[20] it does provide a reasonably accurate description of $D(q)$ for hard-sphere-like dispersions,[65] electrostatically stabilised systems at high volume fraction,[61] and gave qualitatively reasonable results for a wide variety of zero-time (D_S and D_C) and long-time diffusion data, $\langle(\underline{R}_i(0) - \underline{R}_i(t))^2\rangle$ and D_T and scattering functions.[61–64]. For example, the behaviour of $\langle(\underline{R}_i(0) - \underline{R}_i(t))^2\rangle$, D_T, $F_s(q,t)$ and $F(q,t)$ (see Appendices A and J for definitions of all these quantities) was in good qualitative agreement with the experimental results. Even though the method has some theoretical weaknesses it provides a model that qualitatively shows the effect of hydrodynamic interactions on these properties. For example, the calculations clearly show the strong slowing effects of hydrodynamic interactions on single particle motion. A similar approach has been used by Wallrand, Belloni and Drifford[66] for electrostatically stabilised systems.

Dickinson and co-workers[69–75] have also used a two-body tensor modified in order to obtain stable and convergent results, which may be viewed as using a screened tensor with a delta function cut-off rather than a smooth one. They treated a large number of cases involving stability, aggregation, clustering and coagulation and once again provided results which, if not entirely rigorously based, provide a useful qualitative insight into many of these interesting and useful phenomena in concentrated suspensions.

A potentially more theoretically justifiable approach has been developed by Beenakker and Mazur,[20] which gave good results for zero-time properties and is an effective medium theory in which D_{ij} is the Rotne–Prager expression and D_{ii} is D_0 but with a volume-fraction-dependent viscosity background viscosity $\eta(\phi)$ replacing η_f. This form has the advantage of having the correct long-range form for D_{ij} and would be interesting to use in BD simulations at non-zero time.

One problem still to be addressed with any effective two-body approach is the inclusion of lubrication effects, which are particularly important at high volume fraction.

The analysis of experimental results for $D(q)$ by Pusey, Segre and co-workers[76] fraction also lends support the notion of effective two-body tensors.

7.4.2.3 *Mean-field approaches*

An obvious approximation is to assume that the diffusion tensors are diagonal then

$$R_i(t+\Delta t) = R_i(t) + (\partial D_{ii}(t)/\partial r_i)\Delta t + (D_{ii}(t)/k_B T)F_i(t)\Delta t + X_i(\Delta t) \tag{7.26}$$

with,

$$\langle X_i(\Delta t)\rangle = 0 \text{ and}$$

$$\langle X_i(\Delta t)X_i(\Delta t)\rangle = 2D_{ii}(t)\Delta t$$

It should, however, be noted that in general the D_{ii} are still many-body functions, i.e. functions of all the R_{ij}'s.

Van Megen and Snook[62,63] compared the results obtained using their effective two-body tensor with those using only the diagonal elements, D_{ii}, of this tensor to describe self-diffusion in charge-stabilised dispersions and found that there was moderate agreement between the two approaches;[62,63] see Figure 7.1. This method has also been used by Heyes[77] as the basis of a mean-field theory based on the assumption that $\mathbf{D_{ii}} = \mathbf{D_{ii}}(R_{ij})$. He used the theory to describe hard-sphere-like dispersions using an entirely empirical form for $\mathbf{D}_{ii}$ and obtained the parameters by fitting to the experimental values for D_S, which lead to an excellent agreement with the experimental values of the long-time tracer diffusion constant D_T. These results seem to indicate that it may well prove fruitful to extend this approach using a more theoretically justifiable form for $\mathbf{D}_{ii}$ at least to describe self-diffusion. But this method cannot accurately describe properties that depend directly on $\mathbf{D}_{ij}$, e.g. D_c or $D(q)$, at low to moderate values of q, and once again the role of lubrication effects also needs addressing.

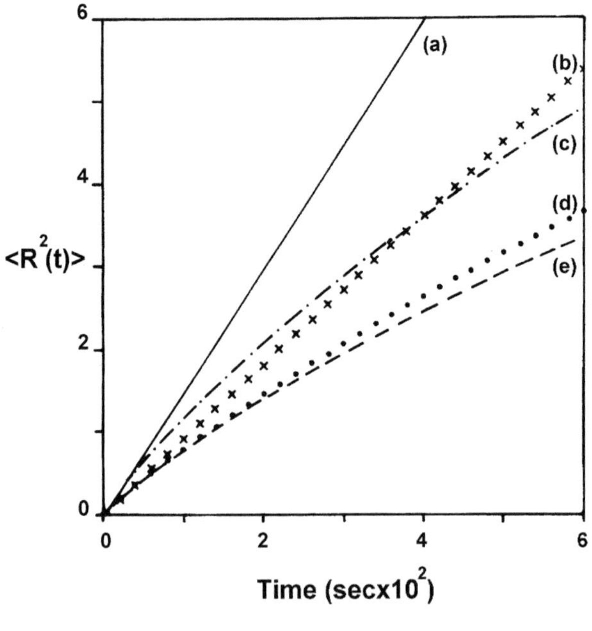

Figure 7.1 Mean square displacement $\langle \Delta R^2(t)\rangle$ versus time t. (a) Free particle; (b) random component; (c) free draining; (d) diagonal component of two-body effective tensor; (e) full two-body tensor (figure redrawn from ref. 63).

7.4.2.4 *Hydrodynamically dilute or free-draining regime*

This approximation is also called "Free Draining" where it is assumed that $\mathbf{D_{ij}} = \boldsymbol{\delta}_{\mathbf{ij}} D_0$ and, thus,

$$R_i(t+\Delta t) = R_i(t) + (D_0/kT)F_i(t)\Delta t + X_i(\Delta t) \tag{7.27}$$

where D_0 is the free particle diffusion coefficient.

$$\langle X_i(\Delta t)\rangle = 0$$

and

$$\langle X_i(\Delta t)X_i(\Delta t)\rangle = 2kT\mu_0\Delta t = 2D_0\Delta t,$$

eq. (7.2) is, of course, the same as eq. (6.45).

This is strictly only applicable to systems where the particle volume fraction is very low but in which the particles are still strongly interacting, e.g. a system of highly charged particles at low particle concentration in a background of very low electrolyte concentration. This algorithm was originally derived by Ermak from the Fokker–Planck equation[78] and has been extensively used to study both systems where it is strictly applicable, e.g. charged polystyrene spheres in ion-exchanged aqueous suspension and for qualitative studies of many other systems where the assumptions are sometimes less well justified.

Gaylor, Snook, van Megen and Watts[79–83] made extensive studies of systems of particles at low volume fraction interacting via a screened coulomb (DLVO type) pair potential and calculated $\langle(\underline{R}_i(0) - \underline{R}_i(t))^2\rangle$, D_T, $G_s(R,t)$, $G_d(R,t)$, $G(R,t)$ $F_s(q,t)$, $F_d(q,t)$ and $F(q,t)$ (see Appendices A and J for definitions of these quantities) as a function of volume fraction. Despite the simplicity of the form of the direct particle interaction excellent agreement with experimental results was obtained[82] for the mean-square displacement; see Figure 7.2 below.

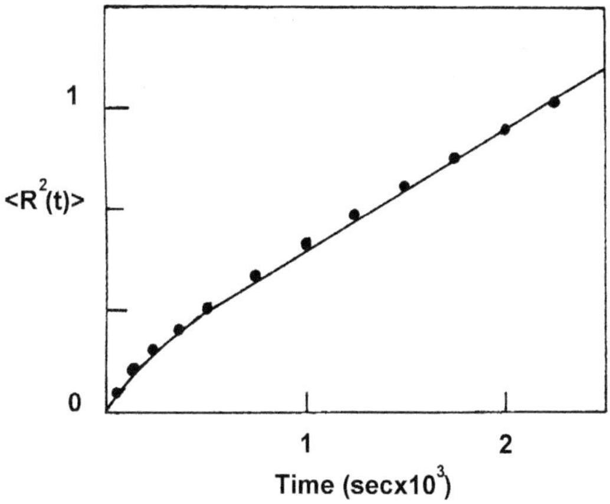

Figure 7.2 Mean square displacement $\langle\Delta R^2(t)\rangle$, continuous line experimental data from Pusey and BD results • (figure redrawn from ref. 82).

Furthermore, it was shown that D_T is strongly influenced by inter-particle forces, that the non-Gausssian effects $\alpha_n(t)$ defined in eq. (A.15), Appendix A are relatively small and that although the Vineyard approximation for $G_s(R,t)$ and $F_s(q,t)$ is not very accurate the delayed Vineyard approximation is excellent although it lacks a theoretical justification.[82]

Despite the fact that this method is only an approximation it has been extensively used to simulate more concentrated systems (see, e.g., refs. 77, 84–91) and has given very useful insights into various colloidal phenomena. Some representative studies are of colloidal diffusion,[84–86,91] glass formation,[87,90] viscous flow[88] and diffusion in porous media.[89]

7.5 BROWNIAN DYNAMICS IN A SHEAR FIELD

From a microscopic starting point this problem has been studied in the case of the GLE; see Chapter 4 and for the limiting case of Brownian motion when hydrodynamic interactions may be neglected, see Section 6.5. Shea and Oppenheim[92] have generalised their results for the Langevin equation for a single Brownian particle in an external field starting from a microscopic description as mentioned in Chapter 4, but no BD algorithm has been derived from this nor has the result yet been generalised to many interacting colloidal particles.

Dickinson and co-workers[69] have developed a BD algorithm for simulating particles in a shear field based on a postulated form of the LE in shear obtained by adding a term to the LE (7.1) to represent the effect of the imposed shear field, i.e. they wrote,

$$M\,\mathrm{d}\underline{V}_i/\mathrm{d}t = \underline{F}_i + \underline{F}_i^F + \underline{F}_i^S \tag{7.28}$$

where $\underline{F}_i$ represents the usual direct interactions, $\underline{F}_i^{\mathrm{F}}$ a frictional force and $\underline{F}_i^{\mathrm{S}}$ a stochastic force. The frictional force $\underline{F}_i^{\mathrm{F}}$ was assumed to have the following form:

$$\underline{F}_i^{\mathrm{F}} = \sum_{j=1}^{n} \zeta_{ij}\cdot[\underline{V}_j + u(\underline{R}_j) + \mathbf{S}_j : \boldsymbol{\rho}] \tag{7.29}$$

where $\boldsymbol{\zeta}_{ij}$ is the normal friction tensor, $u(\underline{R}_j)$ the flow velocity at $\underline{R}_j$, $\mathbf{S}_j$ a configuration-dependent shear tensor and $\boldsymbol{\rho}$ the configuration-dependent rate-of-strain tensor. From eq. (7.28) they obtain a BD algorithm analogous to the Ermak–McCammon algorithm (7.22) which they write as

$$\begin{aligned}\Delta\underline{R}_j &= (k_{\mathrm{B}}T)^{-1}\left[\sum_{j=1}^{n}\underline{F}_j(t)\cdot\mathbf{D}_{ij}(t) + \sum_{j=1}^{n}\partial\boldsymbol{\mu}_{ij}/\partial\underline{R}_j\right]\Delta t \\ &\quad + \underline{X}_i(D_{ij}(t),\Delta t) + \mathbf{S}_i(t)\boldsymbol{\rho}\Delta t + u_j(\underline{R}_j)\Delta t\end{aligned} \tag{7.30}$$

for $i,j = 1$ to n where the random term $\underline{X}_i(D_{ij}(t),\Delta t)$ has mean zero and its mean-square value is

$$\langle \underline{X}_i(D_{ij}(t),\Delta t)\,\underline{X}_i(D_{ij}(t),\Delta t)\rangle = 6\,\mathbf{D}_{ij}(t)\,\Delta t \tag{7.31}$$

This was used to study the rotation and dissociation of colloidal aggregates in a shear flow using a simple two-body tensor representation of $\mathbf{D}_{ij}$ and the lowest order result for $\mathbf{S}_j$ which is a three-body term.[68]

The most complete extension of the BD method of Ermak and McCammon has been given by Brady and Bossis and co-workers[22–27] starting from a macroscopic point of view that essentially generalises Mazur's result based on the Landau–Lifshitz fluctuating hydrodynamics equations.[93] Brady and Bossis[23] have derived a BD algorithm starting with the Newton equation of motion for the combined translational and rotational degrees of freedom as will be discussed in Chapter 8, i.e.

$$\mathbf{M}\cdot \mathrm{d}\mathbf{U}/\mathrm{d}t = \mathbf{F}^{P} + \mathbf{F}^{\mathrm{B}} + \mathbf{F}^{\mathrm{H}} = \mathbf{F} \tag{7.32}$$

where

$$\mathbf{F}^{\mathrm{H}} = -\mathbf{R}_{\mathrm{FU}}\cdot(\mathbf{U}-\mathbf{U}^{\infty}) + \mathbf{R}_{\mathrm{FE}} : \mathbf{E}^{\infty} \tag{7.33}$$

$$\langle \mathbf{F}^{\mathrm{B}} \rangle = 0$$

and

$$\langle \mathbf{F}^{\mathrm{B}}(0)\mathbf{F}^{\mathrm{B}}(t) \rangle = 2k_{\mathrm{B}}T\,\mathbf{R}_{\mathrm{FU}}\delta(t)$$

Generalising the algorithm of Ermak and McCammon[1] (7.22) Brady and Bossis[23] obtain for the change in position $\Delta\mathbf{x}$ of the particle in time step Δt,

$$\begin{aligned}\Delta\mathbf{x} = \mathrm{Pe}\{\mathbf{U}^{\infty} + \mathbf{R}_{\mathrm{FU}}^{-1}\cdot[\mathbf{R}_{\mathrm{FE}} : \mathbf{E}^{\infty} + \gamma^{*-1}\,\mathbf{F}^{\mathrm{P}}]\}\Delta t \\ + \nabla\cdot\mathbf{R}_{\mathrm{FU}}^{-1}\Delta t + \mathbf{X}(\Delta t)\end{aligned} \tag{7.34}$$

where $\gamma^{*} = 6\pi\eta_{\mathrm{f}}a^{2}\gamma/|\mathbf{F}^{\mathrm{P}}|$ and $Pe = 6\pi\eta_{\mathrm{f}}a^{3}\gamma/(k_{B}T) = \gamma\, a^{2}\, D_{0}$

$$\langle \mathbf{X} \rangle = 0$$

and

$$\langle \mathbf{X}(\Delta t)\mathbf{X}(\Delta t) \rangle = 2\,\mathbf{R}_{\mathrm{FU}}{}^{-1}\Delta t$$

Here, $\mathbf{M} = \mathbf{R}_{\mathrm{FU}}{}^{-1}$ is the mobility matrix generalised to include both the translational and rotational effects, the Peclet number Pe measures the relative importance of the shear and Brownian forces whilst the reduced shear rate γ^{*} gives the relative importance of the shear and inter-particle forces.

Phung, Brady and Bossis[27] made an extensive study of the behaviour of a concentrated colloidal dispersion of monodispersed hard spheres in a shear field, which provides an excellent example of what information this method is capable of giving. The diffusion matrices $\mathbf{D}_{ij}$ were calculated by the method outlined in Section 7.2, i.e. a combination of far-field many-body terms and short-ranged two-body lubrication effects. These calculations were carried out for $27 \leq n \leq 123$ in three dimensions at a particle volume fraction of 0.45 for a range of values of the imposed shear field (i.e. for a range of values for the Peclet number *Pe*). The results (see Figure 7.3) showed that at low Peclet numbers the equilibrium structure of the suspension was slightly distorted by the flow and the suspension shear thinned because of the decrease of the direct Brownian contribution to the stress. At moderate values of *Pe* it was found that the hydrodynamic and Brownian forces balanced and the structure ordered into 'strings' arranged in a hexagonal array in the velocity-gradient–vorticity plane; however, the viscosity remained practically constant. At high Peclet numbers this ordered structure gradually "melted" without hysteresis because of the shearing forces overcoming the effect of the Brownian motion that pushed the particles into close contact. After this "melting" the viscosity shear thickened which was attributed to the formation of large non-contact clusters. The long-time self-diffusivities were found to be constant at low *Pe* and then increase at intermediate values of *Pe*. However, when ordering occurs these diffusivities decreased and then grew as "melting" occurs at higher values of Pe.

In order to calculate the bulk stress $\langle \Sigma \rangle$ and hence the rheological properties of the suspension one also needs to calculate two other tensorial quantities as is set out in Appendix J.

7.6 LIMITATIONS OF THE BD METHOD

Numerical techniques such as the BD and MD simulation methods have inherent limitations in addition to those imposed by current computer limitations. Some of those that are inherent to the BD method are as follows:

1. If the particles simulated are neither much larger nor heavier than those of the dispersion medium, it may not be correctly described; see Section 5.5.
2. The detailed coupling between particles of the two species cannot be simulated as the method was derived by coarse graining over the properties of the solvent; see Chapter 5. This may restrict its ability to simulate aggregation and coagulation in which particle surfaces come into close contact.
3. The hydrodynamics used assumes Newtonian viscous behaviour of the background fluid, so particles suspended in a non-Newtonian fluid cannot be simulated.
4. Because of their long range, many-body character and the complexity of their calculation if full hydrodynamic interactions are to be used this imposes serve restrictions on the system size and times which can be simulated.

Of course computer limitations do impose severe restrictions on what type of systems and properties may be simulated even if all the points 1–4 are not relevant, e.g.

1. As with all particle-based simulation techniques there are limitations of length scales that can be simulated, since we use a cubic box in PBCs of relatively small side length, which limits the correlation lengths of phenomena that may be simulated.

2. The use of PBCs introduces a recurrence time t_{rec} because any time-dependent "disturbance", e.g. a density wave will ultimately interact with its periodic image.

Thus, there is a big incentive to find methods to speed up simulations of Brownian systems, such as colloidal suspensions and polymer solutions without significant loss of accuracy and we will now discuss two such methods.

7.7 ALTERNATIVES TO BD SIMULATIONS

One of the major limitations on carrying out BD simulations is the calculation of hydrodynamic interactions. As mentioned in Section 7.3 there have been many schemes devised in order to speed up fluid dynamics calculations in general. Two of these methods, which have been combined with BD-like algorithms, are the LB and the DPD methods, and we shall briefly discuss them in the next two sections.

7.7.1 Lattice Boltzmann approach

In Section 7.3.1 we pointed out that the LB method enables fluid mechanics to be solved much more readily by the use of discrete lattice models than is possible by use of continuum-based methods. The only problem remaining is to allow for the presence of particles

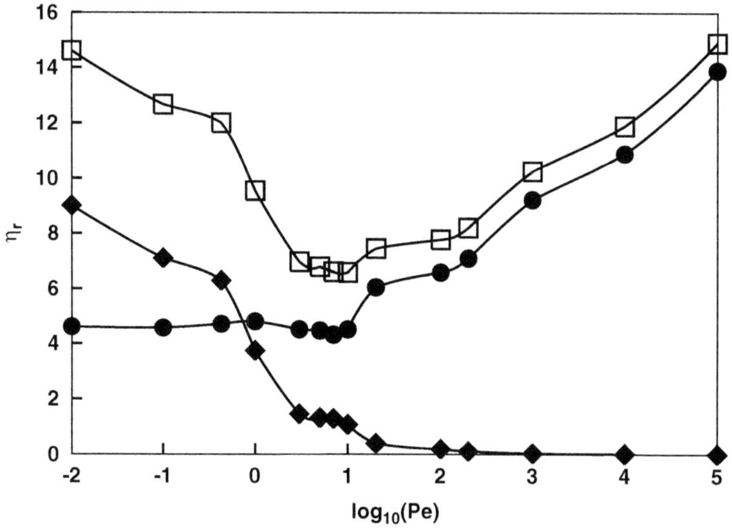

Figure 7.3 The viscosity of a hard-sphere suspension of volume fraction 0.45 as a function of the Peclet number calculated by the BD method.[27] The figure shows the total viscosity □, the hydrodynamic contribution ● and the Brownian contribution ◆.

in the fluid which really amounts to satisfying the correct boundary conditions on the fluid velocity at the particle's surfaces. Several methods have been devised to do this.

Ladd[48,94] represents the colloidal particle by its surface, which cuts some of the links between the lattice sites of the background LB fluid. Then boundary points are placed halfway along these links where the one-particle distribution function representing the fluid is bounced back to mimic no-slip boundaries.

An alternative method due to Lobaskin and Dunweg[95] instead represented the colloidal particles by arrays of point particles connected to one another by springs and these point particles interact with the LB fluid via a frictional force.[96]

A third approach taken by Chatterji and Horbach[97] is essentially a hybrid of these previous two methods, which results in a Langevin equation,

$$M_i\, \mathrm{d}^2 \underline{R}_i/\mathrm{d}t^2 = \underline{F}_c - \zeta_0 \underline{V}(t) + \underline{f}_r \tag{7.35}$$

where $\underline{F}_c$ is the conservative force between the colloidal particles, ζ_0 a friction coefficient and $\underline{f}_r$ the random force due to the solvent, which is taken to be a Gaussian random variable whose Cartesian components α have moments,

$$\langle f_{r,\alpha}(\underline{R},t)\rangle = 0$$

and

$$\langle f_{r,\alpha}(\underline{R},t) f_{r,\beta}(\underline{R}',t')\rangle = 2k_{\mathrm{B}}T\, \zeta_0\, \delta_{\alpha\beta}\, \delta(\underline{R}-\underline{R}')\,\delta(t-t')$$

The more detailed definitions of these forces are given in reference 97 but the method essentially results in a simple equation of motion, and as all the forces involved are short ranged and no many-body hydrodynamic couplings occur (due to the LB representation of the background fluid), the method of solving eq. (7.35) is like those developed in Chapter 6 for the simplest LE equation. This makes this method quite fast and it appears to be a very promising and accurate alternative to a full BD simulation.

These types of methods have generated many interesting results and are many orders of magnitude faster than the full BD method. What is really required now is to test these methods against full BD simulations on the same system in order to assess their accuracy and to help refine them. If this is done, then in future this should lead to their becoming some of the most valuable ways of simulating colloidal and polymer systems.

7.7.2 Dissipative particle dynamics

Dissipative particle dynamics (DPD) represents the whole system of background fluid plus large particles as a collection of frictional "balls" with forces between them or "dissipative particles" representing the various physical effects which would normally be represented by inter-particle forces and hydrodynamic interactions in BD simulations. The original DPD method postulated the following stochastic differential equation,[51]

$$M_i\,\mathrm{d}\underline{V}_i = \sum_{i\neq j} \underline{F}_{ij}^{\mathrm{C}}(\underline{R}_{ij})\mathrm{d}t - \gamma\sum_{i\neq j}\omega(\underline{R}_{ij})(\underline{e}_{ij}\cdot\underline{V}_{ij})\underline{e}_{ij}\mathrm{d}t + \sigma\sum_{i\neq j}\omega^{1/2}(\underline{R}_{ij})\,\underline{e}_{ij}\,\mathrm{d}\underline{W}_{ij} \tag{7.36}$$

and $d\underline{R}_i = \underline{V}_i\,dt$ where $\underline{R}_i$ and $\underline{V}_i$ are the position and velocity of the dissipative particles, M_i their mass, $\underline{F}_{ij}^{\mathrm{C}}$ the conservative, soft, repulsive force between these particles and $\underline{e}_{ij} = \underline{R}_{ij}/|\underline{R}_{ij}|$. The magnitude of the dissipative forces is given by γ, and the similarly σ governs the intensity of the stochastic forces. The functions $\omega(\underline{R}_{ij})$ are weight functions which give the range of interaction of the dissipative particles and are taken to be a short-ranged force, only acting between nearest neighbours and the stochastic forces are represented by $\mathrm{d}\underline{W}_{ij} = \mathrm{d}\underline{W}_{ji}$.

This method is very versatile and has been applied to a wide range of colloidal and polymeric systems and provides a reasonable description of many phenomena.[51,99] It is also extremely fast and thus applicable to large systems. However, there are several problems and shortcomings with the original DPD method, e.g., the method gives poor results for colloids at high volume fractions, it cannot treat entanglements in polymers, the viscosity of the fluid cannot be related to the model parameters and the equation of state is an outcome not an input and the model cannot study energy transport.[51]

Modifications and improvements of the original DPD model based on the postulated eq. (7.36) have been made in order to overcome some of these problems, e.g., Espanol and Revenga[100] have developed a thermodynamically consistent DPD equation which is,[51]

$$M_i\,\mathrm{d}\underline{V}_i/\mathrm{d}t = \sum_j [\mathrm{P}_i/d_i^2 + \mathrm{P}_j/d_j^2]F_{ij}\underline{R}_{ij} - (5\eta/3)\sum_j F_{ij}/(d_i d_j)(\underline{v}_{ij} + \underline{e}_{ij}\underline{e}_{ij}\cdot\underline{V}_{ij}) + \underline{F}_i^{\mathrm{R}} \tag{7.37}$$

$$T_i\mathrm{d}S_i\mathrm{d}t = -2\kappa\sum_j F_{ij}/(d_i d_j)T_{ij} + (5\eta/6)\sum_j F_{ij}/(d_i d_j)\{\underline{v}_{ij}^{\;2} + (\underline{e}_{ij}\cdot\underline{V}_{ij})\} + T_i\,J_i^{\mathrm{R}}$$

and

$$\mathrm{d}\underline{R}_i/\mathrm{d}t = \underline{V}_i$$

where now the temperature and pressure are introduced via the terms T_i and P_i, which are the temperature and pressure of the fluid particle i, $T_{ij} = T_i - T_j$ and S is the entropy. The function $F(R)$ is given by $\nabla W(R) = -\underline{R}\,F(R)$, and $\underline{F}_i^{\mathrm{R}}$ and J_i^{R} are stochastic forces introduced so that the fluctuation–dissipation theorem is obeyed. This equation satisfies conservation of mass, momentum and energy and is consistent with the laws of thermodynamics. It is said to overcome all the limitations of the original DPD equation,[51] e.g., the thermodynamic data and the correct value of the viscosity may be fed in as input. A nice feature of this equation is that if there are no thermal fluctuations then eq. (7.37) is actually a version of SPH,[49] which is very helpful as this method is well studied and well established in the area of fluid mechanics.

Thus, this method and other extensions of the original DPD model are becoming a powerful tool to explore the dynamics of colloidal and polymeric systems. It will be interesting to see if more theoretical work can show the relationship between these new DPD methods and those based on the more traditional and well-founded Langevin-based BD approach. Also rather than just relying on comparisons with experiment to test these methods it would be wise to test them instead against full BD simulations on the same system as was recommended with LB methods.

7.8 CONCLUSIONS

The Ermak–McCammon eq. (7.22) provides an algorithm for calculating the displacement of particle (actually for a component of position R_i) over a time interval Δt for a particle in an n-body system in the diffusive limit to first order in Δt. It cannot be used to find the change in velocity over this interval as a restriction is that the velocity correlations have ceased in the time interval Δt and, thus $\Delta t > 0$ so we cannot take the limit of ΔR_i as $\Delta t \rightarrow 0$. This equation says, quite simply that in time interval Δt the change of the position component $X_i(t)$ of particle is the sum of the effect of direct interactions between the particles, a random term, $X_i(t)$ and a term involving $\partial D_{ij}^0/\partial R_j$. The term $X_i(t)$ is a Gaussian random variable dependent on the diffusion tensor $\mathbf{D_{ij}}$, and the random displacements of each particle are coupled, and the term dependent on $\partial D_{ij}^0/\partial R_j$ is zero if the diffusion tensor $\mathbf{D_{ij}}$ is divergenceless. This basic algorithm has been modified to incorporate the effect of external fields and to treat non-spherical particles (see Section 8.3).

One of the major tasks to perform before this algorithm may be used is to calculate the diffusion tensors $\mathbf{D_{ij}}$, which requires a major mathematical and computational effort as detailed in Section 7.2. These tensors are many-body in character, very long ranged and difficult to calculate, although several schemes exist to do this. This means that the use of the full Ermak–McCammon algorithm is a computationally challenging one and the complexity of calculating $\mathbf{D_{ij}}$ puts a major limit on the number of particles and length of time possible for a simulation. Thus, simpler methods have been sought in order to speed up these BD simulations and these methods fall into two broad categories:

1. Those that use a simplified version of the full BD algorithm which involves simplifying the form of the hydrodynamic interaction tensors used, e.g. by replacing the many-body ones with effective two-body forms.
2. The second type of modification is to use different ways of treating the hydrodynamics compared with those based on the traditional continuum treatments. Two of

these methods, the LB and the DPD methods, are much faster than the full BD method. The main problems remaining with these methods are to assess their accuracy and to understand more about their theoretical grounding.

In conclusion another point that should be made on this topic is that the use of the current BD algorithm could itself possibly be speeded up by developing modifications to the algorithm and improvements in computer codes and its implementation on computers. First the algorithm (7.22) is only of first order in Δt; higher order algorithms could be developed, which would enable a larger time step Δt to be used, multiple time step algorithms might be tried, and neighbour lists and linked lists could be implemented as could the use of multiprocessor techniques. All these improvements will enable the simulation of much larger systems and longer times to be carried out using the BD method.

REFERENCES

1. D.L. Ermak and J.A. McCammon, *J. Chem. Phys.*, **69**, 1352 (1978).
2. J.M. Deutch and I. Oppenheim, *J. Chem. Phys.*, **54**, 3547 (1971).
3. T.J. Murphy and J.L. Aguirre, *J. Chem. Phys.*, **57**, 2098 (1972).
4. H.R. Kruyt (Ed.), *Colloid Science, Volume 1*, Elsevier, Amsterdam, 1952.
5. E.W.J. Verwey and J.Th.G. Overbeek, *Theory of Stability of Lyophobic Colloids,* Elsevier, Amsterdam, 1948.
6. W. van Megen and I. Snook, Equilibrium properties of suspensions, *Adv. Coll. Interf. Sci.*, **21**, 119 (1984).
7. R.J. Hunter (Ed.), *Foundations of Colloid Science*, Oxford University Press, Oxford, 1989.
8. P.N. Pusey, Colloidal suspensions, Course 10. In: *Liquids, Freezing and the Glass Transition* (Eds. J.P. Hansen, D. Levesque and J. Zinn-Justin), Elsevier, B.V., Amsterdam, 1991, pp. 764–942.
9. P. Mazur, *Physica*, **110A**, 128 (1982).
10. J. Happel and H. Brenner, *Low Reynolds Number Hydrodynamics,* Noorhoff, Lyden, 1973.
11. B.U. Felderhof, Physica, **89A**, 373 (1977); R. Schmidtz and B.U. Felderhof, *Physica*, **113A**, 90, 103 (1982) and *Physica*, **116A**, 163 (1982); R.B. Jones and G.S. Burfield, *Physica*, **133A**, 152 (1985); R.B. Jones and R. Schmitz, *Physics*, **149A**, 373 (1988).
12. E. Dickinson, *Chem. Soc. Rev.*, **14**, 421 (1985).
13. P. Ganatos, R. Pfeffer and S. Weinbaum, *J. Fluid Mech.*, **84**, 79 (1978).
14. Q. Hassonjee, P. Ganatos and R. Pfeffer, *J. Fluid Mech.*, **197**, 1 (1988).
15. R.E. Caflisch, C. Lim, J.H.C. Luke and A.S. Sangani, *Phys. Fluids*, **31**, 3175 (1988).
16. P. Mazur and W. van Saarloos, *Physica*, **115A**, 21 (1982).
17. C.W.J. Beenakker and P. Mazur, *Phys. Lett.*, **91**, 290 (1982).
18. C.W.J. Beenakker and P. Mazur, *Phys. Lett.*, **98A**, 22 (1983).
19. C.W.J. Beenakker and P. Mazur, *Physica*, **120A**, 388 (1983).
20. C.W.J. Beenakker and P. Mazur, *Physica*, **126A**, 349 (1984).
21. C.W.J. Beenakker, *Physica A*, **128**, 48 (1984).
22. L. Durlofsky, J.F. Brady and G. Bossis, *J. Fluid Mech.*, **180**, 21 (1987).
23. J. Brady and G. Bossis, *Ann. Rev. Fluid Mech.*, **20**, 111 (1988); *J. Chem. Phys.* **87**, 5437 (1987).
24. R.J. Phillips, J.F. Brady and G. Bossis, *Phys. Fluids,* **31**, 3462 (1988); *ibid.* **31**, 3473 (1973).
25. L. J. Durlofsky and J. F. Brady, *J. Fluid Mech.*, **200**, 39 (1989).

26. J.F. Brady, R.J. Phillips, J.C. Lester and G. Bossis, *J. Fluid Mech.*, **195**, 257 (1988).
27. T.N. Phung, J.F. Brady and G. Bossis, *J. Fluid Mech.*, **313**, 181 (1996).
28. A. Ladd, *J. Chem. Phys.*, **88**, 5051 (1988).
29. A. Ladd, *J. Chem. Phys.*, **90**, 1149 (1989).
30. A. Ladd, *J. Chem. Phys.*, **93**, 3484 (1990).
31. A. Ladd, *Phys. Fluids A*, **5**, 299 (1993).
32. K. Briggs, E.R. Smith, I.K. Snook and W. van Megen, *Phys. Lett.*, **154**, 149 (1991)
33. B.U. Felderhof, Physica A, **159**, 1 (1989); B.U. Felderhof and B. Cichocki, *Physica A*, **159**, 19 (1989).
34. B. Cichocki, B.U. Felderhof, K. Hinsen, E. Wajnryb and J. Blawzdziewicz, *J. Chem. Phys.*, **100**, 3780 (1994).
35. B. Cichocki and K. Hinsen, *Phys. Fluids*, **7**, 285 (1995).
36. K. Hinsen, *Comput. Phys. Commun.*, **88**, 327 (1995).
37. R.W. O'Brien, *J. Fluid. Mech.*, **91**, 17 (1979).
38. C.W.J. Beenakker *J. Chem. Phys.*, **85**, 1581 (1986).
39. E.R. Smith, I.K. Snook and W. van Megen, *Physica*, **143A**, 441 (1987).
40. J. Brady, *Simulations of Particulate Dispersions at Low Reynolds Numbers: What are the Issues?,* CECAM Workshop, *Models and Algorithms for Mesoscopic Simulation of Colloidal Particle Systems,* CECAM, Ecole Normale Superior, Lyon, France, June 1997.
41. A.S. Sangani and G. Mo, *Phys. Fluids*, **6**, 1636 (1994); *ibid.* **8**, 1990 (1996) and A.S. Sangani, *An O(N) Algorithm for Stokes and Laplace Interactions,* CECAM Workshop, *Models and Algorithms for Mesoscopic Simulation of Colloidal Particle Systems,* CECAM, Ecole Normale Superior, Lyon, France, June 1997.
42. J. Higdon, *A P3M Method for Stokes Flow Dynamic Simulations – an O(N) Algorithm,* CECAM Workshop, *Models and Algorithms for Mesoscopic Simulation of Colloidal Particle Systems,* CECAM, Ecole Normale Superior, Lyon, France, June 1997.
43. C.W.J. Beenakker, W. van Saarloos and P. Mazur, *Physica*, **127A**, 451 (1984).
44. C.W.J. Beenakker, and P. Mazur, *Physica*, **131A**, 311 (1985); *Phys. Fluids*, **28**, 3203 (1985).
45. E.R. Dufresne, T.M. Squires, M.P. Brenner and D.G. Grier, *Phys. Rev. Lett.*, **85**, 3317 (2000); R. M. Jendrejack, D.C. Schwartz, M.D. Graham and J.J. de Pablo, *J. Chem. Phys.*, **119**, 1165 (2003); R. M. Jendrejack, D.C. Schwartz, J.J. de Pablo and M.D. Graham, *J. Chem. Phys.*, **120**, 2513 (2004).
46. M. Mareschal and B.L. Holian (Eds.), *Microscopic Simulations of Complex Hydrodynamic Phenomena*, Vol. 292 in NATO ASI Series B, Physics, Plenum Press, New York, 1992; B.M. Boghosian and N.G. Hadjiconstantinou, *Mesoscopic Models of Fluid Dynamics*, *Handbook of Materials Modelling*(Ed. S. Yip), Springer, The Netherlands, 2005, pp. 2411–2414.
47. G. Stumolo and V. Babu, *New Directions in Computational Aerodynamics*, Physics World, August 1997, pp. 45–49; U. Frisch, B. Hasslacher and Y. Pomeau, *Phys. Rev. Lett.*, **56**, 1505 (1986).
48. A.J.C. Ladd and R. Verberg, Lattice–Boltzmann simulations of particle-fluid suspensions, *J. Stat. Phys.*, **104**, 1191–1251 (2001).
49. J.J. Monaghan, Smoothed particle hydrodynamics, *Annual Rev. Astronomy Astrophysics*, **30**, 543 (1992); L.B. Lucy, A numerical approach to the testing of the fission hypothesis, *Astronom. J.*, **82**, 1013 (1977); W.G. Hoover, T.G. Pierce, C.G. Hoover, J.O. Shugart, C.M. Stein and A.L. Edwards, *Comput. Math. Appl.*, **28**, 155 (1994); W.G. Hoover, *Simulating Fluid and Solid Particles and Continua with SPH and PAM*, *Handbook of Materials Modelling* (Ed. S. Yip), Springer, The Netherlands, 2005, pp. 2903–2906.
50. G.A. Bird, *Molecular Gas Dynamics*, Clarendon, Oxford, UK, 1976; G.A. Bird, *Molecular Gas Dynamics and the Direct Simulation of Gas Flows*, Clarendon, Oxford, UK, 1994; F.J. Alexander, *The Direct Simulation Monte Carlo Method: Going Beyond Continuum*

Hydrodynamics, *Handbook of Materials Modelling* (Ed. S. Yip), Springer, The Netherlands, 2005, pp. 2513–2522.
51. P. Espanol, *Dissipative Particle Dynamics*, *Handbook of Materials Modelling* (Ed. S. Yip), Springer, The Netherlands, 2005, pp. 2503–2512.
52. P.J. Hoogerbrugge and J.M.V.A. Koelman, *Europhys. Lett.*, **19**, 155 (1992).
53. E. Dickinson, S.A. Allison and J.A. McCammon, *J. Chem. Soc.*, **81**, 591 (1985).
54. B. Cichocki, *Z. Phys. B*, **66**, 537 (1987).
55. B. Cichocki and K. Hinsen, *Physica A*, **166**, 473 (1990).
56. B. Cichocki and K. Hinsen, *Ber. Bunsenges. Phys. Chem.*, **94**, 243 (1990)
57. K.O.L.F. Jayaweera, B.J. Mason and G.W. Slack, *J. Fluid Mech.*, **20**, 121 (1964).
58. L.M. Hocking, *J. Fluid Mech.*, **20**, 129 (1964).
59. I.K. Snook, K.M. Briggs and E.R. Smith, *Physica*, **240**, 547 (1997).
60. I. Snook, W. van Megen and R.J.A. Tough, *J. Chem. Phys.*, **78**, 5825 (1983).
61. I. Snook and W. van Megen, *J. Coll. Interf. Sci.*, **100**, 194 (1984).
62. W. van Megen and I. Snook, *J. Chem. Soc., Faraday Disc. Chem. Soc.*, **76**, 151 (1983).
63. W. van Megen and I. Snook, *J. Chem. Soc., Faraday Trans. II*, **80**, 383 (1984).
64. W. van Megen and I. Snook, *J. Chem. Phys.*, **88**, 1185 (1988).
65. W. van Megen, R.H. Ottewill, S.M. Owens and P.N. Pusey, *J. Chem. Phys.*, **82**, 508(1985).
66. S. Wallrand, L. Belloni and M. Drifford, *J. Phys.*, **47**, 1565 (1986).
67. J. Bacon, E. Dickinson and R. Parker, *Faraday Trans. Chem. Soc.*, **76**, 165 (1983).
68. J. Bacon, E. Dickerson, R. Parker, N. Anastasiou and M. Lal, *J. Chem. Soc., Faraday Trans. II*, **79**, 91 (1983).
69. E. Dickinson and C. Elvingson, *J. Chem. Soc., Faraday Trans. II*, **84**, 775 (1988).
70. E. Dickinson, S.A. Allison and J.A. McCammon, *J. Chem. Soc., Faraday Trans.*, **81**, 591 (1985).
71. G.C. Ansell and E. Dickinson, *Chem. Phys. Lett.*, **122**, 594 (1985).
72. G.C. Ansell and E. Dickinson, *J. Chem. Phys.*, **85**, 4079 (1986).
73. G.C. Ansell and E. Dickinson, *J. Coll. Interf. Sci.*, **110**, 73 (1986).
74. G.C. Ansell, E. Dickinson and M. Ludvigsen, *J. Chem. Soc., Faraday Trans. II*, **81**, 1269 (1985).
75. E. Dickinson and F. Honary, *J. Chem. Soc., Faraday Trans. II*, **82**, 719 (1986).
76. P.N. Segre, S.P. Meeker, P.N. Pusey and W.C.K. Poon, *Phys. Rev. Lett.*, **75**, 958 (1995); P.N. Segre, O.P. Behrend and P.N. Pusey, *Phys. Rev. E*, **52**, 5070 (1995); P.N. Segre and P.N. Pusey, *Phys. Rev. Lett.*, **77**, 771 (1996).
77. D.M. Heyes, *Mol. Phys.*, **87**, 287 (1996).
78. D.L. Ermak, *J. Chem. Phys.*, **62** 4189 (1975); *ibid* **62** 4197 (1975)
79. K.J. Gaylor, I.K. Snook, W. van Megen and R.O. Watts, *Chem. Phys.*, **43**, 233 (1979).
80. K.J. Gaylor, I.K. Snook, W. van Megen and R.O. Watts, *J.C.S. Faraday II*, **76**, 1067 (1980).
81. K.J. Gaylor, I.K. Snook, W. van Megen and R.O. Watts, *J. Phys. A: Math. Gen.*, **13**, 2513 (1980).
82. K. Gaylor, I. Snook and W. van Megen, *J. Chem. Phys.*, **75**, 1682 (1981).
83. I.K. Snook, W. van Megen, K.J. Gaylor and R.O. Watts, *Adv. Coll. Interface Sci.*, **17**, 33 (1982).
84. C.E. Woodward, B. Jonsson and Akesson, *J. Chem. Phys.*, **89**, 5145 (1989).
85. R. Krause, G. Nagele, D. Karrer, J. Schneider, R. Klein and R. Weber, *Physica A*, **153**, 400 (1988).
86. G. Nagele, M. Medina-Noyola, R. Klein and J.L. Arauz-Lara, *Physica*, **149**, 123 (1988).
87. H. Lowen, J.-P. Hansen and J.-N. Roux, *Phys. Rev. A*, **44**, 1169 (1991);H. Lowen, *Phys. Rep.*, **236**, 249 (1994).
88. D. M. Heyes, *Phys. Lett. A*, **132**, 399 (1988); D. M. Heyes, *J. Non-Newt. Fluid Mech.*, **27**, 47 (1988); D.M. Heyes and P.J. Mitchell, *J. Phys. Condens. Matter*, **6**, 6423 (1994); D.M. Heyes and P.J. Mitchell, *Mol. Phys.*, **84**, 261 (1995).
89. G. Viramontes-Gamboa, J.L. Arauz-Lara and M. Medina-Noyola, *Phys. Rev. Lett.*, **75**, 759 (1995); *Phys. Rev. E*, **52**, 4035 (1995).

90. S. Sood and A.K. Sood, *Phys. Rev. E*, **52**, 4154, 4168 (1995); *Europhys. Lett.*, **34**, 361 (1996); *Progr. Theo. Phys. Suppl.*, **126**, 163 (1997).
91. K. Zahn, J.M. Mendez-Alcaraz and G. Maret, *Phys. Rev. Lett.*, **79**, 175 (1997).
92. J.-E. Shea and I. Oppenheim, *J. Phys.Chem.*, **100**, 19035 (1996).
93. P. Mazur, *Physica*, **110A**, 128 (1982).
94. A.J.C. Ladd, *Phys. Rev. Lett.*, **70**, 1339 (1993); *J. Fluid Mech.*, **271**, 285, 311 (1994).
95. V. Lobaskin and B. Dunweg, *New J. Phys.*, **6**, 54 (2004).
96. W. Kalthoff, S. Schwarzer, G. Ristow and H.J. Herrmann, *Int. J. Mod. Phys. C.*, **7**, 543 (1996); P. Ahlrichs and B. Dunweg, *J. Chem. Phys.*, **111**, 8225 (1999) and P. Ahlrichs, R. Everaers and B. Dunweg, *Phys. Rev. E*, **64**, 040501 (2001).
97. A. Chatterji and J. Horbach, *J. Chem. Phys.*, **122**, 184903 (2005).
98. P. Espanol and P. Warren, *Europhys. Lett.*, **30**, 191 (1995).
99. P.B. Warren, Dissipative particle dynamics, *Curr. Opin. Coll. Interface Sci.*, **3**, 620 (1998).
100. P. Espanol and M. Revenga, *Phys. Rev. E*, **67**, 026705 (2003).

– 8 –

Polymer Dynamics

One of the most significant areas where the GLE approach has proven to be of great value is the study of polymer dynamics. The previous treatments of the GLE method can all be used or extended in some manner to treat such systems. However, there are special problems which arise in treating polymeric systems and we will attempt to outline some of these problems and how GLE's may be used to simulated polymers.

It should first be noted that the area of polymer dynamics is a vast one and there exists an enormous literature on theoretical and experimental studies. Many excellent reviews are available to provide the background in this area and fill in the details for which we have neither the space nor the expertise to do. Good reviews are available[1–14] and these should help the reader to get a much better overview of the field of the theory and simulation of polymer systems.

The application of BD methods to study aspects of this dynamics is similarly a rapidly growing subarea of this field,[14] and so we shall not attempt to summarise all the applications of BD methods to polymers. Instead, we will cover the basic methods involved in applying BD methods to polymer dynamics and merely give a limited number of examples from the literature in order to illustrate the type of results that may be obtained.

If one regards a polymer from an atomic point of view then we may treat the polymer and, if it is a polymer solution, the solvent as a collection of atoms whose dynamics may be described by means of Newton's laws and may, thus, be simulated by means of molecular dynamics. This is at present practical only for rather short-chain polymers either as single chains,[15] in the melt,[16,17] or in solution.[18,19] This does not mean that these methods cannot give useful results for polymeric systems; for example, they can give details of solvent–polymer interactions unavailable to other techniques. Why this limitation on the use of fully atomic method exists will be appreciated by considering the range of length and time scales involved in polymer systems and their dynamics.

Length scales for polymer systems range from naometres if optical and electrical properties are of concern to millimetres or larger for bulk polymer processing. Similarly, relevant time scales vary from femtoseconds for electronic effects to hours for glassy polymers and phase separation. Thus, for phenomena where electronic effects are important, for example optical properties or interatomic interactions, even classical MD methods are not useful and full quantum treatments must be made. So, if a fully theoretical model is desired then one must start with a quantum mechanical method and obtain the interaction energies for atoms within each polymer, and for interactions between each polymer and polymer and solvent molecule. *Ab-initio* MD can also be used, but as in this method one essentially calculates all the interaction energies "on-the-fly" at each MD time step, these methods are limited to treating small systems for short periods.

Alternatively, empirical force fields may be developed, which approximately describe all or some of these interactions.[10–12,14,20,21] These force fields may then be used with MD or GLE methods to simulate the properties of the system. However, as these are based on classical mechanics they may sometimes lead to inaccurate treatments if the details of light-atom dynamics are important and in that case, quantum path integral methods should be used.[22] However, even if quantum dynamics is not important we still have a major limitation inherent in the MD method as the integration time step used must be shorter than the smallest time scale involved, which may be as short as 10^{-15} s. Thus, unless the use of an enormous number of time steps is possible (without significant round-off error occurring), many important phenomena occurring in polymer systems cannot be simulated by MD. This does not even consider the problem of finding force fields for all the many different types of interaction involved and the large range of different length scales often present in such potential simulations. Thus, coarse-graining is forced upon us by the sheer complexity and range of fundamental parameter scales involved.

At the first level of coarse-graining, hydrogen atoms may be combined to form a "united-atom" instead of considering all atoms explicitly.[11] At the next level, many atoms are combined together as ellipsoids, pearls, beads or blobs,[10,11,14] where these elements are supposed to represent a segment of a real chain. These elements are then joined by some means, for example by rigid rods in the bead-rod model or by springs in the bead-spring model. For example, a chain may be represented as a series of beads and springs and the solvent may be treated as having similar short-ranged interactions with other solvent molecules and the beads in the chain as shown in Figure 8.1.

However, even using simplified representations of the polymer chains, solvent molecules and their interactions, these MD-based methods are still limited to treating rather small systems at present.[15–19] Thus, Brownian methods which average over the solvent molecule degrees of freedom must be turned to in order to extend simulations to treat many phenomena in polymeric systems. A concentrated polymer solution represented by a

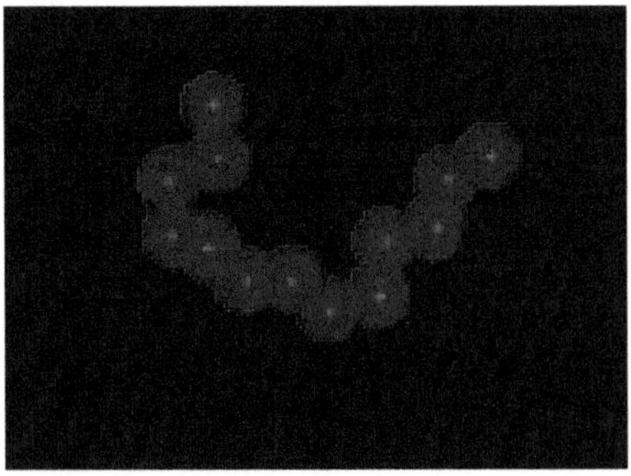

Figure 8.1 A single chain of 13 beads joined by springs used to represent a straight-chain polymer with the solvent being treated as a continuum. (Figure by courtesy of Dr. Peter Daivis, Applied Physics, RMIT.)

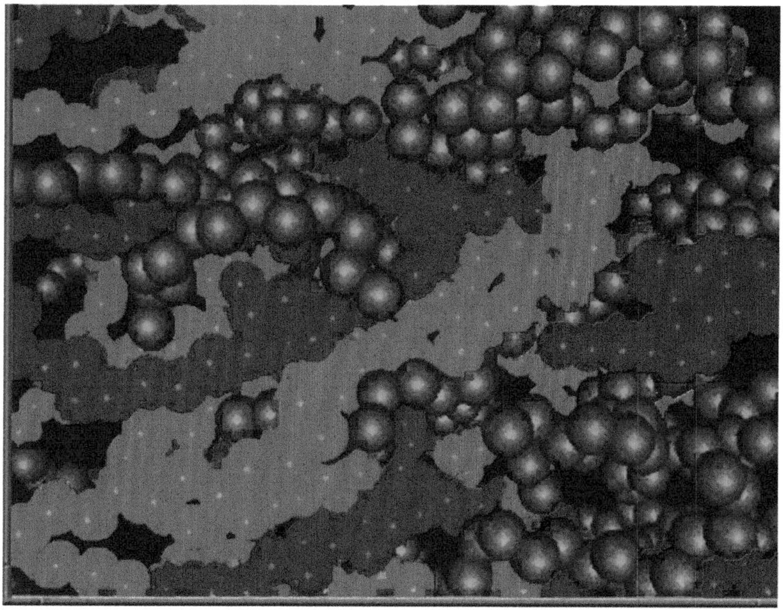

Figure 8.2 A concentrated polymer solution where the polymer chains are represented by a bead-spring model and the solvent atoms are not shown as they are being treated as a continuum. (Figure by courtesy of Dr. Peter Daivis, Applied Physics, RMIT.)

bead–spring model is shown in Figure 8.2, where the solvent molecules are not shown explicitly as they have been replaced by "background" terms in the equations of motion.

8.1 TOXVAERD APPROACH

If we wish to make use of the GLE approach in order to deal with larger systems and longer times, then the approximate GLE's discussed in Chapter 6 may be used. This enables both the velocity as well as position of each atom in the polymers to be obtained as a function of time and, in principle, only the details of the solvent are lost in this approach.

As stated in Chapter 6, it is clear that one can always write an equation of the GLE form, but in practice the problem is to obtain the properties of the memory function and random force. From a theoretical point of view, this amounts to being able to construct a suitable projection operator.

An approximate GLE for a single polymer molecule has been suggested by Bossis *et al.*[23] given by eq. (8.1), and applied by Toxvaerd[24] to treat a single polymer chain consisting of N_b beads with bead–bead interactions being represented by simple LJ12-6 potentials and harmonic forces between neighbouring beads,

$$m_i(\mathrm{d}/\mathrm{d}t)\underline{v}_i(t) = -m_i \int K_v(t-\tau)\underline{v}_i(\tau)\mathrm{d}\tau + \underline{e}_i(t) + \underline{F}_i^{\mathrm{R}}(t) \tag{8.1}$$

for $i = 1$ to N_b, where $\underline{e}_i(t)$ is the sum of the bead−bead forces on the ith bead. Here the random force $\underline{F}_i^R(t)$ is assumed to satisfy the conditions,

$$\langle \underline{v}_i(0) \cdot \underline{F}_i^R(t) \rangle = 0$$

$$(3m_i k_B T) K_v(t) = \langle \underline{F}_i^R(0) \cdot \underline{F}_i^R(t) \rangle + 2m_i \langle \underline{v}_i(0) \cdot \underline{e}_i(0) \rangle \delta(t) \tag{8.2}$$

Toxvaerd[24] performed direct MD simulations for $N_b = 9$ as a free polymer and a polymer in a solvent consisting of 509 LJ12-6 particles at a high density and low temperature corresponding to the liquid state and also BD simulations using eq. (8.1). The forms chosen for $K_v(t)$ in the BD simulations were the Gaussian memory (eq. (3.19)), the Levesque−Verlet memory (eq. (3.50)) and the Lee−Chung memory (eq. (3.51)), and the equation set (8.1) was solved by a combination of MD and the method of Wang and Uhlenbeck (see eq. (6.54)). The BD method was in good agreement with the MD results, but it was decades faster to calculate, thus enabling much longer total times to be simulated. This gain in speed, and hence maximum simulation time, meant that processes which cannot be simulated by direct MD methods can be performed accurately by means of this BD technique.

8.2 DIRECT USE OF BROWNIAN DYNAMICS

One may also use the BD approach based on the many-body Langevin equation as described in Section 7.4.1 to simulate a polymer or polymer solution. This does, however, require some thought as to the approximations being made about the physics of the representation of the polymer and its dynamics. As emphasised in the introduction to this chapter, due to the complexity of the polymer physics involved this means that there are many types of interactions of an intrapolymer, interpolymer and polymer−solvent type operating. Now these types of study treat the solvent as a continuum, Newtonian fluid which is characterised by its density and coefficient of viscosity. If we decide to treat the polymer as consisting of atoms, then we are also assuming that these atoms are of sufficient size and mass to enable the dynamics to be described by the GLE in the Langevin or Brownian limit, that is to use eq. (5.74). Thus, the representation of the units representing the polymer must be of sufficient size and mass to make the LE applicable (see Section 5.5).

Furthermore, if the BD algorithm of Ermak and McCammon[25] (i.e. eq. (7.22)) is used, we are also assuming that we may work in the diffusive limit, that is the velocities of the atoms in the polymer have relaxed to equilibrium before they have undergone significant configurational changes. Thus, we need to make the polymer units large and heavy enough for the conditions required for the normal BD algorithm of Ermak and McCammon[25] to be realised. Such an approach is exemplified by the work of Graham and co-workers,[26–29] which we shall now discuss in some detail in order to illustrate what is involved in such applications and show what may be learnt from this approach. The system studied was a solution of monodispersed, linear polymer chains immersed in an incompressible Newtonian (continuum) solvent. The polymer chain is represented as a sequence of N_b beads connected by N_s springs. A total of N_c chains was simulated in a periodic cubic simulation cell of edge length L and volume V, which means that the system has a total of $N = N_b N_c$ beads per cell at a bulk monomer concentration $c = N/V$. Their notation differs slightly from that used in previous work in Chapter 7 but is in common use in polymer BD work, so we will firstly rewrite the basic equations of motion in a notation which is convenient to use for these polymer simulations. They write the Smoluchowski or stochastic differential equation (see Sections 5.4 and 9.2) representing the dynamics of the system as,

$$\mathrm{d}\mathbf{r} = \{[\kappa \cdot \mathbf{r}(t)] + 1/k_B T[\mathbf{D}\cdot\mathbf{F}^{(\phi)}] + \partial/\partial\mathbf{r}\cdot\mathbf{D}\}\mathrm{d}t + \sqrt{2}\mathbf{B}\cdot\mathrm{d}\mathbf{W} \tag{8.3}$$

Here, $\mathbf{r}$ is a vector containing the $3N$ spatial co-ordinates of the beads, $\mathbf{D}$ is the $3N \times 3N$ diffusion tensor, $\mathbf{F}^{(\phi)}$ is a $3N$ dimensional vector incorporating all the nonhydrodynamic forces, $\boldsymbol{\kappa}$ is a $3N \times 3N$ block diagonal tensor with 3×3 diagonal blocks given by $(\boldsymbol{\nabla}\mathbf{v})^T$, $\mathbf{v}$ the unperturbed solvent flow velocity and $\mathbf{B}$ is a $3N \times 3N$ tensor defined by $\mathbf{B}\cdot\mathbf{B}^T = \mathbf{D}$, and the components of the $3N$ dimensional vector $\mathrm{d}\mathbf{W}$ are obtained from a real-valued Gaussian distribution with mean zero and variance $\mathrm{d}t$.

The hydrodynamic interactions contained in $\mathbf{D}$ were represented by the Rotne–Prager tensor summed by the Ewald technique as discussed in Section 7.2 and given in Appendix M. Where these polymer simulations differ from colloidal simulations is in the description of the nonhydrodynamic interaction. Adjacent beads of a polymer chain interact and in this study these interactions were represented by a worm-like spring model which involves using the force law,

$$\mathbf{F}_{\nu\mu}^{\mathrm{spr}} = k_B T/2b_k[(1-|\mathbf{r}_\mu - \mathbf{r}_\nu|/q_0)^{-2} - 1 + 4|\mathbf{r}_\mu - \mathbf{r}_\nu|/q_0](\mathbf{r}_\mu - \mathbf{r}_\nu)/|\mathbf{r}_\mu - \mathbf{r}_\nu| \tag{8.4}$$

where $\mathbf{r}_\upsilon$ contains the three Cartesian co-ordinates of the position vector of the υth bead, $\mathbf{F}_{\upsilon\mu}^{\mathrm{spr}}$ is the force exerted on bead υ due to connectivity with the bead, b_k is the Kuhn length of the molecule and q_0 is the maximum extension of the spring. The excluded volume interactions were represented by,

$$U_{\nu\mu}^{\mathrm{exv}} = \tfrac{1}{2}\mathrm{v}k_B T N_{k,s}^{\,2}(3/4\pi S_s^2)^{3/2}\exp[-3|\mathbf{r}_\mu - \mathbf{r}_\nu|/4S_s^2] \tag{8.5}$$

and

$$\mathbf{F}_{\nu\mu}^{\mathrm{exv}} = \mathrm{v}k_B T N_{k,s}^{\,2}\pi(3/4\pi S_s^2)^{5/2}\exp[-3|\mathbf{r}_\mu - \mathbf{r}_\nu|/4S_s^2]|\mathbf{r}_\mu - \mathbf{r}_\nu| \tag{8.6}$$

where v is the excluded volume parameter and $S_s^2 = N_{k,s}b_k^2/6$ is the mean square radius of gyration of an ideal chain consisting of $N_{k,s}$ Kuhn segments.

The BD algorithm for the motion of each bead υ used by Graham and co-workers,[26–29] is the Ermak–McCammon first-order algorithm,[25] (eq. (7.22)) which they wrote in the form of,

$$\underline{r}_\nu(t+\Delta t) = \underline{r}_\nu(t) + [\kappa\cdot\underline{r}_\nu(t)]\Delta t + \Delta t/(k_B T)\sum_{\mu=1}^{n}[\mathbf{D}_{\nu\mu}(t)\cdot\underline{F}_\mu(t)] + \sqrt{2}\sum_{\mu=1}^{n}\mathbf{B}_{\nu\mu}(t)\cdot\Delta\mathbf{W}_\mu(t) \tag{8.7}$$

where $\underline{F}_{\mu} = \sum_{\omega \neq \mu} \underline{F}_{\mu\omega}^{\text{exv}} + \underline{F}_{\mu,\mu-1}^{\text{spr}} + \underline{F}_{\mu,\mu+1}^{\text{spr}}$ is the total force on bead μ which consists of an excluded volume interaction with every other bead in the system and a spring interaction with each bead attached to either side of it. An innovation used was to implement a method owing to Fixman[26,30] that enables the Brownian term (the last term on the right-hand side of eq. (8.7)) to be calculated very efficiently, which is essential because of the complexity of the polymer problem. Interestingly, for simulation of infinitely dilute solutions alternative semi-implicit integration schemes were found to be useful, but they were not found to be suitable for these simulations of interacting polymers and the explicit scheme above was therefore used.[29]

The parameters used in these simulations were chosen to represent λ-phage DNA in bulk solution in equilibrium, in shear and in elongational flows using the appropriate boundary conditions (see Section 6.6) in the dilute and semidilute regimes. Properties calculated included mean square radius of gyration, short- and long-time diffusivities, shear viscosity, elongational viscosity, flow direction extension under a variety of conditions (e.g. different concentrations), shear and extensional flow rates. Results for long-time diffusivities as a function of concentration may be seen in Figure 8.3 and for elongational viscosity as a function of extensional rate at different concentrations is shown in Figure 8.4. Shear viscosities were in very good agreement with the experiment and many quantities obtained, for example conformational effects of flow field should aid enormously in interpreting experimental data. Interestingly, ignoring hydrodynamic interactions ("free-draining" calculations) gave reasonable qualitative results, but the effects of hydrodynamic interactions were shown to be a very important factor in determining the magnitude of the quantities calculated. This study shows that it is possible to make realistic simulations on the basis of coarse-grained LE approach of the equilibrium and nonequilibrium properties

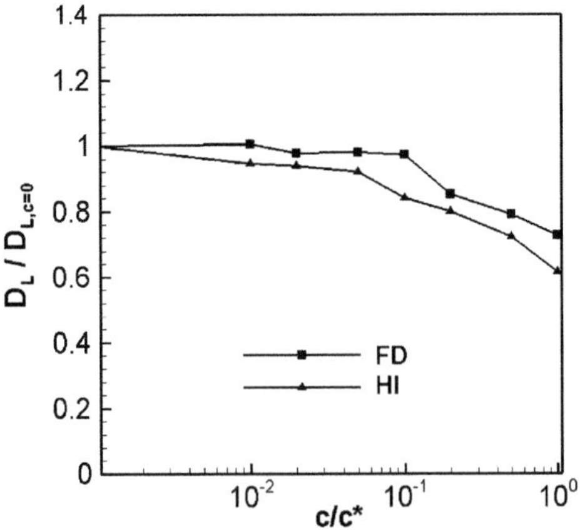

Figure 8.3 Long-time diffusivities of 21 μm DNA system as a function of concentration, c/c^* normalised by the overlap concentration c^*, by BD simulation with and without hydrodynamic interactions. (Figure by kind permission of C. Stoltz, J.J. de Pablo and M.D. Graham, Department of Chemical and Biological Engineering, University of Wisconsin-Madison, WI, USA)

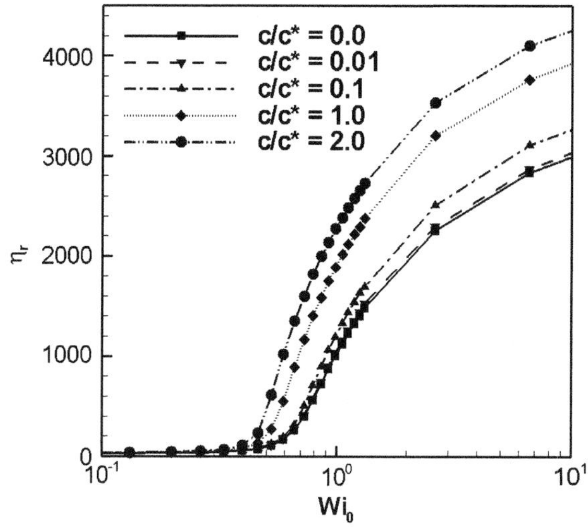

Figure 8.4 Reduced elongational as a function of extensional rate at different normalised concentrations, c/c^* by BD simulation with hydrodynamic interactions. (Figure by kind permission of C. Stoltz, J.J. de Pablo and M.D. Graham, Department of Chemical and Biological Engineering, University of Wisconsin-Madison, WI, USA.)

of polymer solutions under a variety of physical conditions, and also to obtain extra insight into polymer dynamics.

8.3 RIGID SYSTEMS

A relatively simple class of polymers and also colloidal suspensions of nonspherical particles to treat are those that may be regarded as rigid bodies, for example a rigid-rod polymer. One method of describing the dynamics of such systems is to use numerical algorithms based on the GLE or the LE and to apply constraints on the dynamics to maintain the geometry of the whole polymer or sections of it. This may be achieved by a variety of methods, for example by the use of the SHAKE algorithm as introduced by Ryckaert *et al.*[31,32] in MD simulations to maintain rigid geometry of molecules in MD simulation.

Another approach to the problem of simulating the Brownian dynamics of rigid objects is by treating the whole problem by the translational BD method (based on the LE), that is using eq. (7.22) and applying constraints to the rigid bodies or rigid polymer subunits in order to preserve the geometry of these objects by a modification of the SHAKE method mentioned above, and its implementation in BD was originated by Allison and McCammon and termed the SHAKE-HI algorithm.[33] This method applies a translational algorithm, for example (eq. (7.22)) in the case of BD to move all the units whether they are part of a rigid unit or not and subsequently modifies this new configuration until the geometry of the rigid units is preserved. Other methods have been subsequently developed to handle these constraints, for example based on Gauss's principle of least constraint[34] or using very stiff springs.[35] For example, the application of the SHAKE method to a pair of rigid cubic octamer particles using the Rotne–Prager approximation to D_{ij}^T was in good agreement with analytical results for coupling of motions provided the particles were not allowed to approach too closely.[36]

An alternative approach to the treatment of rigid systems is to generalise the BD algorithms, for example eq. (7.22) in the diffusive limit, to treat both the translation and rotation degrees of freedom by starting with the LE for both the translational and rotational degrees of freedom. This is essentially the approach of Brady and Bossis (see Section 7.5) except that here, unlike in their work, we will separate out the translational and rotational degrees of freedom in the equations of motion, the hydrodynamic tensors and the algorithms.

If these degrees of freedom are independent then we may use Ermak–McCammon algorithm[25] (eq. (7.22)) to describe the translational degrees of freedom where now the configuration-dependent friction tensors or diffusion tensors should be labelled with the subscript T to distinguish them from the rotational analogues below.[36] The LE for the rotational degrees of freedom is,

$$I_i(\mathrm{d}\omega_i/\mathrm{d}t) = -\sum_j \zeta_{ij}^{\mathrm{R}}\omega_j + T_i + \sum_j \alpha^{\mathrm{R}}{}_{ij} f_j \tag{8.8}$$

where $1 \le i,j \le 3N$ and I_i is the moment of inertia associated with the index i, ω_i is the angular-velocity component in the direction i, T_i is the sum of the external and interparticle torques acting in the direction i, ζ_{ij}^{R} is an element of the rotational friction tensor and the coefficients $\alpha^{\mathrm{R}}{}_{ij}$ are defined by,

$$\zeta_{ij}^{\mathrm{R}} = (k_{\mathrm{B}}T)^{-1}\sum_l \alpha^{\mathrm{R}}{}_{il}\alpha^{\mathrm{R}}{}_{jl} \tag{8.9}$$

Because there is no coupling between rotation and translation, $\boldsymbol{\zeta}_{ij}^{\mathbf{R}}$ and $\boldsymbol{\zeta}_{ij}^{\mathbf{T}}$ are independent.

The algorithm for the rotational degrees of freedom may be derived from eq. (8.8) by use of the same technique used to derive the BD equation for the translational co-ordinates r_i (see Section 7.4.1),

$$\phi_i(t+\Delta t) = \phi_i(t) + (k_{\mathrm{B}}T)^{-1}\sum_j D_{ij}^{\mathrm{R}}(t)T_j(t)\Delta t + R_i^{\mathrm{R}}(D_{ij}^{\mathrm{R}}(t),\Delta t) \tag{8.10}$$

where $1 \le i,j \le 3N$ and R_i^{R} is a stochastic displacement generated by the same algorithm as for r_i, that is eq. (7.22) with D_{ij}^{R} replacing D_{ij}.

However, if the translational and rotational degrees of freedom are coupled, then instead of simply having diffusion tensors $\mathbf{D}_{ij}^{\mathrm{T}}$ and $\mathbf{D}_{ij}^{\mathrm{R}}$ representing independent translational and rotational effects we have a "super-tensor" or grand diffusion matrix $\mathbf{D}_{\mathbf{ij}}$ with diagonal sub-matrices $\mathbf{D}_{ij}^{\mathrm{T}}$ and $\mathbf{D}_{ij}^{\mathrm{R}}$ and off-diagonal sub-matrices $\mathbf{D}_{ij}^{\mathrm{TR}}$ and $\mathbf{D}_{ij}^{\mathrm{RT}}$ representing this coupling, that is,

$$\mathbf{D}_{\mathbf{ij}} = \begin{bmatrix} \mathbf{D}_{ij}^{\mathrm{T}} & \mathbf{D}_{ij}^{\mathrm{TR}} \\ \mathbf{D}_{ij}^{\mathrm{RT}} & \mathbf{D}_{ij}^{\mathrm{R}} \end{bmatrix} \tag{8.11}$$

where the indices now run from 1 to $3N$ for the translational degrees of freedom and from $3N+1$ to $6N$ for the rotational degrees of freedom, that is the indices of the components of $\mathbf{D}_{ij}{}^{T}$ are now in the range $(1 \le i,j \le 3N)$, for $\mathbf{D}_{ij}{}^{TR}$ in the range $(1 \le i \le 3N, 3N+1 \le j \le 6N)$, $\mathbf{D}_{ij}{}^{RT}$ in the range $(3N+1 \le i \le 6N,\ 1 \le j \le 3N)$ and $\mathbf{D}_{ij}{}^{R}$ in the range $(3N+1 \le i,j \le 6N)$.

The LEs are now,

$$M_i\, \mathrm{d}V_i/\mathrm{d}t = -\sum_{j=1}^{6N}(\zeta_{ij}{}^{T}V_j + \zeta_{ij}{}^{TR}\omega_j) + F_i^{\mathrm{ext}} + \sum_{j=1}^{6N}\alpha_{ij} f_j \tag{8.12}$$

where $1 \le i \le 3N$,

$$I_i(\mathrm{d}\omega_i/\mathrm{d}t) = \sum_{j=1}^{6N}(\zeta_{ij}{}^{RT}V_j + \zeta_{ij}{}^{R}\omega_j) + T_i + \sum_{j=1}^{6N}\alpha_{ij} f_j \tag{8.13}$$

where $3N+1 \le i \le 6N$.

But now unlike using eqs. (7.22) and (8.10), the two Langevin eqs. (8.12) and (8.13) are not independent because the same set of α_{ij} coefficients now occur in both of these equations and,

$$\zeta_{ij} = (K_B T)^{-1}\sum_{1}\alpha_{il}\alpha_{jl} \tag{8.14}$$

where ζ_{ij} is the i,j element of the tensor $\boldsymbol{\zeta}_{\mathbf{ij}}$.

The BD algorithm derived from eqs. (8.12) and (8.13) now becomes,

$$\begin{aligned} r_i(t+\Delta t) = r_i(t) &+ \sum_{j=1}^{6N}(\partial D_{ij}(t)/\partial r_j)\Delta t + (k_B T)^{-1}\sum_{j=1}^{6N} D_{ij}{}^{T}(t)F_j(t)\Delta t \\ &+ (k_B T)^{-1}\sum_{j=1}^{6N} D_{ij}{}^{TR}(t)T_j(t)\Delta t + R_i(D_{ij}(t),\Delta t) \end{aligned} \tag{8.15}$$

where $1 \le i \le 3N$ and,

$$\begin{aligned} \phi_i(t+\Delta t) = \phi_i(t) &+ (k_B T)^{-1}\sum_{j=1}^{6N} D_{ij}{}^{RT}(t)F_j(t)\Delta t + (k_B T)^{-1}\sum_{j=1}^{6N} D_{ij}{}^{R}(t)T_j(t)\Delta t \\ &+ R_i(D_{ij}(t),\Delta t) \end{aligned} \tag{8.16}$$

where $3N+1 \le i \le 6N$ and the divergence term $\partial D_{ij}(t)/\partial r_j$ includes both $D_{ij}{}^{T}$ and $D_{ij}{}^{TR}$.

If we define a set of generalised co-ordinates $\{y_i\}$ which encompasses all translational and rotational co-ordinates and a set of generalised force components F_i acting in direction i which encompasses all the forces F_i and torques T_i, then eqs. (8.15) and (8.16) may be combined into one equation,

$$y_i(t+\Delta t) = y_i(t) + \sum_{j=1}^{6N}(\partial D_{ij}(t)/\partial y_j)\Delta t + (k_B T)^{-1}\sum_{j=1}^{6N} D_{ij}(t)F_j(t)\Delta t + R_i(D_{ij}(t),\Delta t) \tag{8.17}$$

where $(1 \le i, j \le 6N)$.

Examples of the applications of these equations are to enzyme–substrate encounters[37] at the surface of a colloidal particle and to the diffusion of charged rod-like particle suspensions.[38]

The above approach is essentially a generalisation of the Newtonian mechanics method for treating rigid bodies[39]; however, numerical algorithms based on Newton's laws for rigid bodies can result in problems involving singularities, which lead to difficulties in some MD applications. Thus, alternative methods have been developed starting from Newton's equations based on quaternions, which overcome these problems.[32, 39–44] Similar methods have also been applied to BD simulations by Zhang *et al.*[45,46] who simulated self-assembly of various composite nanoparticles such as chains, rings and tetrahedral, and of nanoparticles with tethers attached.

8.4 CONCLUSIONS

The complexity of polymers and polymer solutions means that the direct MD simulation of all the atoms in the system is limited to small systems and often also to the use of very simple interparticle force laws. This means that methods which lead to simplification of the equations of motion by coarse-graining to obtain equations which give a reasonably accurate description of the dynamics of these systems are extremely useful. The approaches considered here average over the properties of the solvent molecules and range from those based on a fully atomic model of the polymer to those representing the polymer as units containing several atoms, for example by using beads linked by springs. BD simulations are one such approach to the problem and have led to many insights into the origins of the dynamical properties of polymers and polymer solutions. Owing to the very large literature on the subject of polymer dynamics, this chapter has only attempted to give a short outline of the basis of the application of BD methods to the investigation of this area.

REFERENCES

1. H. Yamakawa, *Modern Theory of Polymer Solutions*, Harper & Row, New York, 1971.
2. P.-G. DeGennes, *Scaling Concepts in Polymer Physics*, Cornell University Press, Ithaca, New York, USA, 1979.
3. P.J. Flory, *Statistical Mechanics of Chain Molecules*, Hanser, Munich, 1989.
4. M. Doi and S.F. Edwards, *The Theory of Polymer Dynamics*, Clarendon Press, Oxford, 1986.
5. R.B. Bird, C.F. Curtiss, R.C. Armstrong and O. Hassager, *Dynamics of Polymeric Liquids*, Vol. 2, 2nd edn., Wiley, New York, 1987.
6. R.J.J. Jongschaap, Microscopic modelling of the flow properties of polymers, *Rep. Prog. Phys.*, **53**, 1–55 (1990).
7. K. Binder (Ed.), *Monte Carlo and Molecular Dynamics Methods in Polymer Science*, Oxford University Press, UK, 1995.
8. K. Binder (Ed.), *Monte Carlo and Molecular Dynamics Simulations in Polymer Science*, Oxford University Press, UK, 1996.
9. H.-C. Ottinger, *Stochastic Processes in Polymeric Fluids*, Springer, Berlin, 1996.
10. K. Kremer and F. Muller-Plathe, Multiscale simulation in polymer science, *Mol. Simul.*, **28**, 729–750 (2002).

11. S.C. Glotzer and W. Paul, Molecular and mesoscopic simulation methods for polymer materials, *Annu. Rev. Mater. Res.*, **32**, 401–436 (2002).
12. *Macromolecular Theory and Simulations*, Wiley-Interscience Journal, Print ISSN: 1022–1344, Oneline ISSN: 1521–3919.
13. N. Attig, K. Binder, H. Grubmuller and K. Kremer (Eds), *Computational Soft Matter: From Synthetic Polymers to Proteins*, Lecture Notes, John von Neumann Insitut fur Computing, NIC Series Vol. 23, 2002.
14. P.S. Doyle and P.T. Underhill, Brownian dynamics simulations of polymers and soft matter. In: *Handbook of Materials Modeling* (Ed. S. Yip), Springer, The Netherlands, 2005, pp. 2619–2630.
15. B. Dunweg and K. Kremer, *Phys. Rev. Lett.*, **66**, 2996 (1991).
16. M.L. Matin, *Molecular Simulation of Polymer Rheology*, PhD Thesis, RMIT University, Melbourne, Victoria, Australia, 2001.
17. P.J. Daivis, M.L. Matin and B.D. Todd, *J. Non-Newtonian Fluid Mech.*, **111**, 1 (2003).
18. T. Kairn, P.J. Daivis, M.L. Matin and I.K. Snook, *Polymer*, **45**, 2453 (2004).
19. T. Kairn, P.J. Daivis, M.L. Matin and I.K. Snook, *Thermophysics*, **25**, 1075 (2004).
20. V.A. Ivanov, M.R. Stukan, V.V. Vasilevskaya, W. Paul and K. Binder, *Macromolecular Theory and Simulations*, **9**, 488 (2000).
21. P.T. Underhill and P.S. Doyle, On the coarse-graining of polymers into bead-spring chains, *J. Non-Newtonian Fluid Mech.*, **122**, 3 (2004).
22. R. Martonak, W. Paul and K. Binder, *J. Chem. Phys.*, **106**, 8918 (1997).
23. G. Bossis, B. Quentrec and J.P. Boon, *Mol. Phys.*, **45**, 191 (1982).
24. S. Toxvaerd, *J. Chem. Phys.*, **86**, 3667 (1987).
25. D.L. Ermak and J. A. McCammon, *J. Chem. Phys.*, **69**, 1352 (1978).
26. R.M. Jendrejack, M.D. Graham and J.J. de Pablo, *J. Chem. Phys.*, **113**, 2894 (2000).
27. R.M. Jendrejack, J.J. de Pablo and M.D. Graham, *J. Chem. Phys.*, **116**, 7752 (2002).
28. R.M. Jendrejack, D.C. Schwartz, M.D. Graham and J.J. de Pablo, *J. Chem. Phys.*, **119**, 1165 (2003).
29. C. Stoltz, J.J. de Pablo and M.D. Graham, J. Rheol., **50,** 137 (2006).
30. M. Fixman, *Macromolecules*, **19**, 1204 (1986).
31. J.P. Ryckaert, G. Cicotti and H.J.C. Berendsen, *J. Compt. Phys.*, **23**, 327 (1977); W.F. van Gunsteren and H.J.C. Berendsen, *Mol. Phys.*, **34**, 1311 (1977).
32. M.P. Allen and D.J. Tildesley, *Computer Simulation of Liquids*, Oxford Science Publications, Oxford, 1987, p. 92.
33. S.A. Allison and J.A. McCammon, *Biopolymers*, **23**, 167 (1984).
34. R. Edberg, D.J. Evans and G.P. Morriss, *J. Chem. Phys.*, **84**, 6933 (1986); A. Baranyai and D.J. Denis, *Mol. Phys.*, **70**, 53 (1990).
35. D.C. Morse, Theory of constrained Brownian motion, *Adv. Chem. Phys.*, **128**, 65 (2004).
36. E. Dickinson, S.A. Allison and J.A. McCammon, *J. Chem. Soc., Faraday Trans.* 2, **81**, 591 (1985).
37. E. Dickinson and F. Honary, *J. Chem. Soc., Faraday Trans.* 2, **82**, 719 (1986).
38. A.C. Branka and D.M. Heyes, *Phys. Rev. E*, **50**, 4810 (1994).
39. H. Goldstein, *Classical Mechanics*, 2nd edn., Addison-Wesley, Reading, MA, 1980.
40. D.J. Evans, *Mol. Phys.*, **34**, 317 (1977).
41. D.J. Evans and S. Murad, *Mol. Phys.*, **34**, 327 (1977).
42. J.G. Powles, W.A.B. Evans, E. McGrath, K.E. Gubbins and S. Murad, *Mol. Phys.*, **38**, 893 (1979).
43. D. Fincham, *CCP5 Quart.*, **2**, 6 (1981).
44. M.P. Allen, *Mol. Phys.*, **52**, 717 (1984).
45. Z. Zhang, M.A. Horsch, M.H. Lamm and S.C. Glotzer, *Nano Letters*, **3**, 1341 (2003).
46. Z. Zhang and S.C. Glotzer, Nano Letters, **4**, 1407 (2004).

– 9 –

Theories Based on Distribution Functions, Master Equations and Stochastic Equations

In previous chapters we have taken the approach that an equation of motion for the phase space variables describing the physical state of an N-body system is the fundamental starting point for describing the dynamics and properties of such a system. Thus, we based our approach on equations such as Newton's laws of motion, the Gaussian Isokinetic equations and the SLLOD-like equations. If these equations may be solved then this will lead to a complete description of the appropriate properties of the system. However, in order to be able to deal with some systems, for example, those involving processes acting on different time scales these fundamental dynamical, differential equations were then transformed by coarse-graining into new differential equations, GLEs, which contain terms which must, of necessity, be treated as random (or Stochastic) variables.[1,2] There are, however, alternative approaches to describing and simulating many-body systems by means of coarse graining and the introduction of stochastic elements.

An alternative approach is to derive differential equations for N-body distribution functions[1–3] for example starting from the Liouville equation which upon coarse-graining results in a Fokker–Planck (F–P) equation.[3] It should be emphasised that the approach of Deutch and Oppenheim also leads to the F–P equation, see Section 5.4 and so the two approaches are entirely equivalent. However, it does appear that the F–P equation is a better starting point to derive equations in the diffusive limit where the correlations involving velocities have decayed to virtually zero before the particle's spatial configuration has changed significantly. This point will be discussed in Section 9.2. Another area where the F–P equation, or its limiting Smoluchowski form, is very useful is in solving the N-body stationary state Schrodinger equation where stochastic Quantum Monte Carlo methods are proving very valuable in providing very accurate numerical solutions to this equation. This will be discussed in Section 9.3.

Now one of the main purposes of deriving the stochastic differential equations, GLE's, was to be able to deal with some systems whose dynamics involved processes acting on different time scales. However, there are other stochastic methods which have been developed for this purpose and are based on a different starting point and, thus, some explanation as to their motivation is required. The general principle is still that in order to study certain processes it is necessary to ignore the details of some others which occur on a much shorter time scale, as many interesting processes are inherently slow on a molecular time scale. Some of these processes are simply very slow but still need their details following for times which are long. Other processes involve transition events which are relatively fast on an atomic time scale but do not necessarily occur very often and there is a large "waiting time" between events: they are described as "rare events".

Processes which are simply very slow compared to those occurring on an atomic scale, for example the propagation of a crack in a solid ceramic, may take seconds to occur and involve a very wide range of elementary processes, time scales and length scales.[4] These cannot at present be treated by entirely atomic methods and such phenomena are best treated theoretically by the so-called multi-scale modelling techniques in which different contributing processes are treated by different methods ranging from atomic level to the macroscopic scale. The results of each step must then be fed seamlessly into the next step.[4]

In contrast, in rare atomic processes one has to wait an enormous amount of time in order for the system to be transformed into a state from which such processes can occur (ever thought the event itself is capable of being (and must be) simulated by atomic methods once it commences). A beautiful example of this is given by Dellago[5] who points out that although a water molecule in liquid water will dissociate into ionic components in about 10^{-13} s this event only happened every 10 hours! Such processes generally involve an energy barrier which must be overcome and are then described as thermally activated processes.

Now the GLE formalism discussed previously and related distribution function based methods discussed in Section 9.1 are able, in principle, to extend the time scales over which numerical methods may be applied. However, in order to apply such methods one needs to know the form of the memory function $K_A(t)$ and the random force, $\underline{F}^R(t)$. If one knew all the individual dynamical processes (elementary processes), be they fast, slow or rare, and a mathematical description of them then one could construct $K_A(t)$ and $\underline{F}^R(t)$ and solve the GLE. These elementary processes might be studied by using MD methods of sub-systems of the system of interest, using physical models or approximate theories, or some combination of these techniques. However, to enumerate all the relevant microscopic processes and to obtain equations for their contribution to $K_A(t)$ and $\underline{F}^R(t)$ is a difficult task and to even fully unravel the elementary processes occurring for an MD simulation often proves to be difficult.

An alternative approach is based on a kinetic equation usually termed a master equation which describes the system's evolution from one state to the next in terms of probabilities of occurrence of the elementary processes which can occur. The system is thought to pass from one state to another, for example, one spatial configuration to the next, by a series of processes whose relative probabilities are specified. Related to this approach are methods based on stochastic differential equations which provide an equivalent approach.[6] Thus, these approaches introduce stochastic (random) terms by choice rather than by necessity and shift the problem from a detailed description of the dynamics of each elementary process occurring to one of calculating their relative probabilities.

In this chapter we will outline stochastic methods other than the GLE approach which may be based on using distribution functions, master equations or stochastic differential equations.

9.1 FOKKER–PLANCK EQUATION[1–3,6–11]

As outlined in Section 5.4 the F–P equation for reduced distribution function $f^{(n)}(\Gamma_1,t)$ which describes the probability of finding the large particles in the phase space point Γ_1 at time t may be deduced from the Langevin equation approach of Mazur, Deutch and Oppenheim. An equivalent derivation was given by Murphy and Aguirre[3] who started with the Liouville eq. (1.17) for $f^{(N+n)}(\Gamma,t)$ and some arbitrary initial state of the system at time $-t_I$ and chose a state in which the bath particles are in thermal equilibrium in the instantaneous field of the Brownian particles, that is

$$f^{(N+n)}(\Gamma_0,\Gamma_1,-t_I) = f^{(n)}(\Gamma_1,-t_I)\, f^{(N)}{}_{\mathrm{eq}}(\Gamma_0) \tag{9.1}$$

and following a similar argument to and making the same assumptions as Deutch and Oppenheim[2] they obtained the F–P equation for $f^{(n)}(\Gamma_1,t_I)$, that is

$$\begin{aligned} &\partial f^{(n)}/\partial t + \sum_{i=1}^{n} (\underline{P}_i/M_i \cdot \partial/\partial \underline{R}_i + \langle \underline{F}_i \rangle_0 \cdot \partial/\partial \underline{P}_i) f^{(n)} \\ &= k_{\mathrm{B}}T \sum_{i,j=1}^{n} \partial/\partial \underline{P}_i \cdot \zeta_{ij} \cdot (\partial/\partial \underline{P}_j + \underline{P}_j/M_j kT) f^{(n)} \end{aligned} \tag{9.2}$$

where,

$$\zeta_{ij} = (1/k_{\mathrm{B}}T) \int_0^\infty \langle \underline{F}_i(0)[\underline{F}_j(-t') - \langle \underline{F}_j \rangle] \rangle_0 \, \mathrm{d}\tau \tag{9.3}$$

and $\langle \underline{F}_j \rangle$ is the solvent averaged force on large particle j defined by eq. (5.56).

This confirms once again that the F–P approach gives an equivalent description to the coupled Langevin equations. As mentioned in Section 7.4.1 this equation may also be used to derive the BD algorithm of Ermak and McCammon.[7]

To re-iterate remarks made in Chapter 5 about the Langevin equation we could start with the Liouville equation and not make any other assumptions and obtain an exact F–P equation for $f^{(n)}$. Using the Kawasaki formalism as in Chapter 4 one need not even assume that the bath particles are at equilibrium.

9.2 THE DIFFUSIVE LIMIT AND THE SMOLUCHOWSKI EQUATION[3,6–11]

As discussed in Sections 5.4, 6.2.5 and 7.4.1 often the momenta (velocities) of colloidal particles relax to equilibrium in a time over which the particles have moved only a negligible distance which is the so-called Diffusive or Over-damped limit. More strictly all velocity correlations are assumed to be negligible in this limit and then the F–P eq. (9.2) may be further simplified to the Smoluchowski equation or generalised diffusion equation. Murphy and Aguire[3] (see also Wilemski[8]) do by using the n-particle continuity equation,

$$\partial n^{(n)}/\partial t = \sum_{i=1}^{n} \partial/\partial \underline{R}_i \cdot \langle \underline{V}_i \rangle \tag{9.4}$$

where $\langle \underline{V}_i \rangle = \int f^{(n)}(\Gamma_1)\,(\underline{P}_i/M_i)\,\mathrm{d}\underline{P}^{(n-1)}$.

Using the assumption that the large particle momentum variables are at their equilibrium values are, more precisely, described by an equilibrium distribution function, then

$$n^{(n)}(\underline{R}^{(n)},t)=\int f^{(n)}(\Gamma_1,t)\mathrm{d}\underline{P}^{(n)} \tag{9.5}$$

and

$$\partial f^{(n)}/\partial t=(\partial f^{(n)}/\partial n^{(n)})\partial n^{(n)}/\partial t$$

Using eq. (9.5) the F–P eq. (9.2) becomes,

$$\begin{aligned}&\partial f^{(n)}/\partial n^{(n)}\left(-\sum_{i=1}^{n}\partial/\partial\underline{R}_i\cdot\langle\underline{V}_i\rangle\right)+\sum_{i=1}^{n}(\underline{P}_i/M_i\cdot\partial/\partial\underline{R}_i+\langle\underline{F}_i\rangle\cdot\partial/\partial\underline{P}_i)f^{(n)}\\&=k_\mathrm{B}T\sum_{i,j=1}^{n}\partial/\partial\underline{P}_i\cdot\zeta_{ij}\cdot(\partial/\partial\underline{P}_j+\underline{P}_j/M_jkT)f^{(n)}\end{aligned} \tag{9.6}$$

If we then assume that all gradients $\partial/\partial\underline{R}_i$ are small over a mean free path between a large particle and a bath particle then the interaction between the large particles will change negligibly over this path and, thus, the interaction between two large particles will change negligibly over this distance. Thus, retaining only terms to first order in gradients, indicated by the subscript k, Murphy and Aguire obtained,[3]

$$\begin{aligned}&\sum_{i,j=1}^{n}\partial/\partial\underline{P}_i\cdot\zeta_{ij}\cdot(\partial/\partial\underline{P}_j+\underline{P}_j/M_jk_\mathrm{B}T)f_0^{(n)}-\partial f^{(n)}/\partial n^{(n)}\,\partial/\partial\underline{R}_k\cdot\langle\underline{V}_{k,0}\rangle)\\&+(\underline{P}_k/M_k\cdot\partial/\partial\underline{R}_k+\langle\underline{F}_k\rangle\cdot\partial/\partial\underline{P}_k)\,f_0^{(n)}\\&=\sum_{i,j=1}^{n}\partial/\partial\underline{P}_i\cdot\zeta_{ij}\cdot(\partial/\partial\underline{P}_j+\underline{P}_j/M_jk_\mathrm{B}T)f_k^{(n)}\\&=0,\end{aligned} \tag{9.7}$$

where $k=1$ to n,

$$\langle\underline{V}_{k,m}\rangle=\int f_m^{(n)}(\underline{P}_k/M_k)\mathrm{d}\underline{P}^{(n-1)}$$

As the momenta are assumed to be "at equilibrium" then they choose the solution to eq. (9.7) to be

$$f_0^{(n)}=n^{(n)}(\underline{R}^{(n)};t)\prod_{i=1}^{n}(2\pi M_ik_\mathrm{B}T)^{-3/2}\exp[-\underline{P}_i^2/(2\pi M_ik_\mathrm{B}T)] \tag{9.8}$$

which makes $\langle \underline{V}_{k,0} \rangle = 0$ and eq. (9.7) becomes,

$$(\underline{P}_k/M_k \cdot \partial/\partial \underline{R}_k + \langle \underline{F}_k \rangle/(k_B T))n^{(n)}(\underline{R}^{(n)};t)$$
$$\times \prod_{i=1}^{n} (2\pi M_i k_B T)^{-3/2} \exp[-P_i^2/(2\pi M_i k_B T)]$$
$$= k_B T \sum_{i,j=1}^{n} \partial/\partial \underline{P}_i \cdot \zeta_{ij} \cdot (\partial/\partial \underline{P}_j + \underline{P}_j/M_j k_B T) f_k^{(n)}$$

which upon multiplication by $\underline{P}_1$ and integration over momenta results in,

$$(k_B T)^{-1} \sum_{i=1}^{n} \zeta_{ij} \cdot \langle \underline{V}_{i,k} \rangle = -(\underline{\nabla}_l - \langle \underline{F}_l \rangle_0/(k_B T)) n^{(n)} \delta_{lk} \tag{9.9}$$

where $\underline{\nabla}_l = \partial/\partial \underline{R}_l$.

Murphy and Aguire[3] then interpret the left hand side of eq. (9.9) to be $(k_B T)^{-1}$ times the matrix product of the friction tensor matrix $\boldsymbol{\zeta}$ whose elements are the tensors $\boldsymbol{\zeta}_{ij}$ with the column vector whose elements are the vectors $\langle \underline{V}_{i,k} \rangle$. This makes the right hand side of eq. (9.9) a column vector whose elements are all zero except the kth element (because of the term δ_{lk}) which is $(\underline{\nabla}_1 - \langle \underline{F}_1 \rangle_0/k_B T)n^{(n)}$. If we define the diffusion matrix $\mathbf{D} = k_B T \boldsymbol{\zeta}^{-1}$ with elements the tensors $\mathbf{D_{ij}}$ and where $\boldsymbol{\zeta}^{-1}\boldsymbol{\zeta} = \mathbf{I}$ with $\mathbf{I}$ having diagonal elements being the unit tensor and off-diagonal elements being zero (a unit "super tensor"). Multiplying eq. (9.9) by $\mathbf{D}$ to get $\langle \underline{V}_{i,k} \rangle = -\mathbf{D_{ik}} \cdot (\underline{\nabla}_1 - \langle \underline{F}_1 \rangle_0 /(k_B T))n^{(n)}$ and using $\langle \underline{V}_{i,0} \rangle = 0$ then eq. (9.9) becomes,

$$\langle \underline{V}_i \rangle = \sum_{i=1}^{n} \langle \underline{V}_{i,j} \rangle = -\sum_{i=1}^{n} \mathbf{D_{ik}} \cdot (\underline{\nabla}_l - \langle \underline{F}_l \rangle_0/(k_B T)) n^{(n)} \tag{9.10}$$

and substituting eq. (9.10) into eq. (9.5) we have,

$$\partial n^{(n)}/\partial t = \sum_{i,j=1}^{n} + \underline{\nabla}_i \cdot \mathbf{D_{ij}} \cdot (\underline{\nabla}_j - \langle \underline{F}_j \rangle_0/(k_B T)) n^{(n)} \tag{9.11}$$

Eq. (9.11) is Murphy and Aguire's n-body Smoluchowski equation or n-body diffusion equation and it should be noted that the solution of eq. (9.11) gives only the positions of the colloidal particles as a function of time, $(\underline{R}^{(N)}(t))$ and not their momenta.

9.2.1 Solution of the *n*-body Smoluchowski equation

The solution to eq. (9.11) to first order in Δt and resolving $\underline{R}_i$ and $\mathbf{D_{ij}}$ into components as in Sections 7.1 and 7.4.1 is[7] a multivariate Gaussian distribution function defined by the moments,

$$\langle \Delta R_i(\Delta t) \rangle = \sum_{j=1}^{6n} [\partial D_{ij}(t)/\partial R_j + D_{ij}(t)/(k_B T)] \Delta t \tag{9.12}$$

and

$$\langle \Delta R_i(\Delta t)\,\Delta R_i(\Delta t)\rangle = 2D_{ij}(t)\,\Delta t$$

Thus, for short times the Green's function for the differential n-Body Smoluchowski equation is a Gaussian and this may be used as an alternative way to derive BD algorithms of the Ermak-McCammon type[7].

9.2.2 Position-only Langevin equation

The derivation of the equivalent position-only Langevin equation valid in the diffusive limit is more difficult and subtle than the derivation of the n-Body Smoluchowski equation and is discussed at some length in refs. 9–11. For example, Tough et al.[11] derive the following equation:

$$\mathrm{d}R_i/\mathrm{d}t = \sum_{j=1}^{3N} (D_{ij}/kT\, F_j^{\mathrm{ext}} - \partial/\partial R_j\, D_{ij}) + \sum_{j=1}^{3N} \sigma_{ij}\, f_j(t) \tag{9.13}$$

where $f_j(t)$ obey eq. (7.2)

$$D_{ij} = \sum_{k=1}^{3N} \sigma_{ik}\sigma_{jk}$$

A similar equation was developed by Zwanzig[9] which is,

$$\partial R_i/\partial t = -\sum_{j}^{3N} (1/k_{\mathrm{B}}T) D_{ij} \partial U/\partial R_j - \sum_{m,j}^{3N} (\partial/\partial R_j \sigma_{im})\sigma_{jm} + \sum_{j}^{3N} \sigma_{ij}\, f_j(t) \tag{9.14}$$

and Hess and Klein[10] also developed a position-only Langevin equation which reads,

$$\partial R_i/\partial t = -\sum_{j}^{3N} (1/k_{\mathrm{B}}T) D_{ij} \partial U/\partial R_j + \sum_{m,j}^{3N} \sigma_{mj}(\partial/\partial R_j \sigma_{jm}) + \sum_{j}^{3N} \sigma_{ij}\, f_j(t) \tag{9.15}$$

It should be noted that the solution of equations (9.13), (9.14) or (9.15) gives only the positions of the colloidal particles as a function of time, ($\underline{R}^{(N)}(t)$) and not their momenta.

There has been concern expressed as to which equation of the above eqs. (9.13), (9.14) or (9.15), if any, is the "correct" form for the position-only Langevin equation. There does not yet seem to be a definite answer to this question at present.

A further concern surrounds the physical and mathematical interpretation of such equations. There are two standard ways of interpreting stochastic differential equations the Ito and Stratonovich interpretations.[11] However, in practice the answers to such questions are probably of greatest interest those concerned with mathematical rigour and the questions of the meaning of physical theories. If we are only concerned with deriving numerical algorithms, for example, in order to carry out BD simulations, then a safe approach is probably to make sure that the algorithms derived from a position-only Langevin equation are in agreement with those obtained from the Smoluchowski equation (or from the F–P equation or Langevin equation by taking the diffusive limit) for a typical example of a class of the problems being studied. Then, as the position-only Langevin equation appears to be rather easier to work with than the other equations it can be confidently used for other cases of this class of problems. Otherwise the safest approach is always to use the F–P equation, Langevin equation or Smoluchowski equation.

In order to illustrate this, Figure 9.1 shows a modified version of B. Ackerson's famous "Road Map" which he presented at the Faraday Discussion, "*Concentrated Colloidal Dispersions*", The Royal Society of Chemistry, London, No. 76 in 1983 which he used to illustrate the various theoretical approaches to colloidal dynamics and their relationships. This has been re-drawn (with apologies) to illustrate the above discussions about how the various methods are derived and their relationships. The main new feature on the road map is the detour from the Langevin approach to the F–P equation before proceeding to the diffusive limit and the Smouchowski equation which I prefer to the rather dangerous route from the momentum Langevin equation to the position-only Langevin equation.

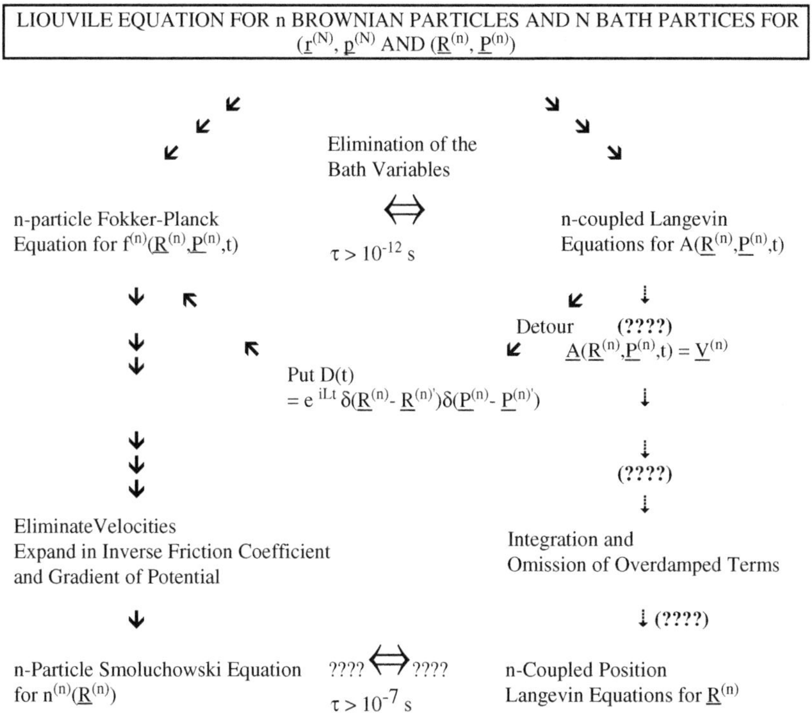

Figure 9.1 A modified "Ackerson Road Map" redrawn with poetic license from the original by B.J. Ackerson, Faraday Discussion, **76**, 237 (1983) with apologies; ⇔ indicates Stochastic Equivalence and (????) indicates that this may not be a possible route.

9.3 QUANTUM MONTE CARLO METHOD[12–18]

The solution of the time-independent or stationary state Schrodinger equation to obtain the wavefunction and/or expectation values such as energy is one of the central problems in electronic structure theory. However, just like the solution of Newton's equations this can only be accomplished analytically for very simple systems. Thus, numerical schemes must be sought to solve the stationary-state Schrodinger equation for realistic systems. This is usually done by using the Quantum Mechanical Variational Principle, Quantum Mechanical Perturbation Theory or some combination of the two. All these methods start with an approximate wavefunction and involve some type of expansion, for example, in terms of a series of determinantal functions constructed from a set of one-electron functions ("orbitals") or some combination of determinants and explicit functions of inter-electronic co-ordinates.

A desirable alternative to this traditional approach is to integrate the stationary-state Schrodinger equation directly. One such approach still uses an approximate wavefunction but evaluates all integrals numerically by Monte Carlo sampling and also performs the evaluation of the unknown parameters (e.g. by use of the Variational Principle or minimization of the variance of the energy) appearing in the approximate wavefunction, entirely numerically. This is known as the Variational Monte Carlo (VMC) method[16,18] but the accuracy of the computed quantities is still limited by assumed form of the wavefunction used.

An alternative numerical approach, which is in principle exact, starts with the time-dependent Schrodinger equation rather than the time-independent one. The problem may be cast in a form such that this equation may be expressed as an N-body diffusion equation based on a distribution function. Now if the Hamiltonian operator **H** does not contain time the solution of this equation starting from some arbitrary state will approach the exact ground state, that is the wavefunction generated will evolve to the exact ground state wavefunction[16,18](see below for the proof). Thus, if we write the imaginary time Schrodinger equation for N electrons and n nuclei as,

$$-\partial\Phi(\underline{r}^{(N)},t)/\partial t = (\mathbf{H} - E_{\mathrm{T}})\Phi(\underline{r}^{(N)},t) \tag{9.16}$$

where t is a real variable which measures the progress in imaginary time, $\Phi(\underline{r}^{(N)},t)$ is a time-dependent function, E_T is a so-called energy offset and the electronic Hamiltonian operator **H** in the Born-Oppenheimer approximation is given by,

$$\mathbf{H} = -1/2\sum_{i=1}^{N}\nabla_i^2 + \mathbf{V}(\underline{r}^{(N)}, \underline{R}^{(n)})$$

where $\underline{r}^{(N)} = (\underline{r}_1,\underline{r}_2,\ldots,r_N)$ and $\underline{R}^{(n)} = (\underline{R}_1,\underline{R}_2,\ldots,\underline{R}_n)$ are the collection of the co-ordinates of the electrons and nuclei respectively and $\mathbf{V}(\underline{r}^{(N)},\underline{R}^{(n)})$ is a potential energy operator and $\Phi(\underline{R}^{(N)},t)$ only depends parametrically on $\underline{R}^{(n)}$. If we can solve eq. (9.16) then as $\tau \rightarrow \infty$ we will show that starting from any arbitrary state $|\Phi(t = 0)>$ the solution to eq. (9.16) will tend to the lowest eigen-state $\Psi_0(\underline{r}^{(N)})$.

Now eq. (9.16) may be written in integral form as,

$$\Phi(\underline{r}^{(N)},t+\tau)=\int \mathrm{G}(\underline{r}^{(N)}\leftarrow\underline{r}^{(N)'},\tau)\Phi(\underline{r}^{(N)'},t)\mathrm{d}\underline{r}^{(\mathrm{n})'} \tag{9.17}$$

where $G(\underline{r}^{(N)} \leftarrow \underline{r}^{(N)'},\tau)$ is a Green's function that obeys the differential equation,

$$-\partial G(\underline{r}^{(N)}\leftarrow\underline{r}^{(N)'},t)/\partial t=(\mathbf{H}-E_{\mathrm{T}})G(\underline{r}^{(N)}\leftarrow\underline{r}^{(N)'},t) \tag{9.18}$$

which may be seen to be of an identical form to that obeyed by $\Phi(\underline{r}^{(N)},t)$ with initial condition,

$$G(\underline{r}^{(N)}\leftarrow\underline{r}^{(N)'},0)=\delta(\underline{r}^{(N)}\leftarrow\underline{r}^{(N)'})$$

We have,

$$G(\underline{r}^{(N)}\leftarrow\underline{r}^{(N)'},\tau)=\langle\underline{r}^{(N)}\,|\exp[-\tau(\mathbf{H}-E_{\mathrm{T}})]|\,\underline{r}^{(N)'}\rangle \tag{9.19}$$

and using,

$$\exp[-\tau\mathbf{H}]=\sum_i|\Psi_i\rangle\exp(-\tau E_i)\langle\Psi_i|$$

so,

$$G(\underline{r}^{(N)}\leftarrow\underline{r}^{(N)'},\tau)=\sum\Psi_i(\underline{r}^{(N)})\exp[-\tau(E_i-E_{\mathrm{T}})]\Psi_i(\underline{r}^{(N)'}) \tag{9.20}$$

where we have used the complete set of eigenfunctions $\{\Psi_i\}$ with eigenvlaues $\{E_i\}$ of $\mathbf{H}$ as a basis. Now as $\tau\rightarrow\infty$ $\exp[-\tau(\mathbf{H}-E_T)]$ projects out the lowest eigen-state$|\Psi_0>$ that has non-zero overlap with $|\Phi(t=0)> = |\Phi_{\mathrm{int}}>$ since,

$$\begin{aligned}
&\lim_{\tau\rightarrow\infty}\langle\underline{r}^{(N)}\,|\exp[-\tau(\mathbf{H}-E_{\mathrm{T}})]|\,\Phi_{\mathrm{int}}\rangle\\
&=\lim_{\tau\rightarrow\infty}\int G(\underline{r}^{(N)}\leftarrow\underline{r}^{(N)'},\tau)\Phi_{\mathrm{int}}(\underline{r}^{(N)'})\mathrm{d}\underline{r}^{(N)'}\\
&=\lim_{\tau\rightarrow\infty}\sum_i\Psi_i(\underline{r}^{(N)})\exp[-\tau(E_i-E_{\mathrm{T}})]\langle\Psi_i|\,\Phi_{\mathrm{int}}\rangle\\
&=\lim_{\tau\rightarrow\infty}\Psi_0(\underline{r}^{(N)})\exp[-\tau(E_0-E_{\mathrm{T}})]\langle\Psi_0|\,\Phi_{\mathrm{int}}\rangle
\end{aligned} \tag{9.21}$$

and hence by adjusting the offset energy E_T to equal the energy of the ground state E_0 then the exponential in eq. (9.21) is constant while making all the other exponential terms damped as for them $E_i > E_0$.

Thus, we have to devise a scheme for numerically solving eq. (9.16) and finding the solution as $\tau\rightarrow\infty$, in order to see how to do this let us re-write eq. (9.16) as,

$$-\partial\Phi(\underline{r}^{(N)},t)/\partial t=\left[1/2\sum_{i=1}^{N}\nabla_i^2\,\Phi(\underline{r}^{(N)},t)\right]+[(\mathbf{V}(\underline{r}^{(N)},\underline{R}^{(n)})-E_{\mathrm{T}})\Phi(\underline{r}^{(N)},t)] \tag{9.22}$$

if we ignore the second term on the right-hand side we would have an N-body diffusion equation of the Smoluchowski type and if instead we ignore the first term we would have a first-order rate process or branching process with rate constant $(\mathbf{V}(\underline{r}^{(N)},\underline{R}^{(n)}) - E_T)$. As we know how to solve these types of equations we need to re-cast the Schrodinger equation in a form which embodies both these processes.

The Green's function for the processes if $(\mathbf{V}(\underline{r}^{(N)},\underline{R}^{(n)}) - E_T) = 0$ would be,

$$G(\underline{r}^{(N)} \leftarrow \underline{r}^{(N)'}, \tau) = (2\pi\tau)^{-3N/2} \exp[-(\underline{r}^{(N)} - \underline{r}^{(N)'})^2/2\tau] \tag{9.23}$$

so this is the basis of a stochastic, Brownian scheme. However, the effect of **H** must be included and so we have to modify eq. (9.23) to allow for interactions to obtain an approximation to the Green's function for the full Schrodinger eq. (9.16). To include the effect of the potential terms in **H** we use the Trotter-Suzuki formula, that is[16,18]

$$\exp\{-\tau(\mathbf{A}+\mathbf{B})\} = \exp(-\tau\mathbf{B}/2)\exp(-\tau\mathbf{A})\exp(-\tau\mathbf{B}/2) + O(\tau^3) \tag{9.24}$$

where $A = -\frac{1}{2}\sum_{i=1}^{N}\nabla_i^2$ and $\mathbf{B} = \mathbf{V} - E_T$ then we have using eqs. (9.19) and (9.24),

$$\begin{aligned} &G(\underline{r}^{(N)} \leftarrow \underline{r}^{(N)'}, \tau) \\ &\approx \exp\{-\tau[\mathbf{V}(\underline{r}^{(N)}, \underline{R}^{(n)}) - E_T]/2\}\left\langle \underline{r}^{(N)} \left| \exp\left[-1/2\tau\sum_{i=1}^{N}\nabla_i^2\right] \right| \underline{r}^{(N)'} \right\rangle \\ &\times \exp\{-\tau[\mathbf{V}(\underline{r}^{(N)'}, \underline{R}^{(n)}) - E_T]/2\} \end{aligned} \tag{9.25}$$

Thus, from eqs. (9.23) and (9.25) we have,

$$\begin{aligned} &G(\underline{r}^{(N)} \leftarrow \underline{r}^{(N)'}, \tau) \\ &\approx (2\pi\tau)^{-3N/2} \exp[-|\underline{r}^{(n)} - \underline{r}^{(n)'}|/2\tau] \exp\{-\tau[\mathbf{V}(\underline{r}^{(N)}, \underline{r}^{(n)}) \\ &+ \mathbf{V}(\underline{r}^{(N)'}, \underline{r}^{(n)}) - 2E_T]/2\} + O(\tau^3) \\ &= P(2\pi\tau)^{-3N/2} \exp[-|\underline{r}^{(n)} - \underline{r}^{(n)'}|/2\tau] \end{aligned} \tag{9.26}$$

and this means that we now have a factor $P = \exp\{-\tau[\mathbf{V}(\underline{r}^{(N)},\underline{R}^{(n)}) + \mathbf{V}(\underline{r}^{(N)'},\underline{R}^{(n)}) - 2E_T]/2\}$ which acts as a time-dependent renormalization or re-weighting of the diffusion Green's function. If we ignore the diffusive term in eq. (9.26) we see that $G(\underline{r}^{(N)} \leftarrow \underline{r}^{(N)'}, \tau)$ would describe a first-order rate process or branching process with rate constant $\{-\tau[\mathbf{V}(\underline{r}^{(N)},\underline{R}^{(n)}) + \mathbf{V}(\underline{r}^{(N)'},\underline{R}^{(n)}) - 2E_T]/2\}$.

Several schemes, broadly termed Diffusion Monte Carlo (DMC) methods, have been developed using the above or related ideas but, in practice, there are some important practical problems to be overcome before accurate results may be obtained from this approach.[13–18] For example, using obvious methods like a configuration ("walker") evolution scheme or a branching or birth/death algorithm in which P determines the number of walkers that survive to the next step of the process is very inefficient if eq. (9.26) is used due to the potentially unbounded nature of $\mathbf{V}(\underline{r}^{(N)}, \underline{R}^{(n)}) + \mathbf{V}(\underline{r}^{(N)}, \underline{R}^{(n)'})$.[16,18] A more efficient scheme is to incorporate importance sampling (see Appendix K) into the procedure using a "trial" or "guiding" wave function $\Psi_T(\underline{R}^{(n)})$. This is done by multiplying eq. (9.16) by $\Psi_T(\underline{R}^{(N)})$ and introducing the function $f(\underline{r}^{(N)},t) = \Phi(\underline{r}^{(N)})\,\Psi_T(\underline{r}^{(N)})$ which leads to,

$$\partial/\partial t\ f((\underline{r}^{(N)},t) = -1/2\,\underline{\nabla}^2 f(\underline{r}^{(N)},t) + \underline{\nabla}\cdot\left[v_D(\underline{r}^{(N)}) f(\underline{r}^{(N)},t)\right] + E_L(\underline{r}^{(N)}) - E_T]\,f(\underline{r}^{(N)},t) \tag{9.27}$$

where $\underline{\nabla} = (\nabla_1, \nabla_2, ..., \nabla_N)$ is the 3-N dimensional gradient operator and $v_D(\underline{r}^{(N)},t)$ is called the 3-N dimensional drift velocity given by,

$$v_D(\underline{r}^{(N)},t) = \underline{\nabla}\ln|\Psi_T(\underline{r}^{(N)})| = \Psi_T(\underline{r}^{(N)})^{-1}\,\nabla\Psi_T(\underline{r}^{(N)})$$

and the "local" energy is given by,

$$E_T = \Psi(\underline{r}^{(N)})^{-1}\,\mathbf{H}\Psi(\underline{r}^{(N)})$$

Now instead of eq. (9.17) we have,

$$f(\underline{r}^{(N)},t+\tau) = \int G'(\underline{r}^{(N)} \leftarrow \underline{r}^{(N)'},\tau)\ f(\underline{r}^{(N)} \leftarrow \underline{r}^{(N)'})\,\mathrm{d}\underline{r}^{(N)'}$$

where we have introduced the modified Green's function $G'(\underline{r}^{(N)} \leftarrow \underline{r}^{(N)'},\tau) = \Psi_T(\underline{r}^{(N)})G(\underline{r}^{(N)} \leftarrow \underline{r}^{(N)'},\tau)\ \Psi_T(\underline{r}^{(N)})^{-1}$ and the short-time approximation to this is,

$$G'(\underline{r}^{(N)} \leftarrow \underline{r}^{(N)'},\tau) \approx G_d(\underline{r}^{(N)} \leftarrow \underline{r}^{(N)'},\tau)\,G_b(\underline{r}^{(N)} \leftarrow \underline{r}^{(N)'},\tau) \tag{9.28}$$

where

$$G_d(\underline{r}^{(N)} \leftarrow \underline{r}^{(N)'},\tau) = (2\pi\tau)^{-3N/2}\exp\left|-\{\underline{r}^{(N)} - \underline{r}^{(N)'} - \tau v_D(\underline{r}^{(N)},t)\}^2/2\tau\right|$$
$$G_d(\underline{r}^{(N)} \leftarrow \underline{r}^{(N)'},\tau) = \exp\left|-\tau[E_L(\underline{r}^{(N)}) + E_L(\underline{r}^{(N)'}) - 2E_T]/2\right|$$

This eq. (9.28) leads to more efficient and stable numerical schemes than eq. (9.26) and is the basis of methods based on the biased random walk method called the Diffusion Monte Carlo method.[13–18] The methods based on eq. (9.28) are much more stable than those based on eq. (9.26) principally because, unlike $\{\mathbf{V}(\underline{r}^{(N)}) + \mathbf{V}(\underline{r}^{(N)'}) - 2E_T\}$ the term which replaces it, that is $\{E_L(\underline{r}^{(N)}) + E_L(\underline{r}^{(N)'}) - 2E_T\}$ becomes small as the solution approaches the correct ground-state one. To implement such schemes a choice of the form for the trial wavefunction $\Psi_T(\underline{r}^{(N)})$ must first be made and several successful approaches to implementing DMC have been developed. These different approaches use a variety of numerical methods, for example, to whether to use importance sampling or not.[12–18]

What ever numerical scheme is used once the system has evolved to the ground state then expectation values, such as the energy may be accumulated by continuing the DMC process and, in principle, the values obtained are exact.

However, no matter what scheme is used, several common problems are encountered in practice,

1. The wavefunction for a fermionic system may be positive or negative and on the planes separating positive and negative regions (the nodes of the wavefunction) the wavefunction will be zero. This is unlike the densities appearing in a Smoluchowski equation which are always positive. Thus, one must handle this nodal problem in any practical scheme and this is usually done by using the trial wavefunction $\Psi_T(\underline{r}^{(N)})$ to define the nodes which are usually fixed during the simulation ("Fixed Nodes Approximation"). Apart from this the method may be shown to give the exact ground state energy for a wavefunction having the chosen nodal surface, that is, the wavefunction will be exact between the nodes but the nodes may not be exact.[18] However, the error in the fixed nodes energy should be second order in the error in the position of the nodal surfaces.[18]
2. Ignoring the fixed nodes error we must generate enough configurations in order to obtain accurate numerical approximations to expectation values and to check for this the uncertainties in expectation values should be calculated as the results are often affected by serial correlation.
3. Lastly the results may depend on the size of the time step τ used in the numerical procedure and results should be extrapolated to $\tau = 0$ to eliminate this effect.

To illustrate points 2 and 3 Figures 9.2 and 9.3 show the results for the energy versus time step and energy versus the number of configurations for a fixed nodes DMC calculation of the electronic energy of the H_2O dimer and of solid Ni, respectively.

9.4 MASTER EQUATIONS

There are many schemes based on master equations and stochastic differential equations and applications of them to problems in chemistry and physics.[6] We will not attempt to cover this vast field here but only indicate how these methods may be thought of as related to the approach used in this book.

As discussed in the introduction to this chapter the approach in this book has been to start with a fundamental equation of motion for atoms and to average over certain degrees of freedom to obtain a more coarse-grained equation in which some information about certain degrees of freedom (i.e. some physical processes) is lost. This introduces stochastic terms, that is random effects, by necessity. By contrast one may chose to postulate equations which contain some stochastic terms and accept this loss of information as a basic tenet of the theoretical approach. This is the approach based on stochastic differential equations and master equations. Many of these approaches are discussed in the literature and the book by van Kampen[6] is an excellent standard work in this area and for detailed background, definitions and discussions of these general techniques provides an ideal starting point.

One central concept in this area is that of a Markov process. Broadly, if we know a state of the system say, $y_{n-1} = (\underline{r}^{(N)}{}_{n-1}, \underline{p}^{(N)}{}_{n-1})$ at time t_{n-1} then the state of the system at a later

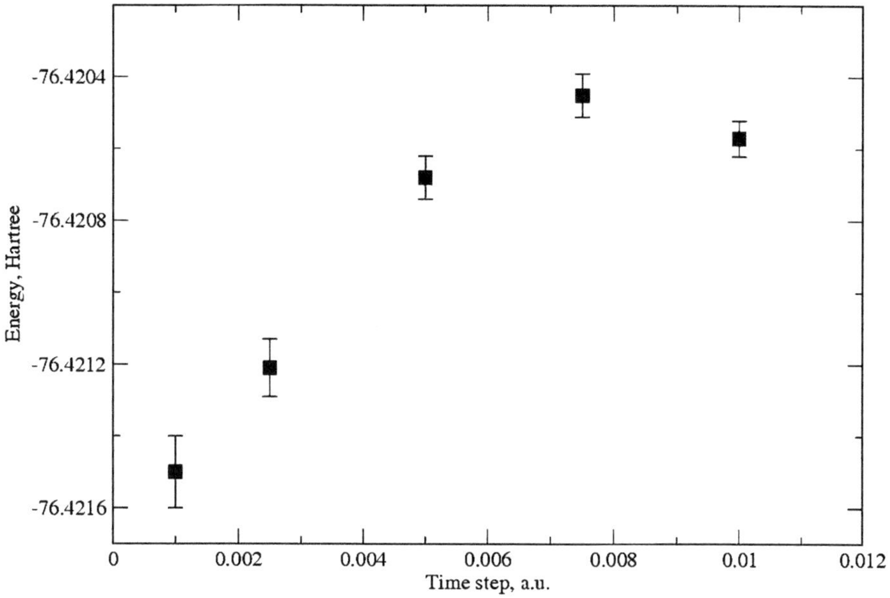

Figure 9.2 The results of the energy versus the size of the time step for a fixed nodes DMC calculation of the electronic energy of the H_2O dimer. (Figure courtesy of Ms. Nicole Benedek, RMIT.)

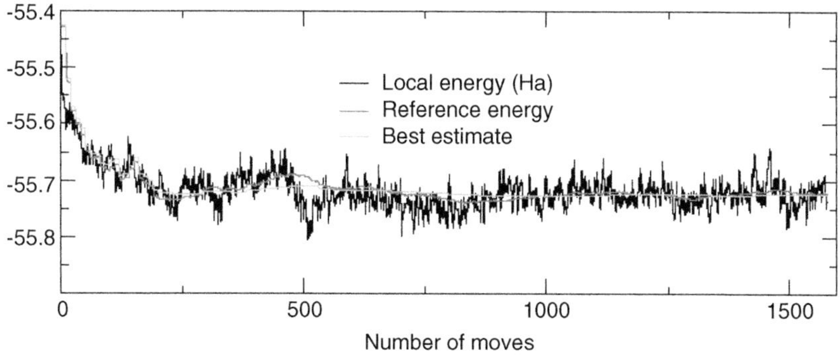

Figure 9.3 The electronic energy of antiferromagnetic NiO crystal versus the number of DMC moves (Figure courtesy of Dr. Mike Towler, Cavendish Laboratory, Cambridge University, UK)

time t_n, $y_n = (\underline{r}^{(N)}{}_n, \underline{p}^{(N)}{}_n)$ only depends on the state at t_{n-1}. More precisely adopting van Kampen's notation[6] he gives the definition of a Markov processes as, *"the conditional probability density at time* t_n*, given the value* y_{n-1} *at time* t_{n-1}*, is uniquely determined and* *is not affected by any knowledge of the values at earlier times.*" Van Kampen[19] expresses this as,

$$P_{1|n-1}(y_n,t_n \mid y_1,t_1;...;y_{n-1},t_{n-1}) = P_{1|n-1}(y_n,t_n \mid y_1,t_1;y_{-1},t_{n-1}) \tag{9.29}$$

where $P_{1|n-1}$ is called the transition probability. As he points out all that is needed to fully determine a Markov process is $P_1(y_1,t_1) = P_{1|n-1}(y_2, t_2 | y_1,t_1)$ since by iteration one may construct all of the P_n's from this information by repeatedly using eq. (9.29).
Using this process we have for $t_1 < t_2 < t_3$,

$$P_3(y_1,t_1;y_2,t_2;y_3,t_3) = P_1(y_1,t_1)P_{1|1}(y_2,t_2 \mid y_1,t_1)P_{1|1}(y_3,t_3 \mid y_2,t_2) \tag{9.30}$$

then integrating eq. (9.30) over y_2 and dividing both sides by $P_1(y_1,t_1)$ we obtain the famous Chapman−Kolmogorov equation which is obeyed by the transition probability of any Markov process,

$$P_{1|1}(y_3,t_3 \mid y_1,t_1) = \int P_{1|1}(y_3,t_3 \mid y_2,t_2)P_{1|1}(y_2,t_2 \mid y_1,t_1)\,dy_2 \tag{9.31}$$

The Chapman−Kolmogorov equation is not very easy to use in practice and so it is usually converted to an equivalent equation called the master equation.[6,20–23] Take $t_3 = t_2 + \Delta t$ where Δt is small and so as $P_{1|1}(y_3,t_3 | y_2,t_2) = \delta(y_3 - y_2)$ then,

$$P_{1|1}(y_3,t_2+\Delta t \mid y_2,t_2) = \delta(y_3 - y_2)[1 - A(y_2)\Delta t] + \Delta t\, W(y_3 \mid y_2) + O(\Delta t^2) \tag{9.32}$$

where $W(y_3|y_2)$ is the transition probability per unit time from y_2 to y_3 and since we require the probability to be normalised,

$$A(y_2) = \int W(y_3 \mid y_2)\,dy_3 \tag{9.33}$$

Substituting eqs. (9.32) and (9.33) into eq. (9.31) we have,

$$\begin{aligned}&[P_{1|1}(y_3,t_2+\Delta t \mid y_1,t_1) - P_{1|1}(y_3,t_2 \mid y_1,t_1)]/\Delta t\\ &= [W(y_3 \mid y_2)P_{1|1}(y_2,t_2 \mid y_1,t_1) - W(y_2 \mid y_3)P_{1|1}(y_3,t_2 \mid y_1,t_1)]\,dy_2\end{aligned}$$

and taking the limit $\Delta t \to 0$,

$$\partial/\partial t\, P_{1|1}(y,t \mid y_0,t_0) = \int [W(y \mid y')P_{1|1}(y',t' \mid y_0,t_0) - W(y' \mid y)P_{1|1}(y,t \mid y_0,t_0)]\,dy'$$

or

$$\partial/\partial t\, P(y,t) = \int [W(y \mid y')P(y',t') - W(y' \mid y)P(y,t)]\,dy' \tag{9.34}$$

Eq. (9.34) is the so-called master equation which if the range of the variables is discrete is usually written as,[6,20–22]

$$\partial/\partial t\, P_n(t) = \sum_{n'} [W_{nn'}(t)P_{n'}(t) - W_{n'n}(t)P_n(t)] \tag{9.35}$$

this form shows that it is a balance or gain−loss equation for the probabilities of the states n. The first term is the gain of state n due to transitions from other states n' and the second term is the loss due to transitions from n to the other states.

At this stage it should be remarked that all Markov processes may be described by the Chapman−Kolmogorov equation or equivalently the master equation. However, this does not imply that all physical processes are described by the master equation since not all physical processes are Markov processes. It might even be speculated that all master equations that one can devise do not necessarily describe real physical processes. In fact van Kampen[6] has excellent discussions as to the conditions that the transition probability $W_{n'n}(t)$ must satisfy in various systems, for example, for a closed, isolated physical system.[19]

It may be shown that the F−P equation is a special type of master equation and may be used as an approximate model for a general Markov processes.[6,20–24] To derive the F−P equation from the general master equation one must limit the size of the "jumps" from one state of the system to another state. Van Kampen writes,[6]

$$W(y \mid y') = W(y';r), \quad r = y - y' \tag{9.36}$$

substituting eq. (9.36) into eq. (9.34) we have,

$$\partial/\partial t\, P(y,t) = \int [W(y-r;r)P(y',t')]\mathrm{d}r - P(y,t)\int W(y;-r)P(y,t)\mathrm{d}r \tag{9.37}$$

Assuming that the jumps are small then $W(y';r)$ is a sharply peaked function of r but a slowly varying function of y' so,

$$\begin{aligned} &W(y';r) \approx 0 && \text{for} \mid r \mid > \delta \\ &W(y'+\Delta y;r) \approx W(y';r) && \text{for} \mid \Delta y \mid < \delta \end{aligned}$$

Now assuming that $P(y,t)$ varies slowly with y we may use a Taylor series expansion of the integral $\int[W(y-r;r)\ P(y',t')]\mathrm{d}r$ which to second order gives,[6,20–22]

$$\begin{aligned} &\partial/\partial t\, P(y,t) \\ &= \int [W(y;r)P(y,t)]\mathrm{d}y - \int r\partial/\partial y\{W(y;r)P(y,t)]\mathrm{d}r \\ &\quad + \tfrac{1}{2}\int r^2\partial^2/\partial y^2\{W(y;r) - P(y,t)]\mathrm{d}r - P(y,t)\int W(y;-r)\mathrm{P}(y,t)\mathrm{d}r \\ &= -\int r\partial/\partial y\{W(y;r)P(y,t)]\mathrm{d}r + \tfrac{1}{2}\int r^2\partial^2/\partial y^2\{W(y;r) - P(y,t)]\mathrm{d}r \\ &= \int [W(y;r)P(y,t)]\mathrm{d}y - \int r\partial/\partial y\{W(y;r)P(y,t)]\mathrm{d}r \\ &\quad + \tfrac{1}{2}\int r^2\partial^2/\partial y^2\{W(y;r) - P(y,t)]\mathrm{d}r - P(y,t)\int W(y;-r)P(y,t)\mathrm{d}r \\ &= -\int r\partial/\partial y\{W(y;r)P(y,t)]\mathrm{d}r + \tfrac{1}{2}\int r^2\partial^2/\partial y^2\{W(y;r) - P(y,t)]\mathrm{d}r \end{aligned} \tag{9.38}$$

If we define the jump moments by,[6]

$$a_\upsilon = \int_{-\infty}^{\infty} r^\upsilon W(y;r)\,\mathrm{d}r \tag{9.39}$$

then eq. (9.38) becomes,

$$\partial/\partial t\, P(y,t) = -\partial/\partial y\,\{a_1(y)P(y,t)\} + 1/2\,\partial^2/\partial y^2\,\{a_2(y)P(y,t)\}$$

or

$$\partial P(y,T)/\partial t = -\partial/\partial y\; A(y)P(y,t) + 1/2\partial^2/\partial y^2\; B(y)P(y,t) \tag{9.40}$$

where $A(y)$ and $B(y)$ are real differentiable functions with $B(y) > 0$. This is van Kampen's form of the F−P equation resulting from the master equation.[6,20–23]

As the Langevin equation (or rather Langevin equations for all the relevant dynamical variables) is equivalent to the F−P equation (see Section 5.4) then the above remarks about the F−P equation also applies to the LE equation.

9.4.1 The identification of elementary processes

The first problem to be solved in developing methods based on a master equation (9.35) is to identify the elementary processes involved and to calculate their relative probability, that is, the rate constants of these processes.

Methods based on solving dynamical equations, for example, Newton's equations (i.e. MD methods) have been developed specifically to treat phenomena involving rare events and to find the nature of the elementary events occurring. These methods are termed accelerated dynamics methods are usually based either directly or indirectly on Transition State Theory (TST).[21,25–28] Basically, TST is used to calculate a rate constant for the escape from one state to another and this is evaluated from the equilibrium flux through the dividing surface between these states. Thus, if one can identify the states involved, which will include the barrier (transition state or saddle point) between the two states, the calculation amounts to an equilibrium one. If no correlated events occur this method is exact and is certainly a reasonable one for most rare events, as the system spends so much time in the initial state that it must achieve at least meta-stability or pseudo-equilibrium.

In TST the probability per unit time for the transition to occur is given by[21,25–28],

$$W_{IJ} = \upsilon_{IJ}\exp(-\Delta E_{IJ}/k_\mathrm{B}T) \tag{9.41}$$

where υ_{IJ} is the attempt frequency with which the system attempts the transition and ΔE_{IJ} is the activation barrier for the transition from state I to state J. Often the further approximation of a harmonic transition state, that is, that the vibrational modes in the transition state are harmonic is made but this is not essential. In this quasi-harmonic approximation,

$$\upsilon_{IJ} = \prod_{k=1}^{3N-3} \upsilon_k \Big/ \prod_{k=1}^{3N-4} \upsilon_{k'} \tag{9.42}$$

where υ_k and $\upsilon_{k'}$ are the vibrational eigen-frequencies of the system in the initial configuration and at the saddle point configuration between the states I and J respectively. A example of such a processes is diffusion in a solid or on a solid surface where the two states of the system may be linked by atomic jumps of an atom from site α to site α' which changes the spatial configuration of the solid from I to J.

In most versions of TST the rate constants are computed after the dividing surface is specified, for example, by an exploration of energy pathways by Density Functional Theory (DFT) theory. However, accelerated MD uses TST to find these dividing surfaces and there are several such methods three of which are:

1. *Parallel-Replica Dynamics*[27–29] which is the simplest and most accurate of these methods as it involves the least assumptions, the only one being that the infrequent events obey first-order kinetics, that is, the decay process is exponential. Thus, after a time $t > \tau_{\text{corr}}$ after entering the transition state the probability distribution for the time for the next escape is given by,

$$p(t) = k_{\text{tot}} \mathrm{e}^{-k_{\text{tot}} t} \tag{9.43}$$

where k_{tot} is the rate constant for escape from the state. In practice the system's dynamics is followed on M processors of a multiprocessor computer in which the MD simulation is interrupted at periodic intervals to perform a steepest-descent or conjugate-gradient minimization which, if successful, this procedure will take the system towards the minimum of its current potential basin. If the geometry changes significantly from one minimization to the next then this indicates that a new basin has been entered. This is actually performed on M replicas of the original system which originated from the same state but with a different initial condition of the trajectory taken from this common starting point.

2. *Hyperdynamics*[29–34] which uses importance sampling in the time domain by adding a bias potential $V_b(r)$ to the actual potential energy function and requires the assumptions of TST but not of a harmonic transition state. The bias potential is designed to raise the energy in regions other than the dividing surface and, thus, the problem is to choose a suitable bias potential for which $V_b(r) = 0$ on the dividing surface and $V_b(r) \leq 0$ otherwise.
3. *Temperature accelerated dynamics*[35–37] in which the transitions are sped up by increasing the temperature in the simulation, whilst filtering out the transitions that should not have occurred at the temperature of interest. This is the most approximate method and replies on harmonic TST. An MD run is performed at a higher temperature than desired, T_{high} and the activation energy E_i for this process evaluated by use

of eq. (9.41) then using eq. (9.42) the time at which the processes would have occurred at T_{low} is given by,

$$t_{i,\text{low}} = t_{i,\text{high}} \exp[E_i(1/k_B T_{\text{low}} - 1/k_B T_{\text{high}})] \tag{9.44}$$

The shortest-time event found at the high temperature T_{high} is then taken to be the proper transition at the desired temperature T_{low}.

Other methods have been developed to identify and characterise the rate constants of elementary processes, for example, Quantum Mechanical calculations of the energies of pathways,[38] Transition Path Sampling techniques (TPS)[39] and "Blue Moon" ensemble methods.[40] However, the identification of all the relevant elementary processes which contribute to many overall phenomena, for example, surface diffusion, still remains a significant problem. Interesting cases studies indicate that the elementary processes which underlie even well-researched phenomena may not be as well understood as is often thought.[27,28,38,41,42]

Once a choice of elementary processes has been made and the transition probabilities, that is, relative probabilities or rate constants associated with each of these processes found a scheme for solving the master equation may be implemented and one such scheme, the Kinetic Monte Carlo (KMC) Method is discussed in the next section. Many other methods have also been developed, for example, based on Stochastic Langevin equations or F−P equations.

9.4.2 Kinetic MC and master equations[24,38,43–48]

The idea behind the Kinetic Monte Carlo (KMC) is the connection between the master equation (9.35) and equilibrium Monte Carlo simulations and its origins may be traced back to Kinetic Ising models.[38] There are now an enormous number of techniques which may be classified under the heading of KMC, some which are lattice based, some off lattice and even those based on continuum modelling. The range of topics to which these techniques have been applied is similarly enormous and to name but a few: aggregation in colloids and aerosols,[49–51] chemical reactions in general,[21,22] and in particular for catalytic reactions[52] and combustion,[53] diffusion in solids[54] and at surfaces,[55] many aspects of the growth of crystals, thin films, nanostructures on surfaces and catalysis.[23,38,56–70] The study of surfaces have been a particularly fruitful area of research and such studies encompass a very wide range of problems in surface and materials science for example thin film growth,[56,57] surface adsorption,[58] thin film deposition,[59,60] nanocrystals growth[65–68] and chemical vapour deposition.[69–71]

Other applications include grain growth,[72] semiconductor process simulation,[73,74] electron kinetics in plasmas and semiconductors,[75] crystal growth from solution,[76] phase transitions in solids[77] and phase separation in alloys.[78] This list is certainly not exhaustive and the references quoted contain references to many, many more studies using KMC methods.

A basic KMC procedure may be described as follows. A list of possible processes is made and a constant time increment Δt is chosen to accommodate the fastest transition used. Then at each time step,

1. A transition process is chosen at random from the list of the possible transitions given the current state of the system.
2. The probability of the processes happening during Δt, P_{proc} is calculated.

3. The transition is accepted with probability P_{proc}, it will be accepted if $P_{proc}<R$ otherwise it will be rejected, were R is a random number.
4. The time is incremented by Δt and the procedure is continued from step 1 using the new configuration if the process is accepted otherwise the old configuration is used.

This overall procedure is continued until the phenomena being studied is deemed to have completed.

However, this is a fairly slow procedure and other more efficient ones have been devised which are much faster.[38,48,58,60,63,64,65,71,79–81] These more efficient MC schemes are based on the ideas of Bortz et al.[79] (or closely related ones[80,81]) which gives an algorithm which accepts all moves. The basic procedure is,

1. Calculate the rate constants for the possible processes and form $K = \Sigma k_j$ where j runs over all processes which are possible for the current state of the system.
2. Randomly choose a possible process, i with probability depending on its relative rate that is, use,

$$\sum_{j=1}^{i=1} k_j \leq R_1 K \leq \sum_{j=1}^{i} k_j \tag{9.45}$$

where R_1 is a random number chosen from a uniform distribution on [0,1).

3. Execute process i.
4. Increment the time by Δt given by,

$$\Delta t = -\ln(R_2)/K \tag{9.46}$$

where R_2 is a random number chosen from a uniform distribution on [0,1).

Once again this overall procedure is continued until the processes being studied is deemed to be completed. The difference between this MC procedure (often called N-Fold Way) and the simple KMC procedure is that all configurations generated are accepted, however, they are generated with the correct probability and it may be shown that it is simply a re-arrangement of the simple Metropolis KMC algorithm.[60,79]. In practice his method is much faster than the simple one and is continually being built upon to further improve its efficiency.[27,82,83]

9.4.3 KMC procedure with continuum solids

Some processes involve both large length scales and also long times, for example, electron and ion transport in solids where events may occur over length scales from 10's of nm to μm. Although the use of parallel computers enables large length scales to be simulated the long times involved make such calculations extremely difficult. Thus, for this class of phenomena it is necessary to represent the solid as a continuum and use mean free paths of the elementary processes to control the KMC computation.[83–89] In order to illustrate this type of approach we will describe the simulation of electron transport through a solid which is very similar to the previously outlined Bortz method.

Each electron's individual passage through a solid is simulated by a stochastic Monte Carlo procedure in which a possible sequence of collision events is computed. The path is controlled by values of the scattering cross-sections (or equivalently mean free paths) for the various scattering processes which are possible. The properties dependent on the electron interactions with the solid are then calculated by averaging over a large number of these trajectories. The basic idea for this MC process is that one takes an electron, decides what will be its initial energy, angle to the normal of the surface of the material and on the first type of scattering event to occur. The electron travels a certain distance s until the next scattering event occurs. How far the electron travels through the material until this next scattering event occurs is determined by the mean-free paths λ_i for each type of scattering event which may be calculated from the total cross section σ_i for that type of scattering by,

$$\lambda_i = A/(N_{\mathrm{A}}\,\rho\sigma_i) \tag{9.47}$$

where A = atomic weight, ρ the mass density and N_{A} Avogadro's number.

In order to calculate the total mean free path λ_{T} we write it in terms of the mean free paths of each type of scattering event i, both elastic and inelastic, which are possible, by,

$$1/\lambda_T = \sum_{i=1}^{j} 1/\lambda_i = 1/\lambda_{\mathrm{e}} + 1/\lambda_{\mathrm{in}} \tag{9.48}$$

where λ_{e} is the elastic mean-free path and λ_{in} the inelastic mean-free path.

If we assume that the step length between two successive collision events is a Poisson stochastic process the path length S between scattering events is then given by,

$$S = -\lambda_{\mathrm{T}} \ln(R_1) \tag{9.49}$$

where R_1 is a random number chosen from a uniform distribution on [0,1).

To decide the type of scattering event j which occurs another random number R_2 is chosen from a uniform distribution on [0,1) and the event is selected by use of,

$$\lambda_{\mathrm{T}} \sum_{i=1}^{j} 1/\lambda_{i-1} \le R_2 \le \lambda_{\mathrm{T}} \sum_{i=1}^{j} 1/\lambda_i \tag{9.50}$$

where i labels the type of elementary scattering events.

Once the type of scattering is decided upon the solid scattering angle $\Omega = (\theta,\phi)$, energy loss ΔE (if any is involved in the process chosen) and the generation of secondary electrons (if any are generated in the process chosen) are calculated. The angle of scattering θ is determined from another random number R_3 and the azimuthal angle ϕ is assumed to be uniformly distributed around the direction of travel and this is determined by fourth random number R_4. The energy loss ΔE (if any) are calculated from the partial scattering cross sections for the scattering process involved via the calculation of one more random number R_5.

The process is continued until the electron has either,

1. Left the material by backscattering out of the surface it entered through,
2. Left the material by transmission through the material if it is of semi-infinite extent

or

3. Its energy has fallen below some pre-determined minimum threshold value, E_{min}.

The properties dependent on the electron interactions with the material are then calculated by averaging over a large number of these trajectories. Properties which may be simulated include, electron transmission and backscattering coefficients, the energy distribution of transmitted, backscattered and secondary electrons, the distribution of depths of penetration of electrons, mean depth of penetration of various types of electrons and the secondary electron yield.

Figure 9.4 shows a representative set of trajectories for 2 keV electrons normally incident on Al.

Usually it is assumed that the electron trajectories are independent and so only one electron trajectory is simulated at a time so if many processors are used, each trajectory may be run on a different processor. Similar methods have been used to simulate ion transport through solids.[84,89]

9.5 CONCLUSIONS

In this chapter we gave an outline of some stochastic methods based on distribution functions or on a master equation which provides an alternative starting point to the use of dynamical equations of motion.

As emphasised before, if all particles and all degrees of freedom are explicitly treated then a solution of the appropriate dynamical equations gives a complete description of all the physical processes which these equations are capable of describing and, by suitable averaging, all the macroscopic processes for the N-body system may be obtained. In principle, an appropriate numerical scheme for solving such equations, for example, the EMD and NEMD methods will (ignoring problems of numerical accuracy and the storage of such data) give a sufficiently accurate picture of all processes which may occur. However, this is not always possible as the numerical methods used to solve such equations need an integration step smaller than the fastest atomic processes being simulated and many processes occur on much, much longer time scales than this.

In order to overcome this and treat processes with time scales long on an atomic scale some sort of coarse-graining over the fastest variables is necessary. Previous chapters have developed coarse-gained stochastic differential equations termed generalised Langevin equations (GLE) which in certain limits reduce to the Langevin equation (LE). In Sections 9.1 and 9.2 above we discussed an alternative but equivalent approach based on the N-body distribution function which lead to the F−P equation and in the diffusive limit the Smoulochkowski, N-body diffusion (SE) equation. The F−P equation provides

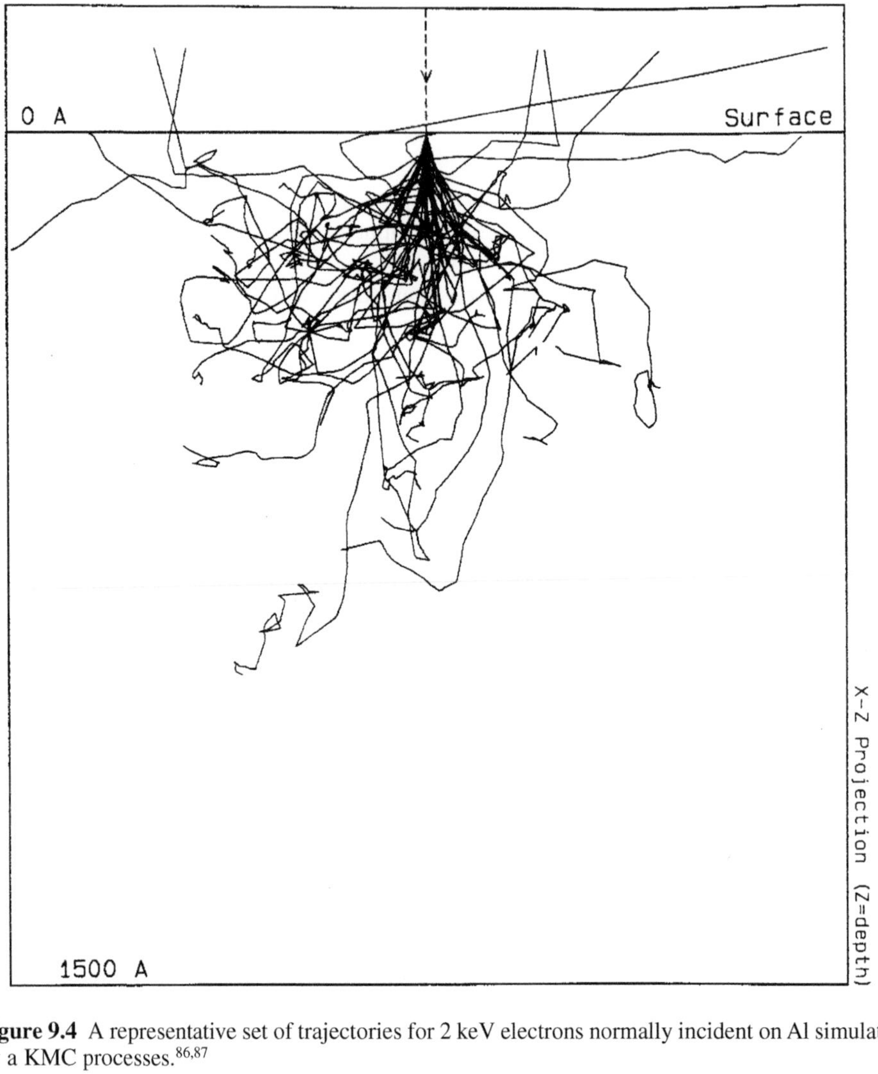

Figure 9.4 A representative set of trajectories for 2 keV electrons normally incident on Al simulated by a KMC processes.[86,87]

an alternative to the LE which can sometimes be a more convenient starting point to discuss dynamics. An illustration of this is that if velocity correlations have relaxed to equilibrium before the system's spatial configuration has changed significantly then the SE provides a convenient description. However, the derivation of a position only LE which should provide an equivalent description to that given by the SE has proved to be difficult to develop.

Another example of the use of the SE is that the time-dependent Schrödinger equation may be cast into a form of an SE which forms the basis of a convenient and accurate stochastic method of obtaining quantum mechanical expectation values termed the Diffusion Quantum Monte Carlo method which was discussed in Section 9.3.

Another alternative approach to coarse-graining is the use of a stochastic description based on the master equation (ME) where one starts with the idea that a reasonable description of many phenomena is an inherently probabilistic one. This approach which was discussed in Section 9.4 may loosely be looked on as the inverse of the approach used in most of this book in which a fundamental dynamical description is coarse grained by necessity. Rather in these theories an ME involving the probability or rates at which processes occur is postulated and an inherently stochastic approach is used from the very beginning. The system is then regarded as passing from one state to another by a set of randomly selected elementary processes whose probability of occurrence is given by a rate law.

The LE or equivalently F−P equation may be shown to only approximate the dynamics described by the general ME and it is also sometimes stated that LEs (or the F−P equations) may be derived by adding random terms to equations of motion. This seems to imply that all LE or F−P descriptions have a somewhat arbitrary basis in physics and are hard to justify and improve. However, as shown earlier the Langevin and F−P equations developed here are based on well-defined approximations to exact equations of motion or rather to equations of motion which exactly describe some physical phenomena. Thus, these examples are not arbitrary and they may be systematically improved by eliminating some of the approximations used or by making better approximations.

The skill in developing a method based on the ME is choosing all the relevant dynamical processes and then calculating their rates which were briefly described in Section 9.4.1. The first problem in carrying these procedures in practice is that it may be quite difficult to determine the form of all the important elementary processes. Secondly it is often a challenge to calculate the rates of all these processes but this is often accomplished by use of Transition State Theory. Once these two tasks are accomplished successfully a kinetic MC process may be used in order to fully describe the overall phenomena to be investigated as described in Sections 9.4.2 and 9.4.3.

Many ME-based methods have lead to much new understanding of important phenomena and the KMC method is now a standard tool in many areas of science and engineering. However, this success and widespread use does not mean that all stochastic models which have been developed are useful. There is a danger of developing models based on somewhat arbitrary equations and unless the physics upon which they are based is well defined then one runs the risk of not really providing any physical understanding or insight into the phenomena being modelled. Even good agreement with experimental data without physical insight seems to be a rather unsatisfactory exercise.[22]

REFERENCES

1. J. Albers, J.M. Deutch and I. Oppenheim, *J. Chem. Phys.*, **54**, 3541 (1971).
2. J.M. Deutch and I. Oppenheim, *J. Chem. Phys.*, **54**, 3547 (1971).
3. T.J. Murphy and J.L. Aguirre, *J. Chem. Phys.*, **57**, 2098 (1972).
4. F.F. Abraham, J.Q. Broughton, N. Bernstein and E. Kaxiras, See for example, Spanning the length scales in dynamic simulations, *Comp. Phys.*, **12**, 538−546 (1998) and F.F. Abraham, N. Bernstein, J.Q. Broughton, and D. Hess, *MRS Bull.*, **25**, 27−32 (2000).

5. C. Delargo, Transition path sampling. In *Handbook of Materials Modelling* (Ed. S. Yip), Springer, The Netherlands, 2005, pp. 1585−1596.
6. N.G. Van Kampen, *Stochastic Processes in Physics and Chemistry*, North-Holland, Amsterdam, 1992.
7. D.L. Ermak and J.A. McCammon, *J. Chem. Phys.*, **69**, 1352 (1978).
8. G. Wilemski, *J. Stat. Phys.*, **14**, 153 (1976).
9. R. Zwanzig, *Adv. Chem. Phys.*, **15**, 325 (1969).
10. W. Hess and R. Klein, *Physica A*, **97**, 71 (1978).
11. R.J.A. Tough, P.N. Pusey, H.N.W. Lekkerkerker and C. van den Broeck, *Mol. Phys.*, **59**, 595 (1986).
12. W.A. Lester Jr. and B.L. Hammond, Quantum Monte Carlo for the electronic structure of atoms and molecules, *Ann. Rev. Phys. Chem.*, **41**, 283−311 (1990).
13. P.J. Reynolds, J. Tobochnik and H. Gould, Diffusion Monte Carlo, *Comp. Phys.*, March/April, 192−197 (1992).
14. M.A. Lee and K.E. Schmidt, Green's function Monte Carlo, *Comp. Phys.*, March/April, 192−197 (1992).
15. B.L. Hammond, M.M. Soto, R.N. Barnett and W.A. Lester Jr., On quantum Monte Carlo for the electronic structure of molecules, *J. Mol. Struct.* (*Theochem*), **234**, 525−538 (1991).
16. *B.L. Hammond, W.A. Lester* Jr. and P.J. Reynolds, "*Monte Carlo Methods in Ab Initio Quantum Chemistry*", World Scientific, Singapore, 1994.
17. W. A. Lester Jr., S.M. Rothstein and S. Tanaka (Eds.), *Recent Advances in Quantum Monte Carlo Methods,* Part I and Part II, World Scientific, Singapore.
18. W.M.C. Foulkes, L. Mitas, R.J. Needs and G. Rajagopal,Quantum Monte Carlo simulations of solids, *Rev. Mod. Phys.*, **73**, 33−83 (2001).
19. N.G. Van Kampen, *Stochastic Processes in Physics and Chemistry*, North-Holland, Amsterdam, 1992, pp. 108−111.
20. H.S. Wio, *An Introduction to Stochastic Processes and Nonequilibrium Statistical Physics*, World Scientific, Singapore, 1994.
21. A. Nitzan, Stochastic theory of rate processes. In *Handbook of Materials Modelling* (Ed. S. Yip), Springer, The Netherlands, 2005, pp. 1635—1672.
22. D.T. Gillespie, Stochastic chemical kinetics. In *Handbook of Materials Modelling* (Ed. S. Yip), The Netherlands, 2005, pp 1735—1752.
23. 18. D.D. Vvedensky, Stochastic equations for thin film morphology. In *Handbook of Materials Modelling* (Ed. S. Yip), Springer, The Netherlands, 2005, pp. 2351−2361.
24. L. Mandreoli, *Density based kinetic Monte Carlo methods*, PhD Thesis, Technischen Universitat, Berlin, Germany, 2005.
25. H. Metiu, Introduction: Rate processes., In *Handbook of Materials Modelling* (Ed. S. Yip), Springer, The Netherlands, 2005, pp. 1567–1571.
26. J.D. Doll, A Modern Perspective on Transition State Theory. In *Handbook of Materials Modelling* (Ed. S. Yip), Springer, The Netherlands, 2005, pp. 1573–1583.
27. A.F. Voter, F. Montalenti and T.C. Germann, Extending the time scale in atomic simulation of materials, *Annu. Rev. Mater. Res.*, **32**, 321 (2002).
28. B.P. Uberuaga, F. Montalenti, T.C. Germann and A.F. Voter, Accelerated molecular dynamics. In *Handbook of Materials Modelling* (Ed. S. Yip), The Netherlands, 2005, pp. 629–648.
29. A.F. Voter, *Phys. Rev. B*, **57**, 13985 (1998).
30. A.F. Voter, *J. Chem. Phys.*, **106**, 4665 (1997).
31. A.F. Voter, *Phys. Rev. Lett.*, **78**, 3908 (1997).
32. D. Moroni, T.S. van Erp and P.G. Bolhuis, *Physica A*, **340**, 395 (2004).
33. X.M. Duan, D.Y. Sun and X.G. Gong, *Compt. Matter. Sci.*, **20**, 151 (2001).

34. X.M. Duan and X.G. Gong, *Comput. Mater. Sci.*, **27**, 375 (2003).
35. M.R. Sorensen and A.F. Voter, *J. Chem. Phys.*, **112**, 9599 (2000).
36. F. Montalenti and A.F. Voter, *Phys. Stat. Sol.*, **226**, 21 (2001).
37. F. Montalenti and A.F. Voter, *J. Chem. Phys.*, **116**, 4819 (2002).
38. K. Reuter, C. Stampfl and M. Scheffler, Ab initio thermodynamics and statistical mechanics of surface properties and functions., In *Handbook of Materials Modelling* (Ed. S. Yip), Springer, The Netherlands, 2005, pp. 149−194.
39. P.G. Bolhuis, D. Chandler, C. Dellago and P.L. Geissler, Transition path sampling: throwing ropes over rough mountain passes, in the dark, *Annual Rev. Phys. Chem.*, **53**, 291−318 (2002).
40. E.A. Carter, G. Ciccotti, J.T. Hynes and R. Kapral, *Chem. Phys. Lett.*, **156**, 472 (1989), and G. Ciccotti, R. Kapral and A. Sergi, Simulating reactions that occur once in a blue moon, In: *Handbook of Materials Modelling* (Ed. S. Yip) Springer, The Netherlands, 2005, pp. 1597–1611.
41. T.T. Tsong, Mechanisms of surface diffusion, *Prog. Surf. Sci.*, **67**, 235 (2001).
42. O.S. Trushin, P. Salo, M. Alatalo and T. Ala-Nissila, *Surf. Sci.*, **482−485**, 365 (2001).
43. D.P. Landau and K. Binder, *A Guide to Monte Carlo Simulations in Statistical Physics*, Cambridge University Press, Cambridge, UK, 2000, see in particular pp. 39−47 and 150−153 and 299–319.
44. K.H. Kehr and K. Binder, Simulation of diffusion in lattice gases and related phenomena., In *Applications of the Monte Carlo Method in Statistical Physics* (Ed. K. Binder), Springer, Berlin, 1987, pp. 181−329.
45. G. Gilmer and S.Yip, Basic Monte Carlo Models: Equilibrium and Kinetics. In: *Handbook of Materials Modelling* (Ed. S. Yip), Springer, The Netherlands, 2005, pp. 613−628 and p. 1941.
46. A.F. Voter, Introduction to the kinetic Monte Carlo method. In: *Radiation Effects in Solids* (Eds. K.E. Sickafus and E.A. Kotomin), Springer, NATO Publishing Unit, Dordrecht, The Netherlands, 2005.
47. T. J. Walls, *Kinetic Monte Carlo Simulations of Perovskite Crystal Growth with Long Range Coulomb Interactions*, Thesis for Bachelor of Science with Honors in Physics, College of William and Mary, VA, USA, 1999.
48. P. Jensen, Growth of nanostructures by cluster deposition: experiments and simple models, *Rev. Mod. Phys.*, **71**, 1695−1735 (1999).
49. P. Meakin, Fractal aggegates, *Adv. Coll. Interf. Sci.*, **28**, 249−331 (1988).
50. D. Mukherjee, C.G. Sonwane and M.R. Zachariah, *J. Chem. Phys.*, **119**, 3391 (2003).
51. P. Sandkuhler, M. Lattuada, H. Wu, J. Sefcik and M. Morbidelli, Further insights into the universality of colloidal aggregation, *Adv. Coll. Interf. Sci.*, **113**, 65−83 (2005).
52. P. Stoltze, *Microkinetic simulation of catalytic reactions*, *Prog. Surf. Sci.*, **65**, 65−150 (2000).
53. M.J. Pilling and S.H. Robertson, Master equation models for chemical reactions of importance in combustion, *Ann. Rev. Phys. Chem.*, **54**, 245−275 (2003).
54. G. Wahnstrom, Diffusion in solids. In *Handbook of Materials Modelling*(Ed. S. Yip), Springer, The Netherlands, 2005, pp. 1787–1796.
55. G.L. Kellog, Field ion microscope studies of single-atom surface diffusion and cluster nucleation on metal surfaces, *Surf. Sci. Rep.*, **21**, 1−88 (1994); A.G. Naumovets, Collective surface diffusion: an experimentalist's view, *Physica A*, **357**, 189−215 (2005).
56. G.H. Gilmer and K.A. Jackson, Computer simulation of crystal growth. In: *1976 Crystal Growth and Materials* (Eds. E. Kaldis and H.J. Scheel), North-Holland, Amsterdam, 1977, pp. 80–114.
57. G.H. Gilmer, Computer simulation of crystal growth, *J. Crystal Growth*, **42**, 3 (1977).
58. J.W. Evans, Kinetic Monte Carlo simulation of non-equilibrium lattice-gas models: Basic and refined algorithms applied to surface adsorption processes. In: *Handbook of Materials Modelling* (Ed. S. Yip), Springer, The Netherlands, 2005, pp. 1753–1767.

59. H. Huang, Texture evolution during thin film deposition. In: *Handbook of Materials Modelling* (Ed. S. Yip), Springer, The Netherlands, 2005, pp. 1039–1049.
60. C. Battaille, Monte Carlo methods for simulating thin film deposition. In: *Handbook of Materials Modelling* (Ed. S. Yip), Springer, The Netherlands, 2005, pp. 2363−2377.
61. B.I. Lundquist, A. Bogicevic, S. Dudiy, P. Hyldgaard, S. Ovesson, C. Ruberto, E. Schroder and G. Wahbstrom, Bridging the gap between micro- and macroscales of materials by mesoscopic models, *Compt. Mater. Sci.*, **24**, 1 (2002).
62. M. Giesen, *Step and island dynamics at solid/vacuum and solid/liquid interfaces, Prog. Surf. Sci.*, **68**, 1−153 (2001).
63. T.J. Walls, *Kinetic Monte Carlo Simulations of Perovskite Crystal Growth with Long Range Coulomb Interactions*, Thesis for Bachelor of Science with Honors in Physics, College of William and Mary, VA, USA, 1999.
64. L. Mandreoli, *Density Based Kinetic Monte Carlo Methods*, PhD Thesis, Technischen Universitat, Berlin, Germany, 2005.
65. P. Jensen, "Simple models for nanocrystal growth", In *Handbook of Materials Modelling* (Ed. S. Yip), Springer, The Netherlands, 2005, pp. 1769–1785.
66. P. Jensen, Growth of nanostructures by cluster deposition: experiments and simple models, *Rev. Mod. Phys.*, **71**, 1695−1735 (1999).
67. P. Jensen and N. Combe, Understanding the growth of nanocluster films, *Compt. Matter. Sci.*, **24**, 78 (2002).
68. H. Mizuseki, K. Hongo, Y. Kawazoe and L.T. Wille, Multiscale simulation of cluster growth and deposition process by hybrid model based on direct simulation Monte Carlo method, *Compt. Mater. Sci.*, **24**, 88 (2002)
69. H.N.G. Wadley, X. Zhou, R.A. Johnson and M. Neurock, Mechanisms, models and methods of vapour deposition, *Prog. Mater. Sci.*, **46**, 329−377 (2001).
70. C.C. Battaile and D.J. Srolovitz, Kinetic Monte Carlo simulation of chemical vapor deposition, *Ann. Rev. Mater. Res.*, **32**, 297−319 (2002).
71. C.C. Battaile, D.J. Srolovitz and J.E. Butler, A kinetic Monte Carlo Method for the atomic-scale simulation of chemical vapour deposition: application to diamond, *J. Appl. Phys.*, **82**, 6293 (1997).
72. Q. Yu and S.K. Esche, A Monte Carlo algorithm for single phase normal grain growth with improved accuracy and efficiency, *Compt. Matter. Sci.*, **27**, 259 (2003); G.H. Gilmer and J.Q. Broughton, Computer modelling of mass transport along surfaces, *Ann. Rev. Mater. Sci.*, **16**, 487−516 (1986).
73. A. Touzik, H. Hermann and K. Wetzig, General-purpose distributed software for Monte Carlo simulations in materials design, *Compt. Matter. Sci.*, **28**, 134 (2003).
74. La Magna, S. Coffa, S. Libertino, M. Strobel and L. Colombo, Atomistic simulations and the requirements of process simulation for novel semiconductor devices, *Compt. Matter. Sci.*, **24**, 213 (2002).
75. V.I. Kolobov, Fokker-Planck modelling of electron kinetics in plasmas and semiconductors, *Compt. Matter. Sci.*, **28**, 302 (2003).
76. S. Piana, M. Reyhani and J.D. Gale, Simulating micrometre-scale crystal growth from solution, *Nat. Lett.*, **438**, 70 (2005).
77. G. Martin and F. Soisson, Kinetic Monte Carlo method to model diffusion controlled phase transformations in the solid state. In: *Handbook of Materials Modelling* (Ed. S. Yip), Springer, The Netherlands, 2005, pp. 2223–2248.
78. R. Weinkamer, P. Fratzl, H.S. Gupta, O. Penrose and J.L. Lebowitz, Using kinetic Monte Carlo simulations to study phase separation in alloys, *Phase Transit.*, **77**, 433 (2004).
79. A.B. Bortz, M.H. Kalos and J.L. Lebowitz, *J. Comp. Phys.*, **17**, 10 (1975).

80. D.T. Gippespie, *J. Phys. Chem.*, **81**, 2340 (1977).
81. A.F. Voter, *Phys. Rev. B*, **34**, 6819 (1986).
82. A. La Magna and S. Coffa, *Compt. Mater. Sci.*, **17**, 21 (2000).
83. G. Henkelman and H. Jonsson, *J. Chem. Phys.*, **115**, 9657 (2001).
84. S. Valkealahti and R.M. Nieminen, *Appl. Phys.*, **32**, 95 (1983) and **35**, 51 (1984).
85. S. Valkealahti, *Monte Carlo and Molecular Dynamics Simulations of Near-Surface Phenomena*, PhD Thesis, University of Jyvaskyla, Jyvaskyla, Finland, 1987.
86. K.L. Hunter, I.K. Snook and H.K. Wagenfeld, *Phys. Rev. B*, **54**, 4507 (1996).
87. K.L. Hunter, *Secondary Electron Emission from Solid Surfaces*, PhD Thesis, RMIT University, Melbourne, Victoria, Australia, 1991.
88. R. Shimizu and D. Ze-Jun, Monte Carlo modelling of electron-solid interactions, *Rep. Prog. Phys*, 487–531 (1992).
89. See for example, J.F. Ziegler, J.P.Biersack and U. Littmark, *The Stopping and Range of Ions in Solids*, Pergamon Press, New York, 1985 which was updated in 1996 and R.E. Johnson and B.U.R. Sundqvist, Electronic sputtering: from atomic physics to continuum mechanics, Phys. Today, **1**, 28–36 (1992).

– 10 –

An Overview

In this book I have tried to give a general, overall treatment of the physics of many-body systems of interacting particles by means of generalised Langevin equations (GLEs). The starting points used were the fundamental dynamical equations for the atoms in the system such as Newton's laws and the generalised SLODD equations of non-equilibrium statistical mechanics. Although an outline was given of the background to these basic dynamical equations no attempt was made to justify them or to discuss their applications in any detail.

Starting from these dynamical equations for atomic motion, we developed a variety of coarse-grained dynamical equations by averaging over certain degrees of freedom by the use of projection operators. This enabled us to obtain equations of motion in which the variables we are not interested may be treated in some average way. For example we might choose to eliminate the details of the dynamics of "fast variables" and to study only the "slow" ones in detail. This led to several fundamental equations of motion of the GLE type.

1. Equations for a single dynamical variable or a set of such variables for representative particles for a system in equilibrium (Chapter 2).
2. Equations for a single dynamical variable or a set of such variables for representative particles for a system in a non-equilibrium state generated by the generalised SLLOD equations of motion. The particular case of the non-equilibrium steady state was looked at in detail (Chapter 4).
3. Equations for a group of particles selected from the other particles in the system (Chapter 5).

The limiting case of this later type of equation, when the mass of the selected particles was much larger than that of the other particles ("The Brownian Limit"), was considered in detail. This led to the Langevin equation for any arbitrary function of the dynamical variables of the large particles and the Fokker-Planck equation for the N-body distribution function for these particles (Chapter 5).

From these equations of motion numerical algorithms ("Brownian Dynamics" methods) were developed in a variety of cases which, when combined with suitable boundary condition, enable the properties of many-body systems to be simulated (Chapters 6 and 7).

When the algorithms of this type arising from the Langevin Equation are used in their most general form indirect, solvent mediated hydrodynamic interactions occur between the particles. These are complicated many-body interactions of very long range which play a key role in determining the properties of all but very dilute systems. Therefore, we spent some time outlining methods to calculate these interactions. Unfortunately, the direct calculation of these hydrodynamic effects from the low Reynolds number ("Creeping Flow") equations of hydrodynamics is very complicated and requires large amounts of computer time. Thus, new methods have been sought in order to overcome this problem, e.g. Smoothed Particle Hydrodynamics, Lattice Boltzmann methods and Dissipative Particle Dynamics (Chapter 7), all of which show promise in making calculations much faster whilst attempting not to lose accuracy. However, further tests of these methods against accurate BD calculations using traditional methods to calculate the hydrodynamic interactions would be useful to help judge the accuracy of these methods and allow for their refinement.

It should be remarked that general forms of the Langevin equation for any function of the dynamical variables characterising the state of the large particles, i.e. eqs. (5.58) and (5.66), are exact. Thus, we could use one of these equations which would not involve making the assumptions and taking the limits involved in arriving at the many-body Langevin equation (5.68). Even the assumption that the bath particles are at equilibrium could be eliminated by use of the Kawasaki formalism as in Chapter 4 and redefining the projection operator used, i.e. eq. (5.51). The main limitation of such equations is that in arriving at them by coarse-graining we have lost information. Thus, if we wish to use these equations without adding extra information not contained in them we are forced by necessity to treat some terms appearing in them by stochastic methods, i.e. by the use of random variables.

In Section 5.5, we discussed the Brownian limit and pointed out that a basic requirement for this to be valid is that all dynamical processes should have ceased in the bath system in the presence of the Brownian particle before it has changed its dynamical state significantly. This does not, however, imply that the dynamical processes involving the velocity of the Brownian particle have ceased before the Brownian particle has significantly changed its spatial position. When this extra assumption is reasonable we may take the Diffusive (or "Overdamped") Limit which is often a very good assumption for colloidal particles and units of polymers. If we start with the Fokker–Planck equation and take the Diffusive Limit then this results in an equation called the Smoluchowski or many-body diffusion equation. This is much easier to use than the Langevin or FP equations and, thus, is often used in practice, especially to derive BD algorithms. Such methods, their limitations and typical results obtained are discussed in Chapters 6, 7 and 8. An interesting application of the N-body Diffusion or Smoluchowski equation is to develop Diffusion Monte Carlo methods to solve the Schrodinger equation by numerical methods as shown in Section 9.3.

As stated above, one of the main purposes of deriving GLEs was to be able to deal with some systems whose dynamics involves processes acting on different time scales. However, there are other stochastic methods which have been developed for this purpose which are particularly useful to treat processes that are "rare events". These methods are usually based on postulating a kinetic Master Equation for a particular overall phenomenon which describes the system's evolution from one state to the next in terms of probabilities of occurrence of the elementary processes that determine the phenomenon. Often these Master Equations are solved by means of a Kinetic Monte Carlo procedure. Chapter 9 discusses this approach.

The LE or equivalently FP equation may be shown to only approximate the dynamics described by the general ME, and it is also sometimes stated that LEs (or the F–P equations) are be derived by adding random terms to equations of motion. This seems to imply that all LE or FP descriptions have a somewhat arbitrary basis in physics and are hard to justify and improve. However, as shown in Chapter 5, the Langevin and Fokker–Planck equations developed here are based on well-defined approximations to exact equations of motion or rather to equations of motion which exactly describe some physical phenomena. Thus, these examples are not arbitrary and they may be systematically improved by eliminating some of the approximations used or by making better approximations.

As discussed in Chapter 9, many ME-based methods and KMC methods have led significant advances in understanding of an enormously wide range of important phenomena, and the KMC method is now a very widely used technique. However, several points should be emphasised about the use of master equations.

1. Unless the physics upon which they are based is well defined then one runs the risk of not really providing any physical understanding or insight into the phenomena being modelled.
2. The general Master Equations describes a general Markov process, but not all physical processes may be described as a Markov process.
3. The Langevin and FP equations are, in general, approximations to the Master Equation. This does not mean that all Master Equations give a more accurate description of all physical processes than that given by the Langevin and FP equations.

Finally, it is to be hoped that this book may prove useful to some in their endeavours to understand the properties of many-body systems of particles.

Appendix A

Expressions for Equilibrium Properties, Transport Coefficients and Scattering Functions

A.1 EQUILIBRIUM PROPERTIES

The numerical evaluation of any property B of a system in equilibrium may be obtained from time averages ($\langle\langle B(\Gamma)\rangle\rangle$) over a trajectory using eq. (1.2). In practice this means approximating the integral over an infinite time interval given by eq. (1.2) by a finite sum of values of B evaluated at N_T discrete, time-ordered values of $\Gamma_i\,(t_i)$, i.e.

$$\langle B\rangle = \lim_{T\to\infty} 1/T \int_{t_0}^{t_0+T} B(t)\,\mathrm{d}t \approx 1/T\sum_{i=1}^{N_T} B(\Gamma_i(t_i)) \tag{A.1}$$

Alternatively one may use the ensemble average expression (1.3) for $\langle B\rangle$ and approximate this integral by a finite sum of N_c values of Γ_i, i.e.

$$\begin{aligned}\langle B\rangle &= \int B(\Gamma)\,f^{\mathrm{eq}}(\Gamma)\,\mathrm{d}\Gamma \\ &\approx \sum_{i=1}^{N_c} B(\Gamma_i)\,f^{\mathrm{eq}}(\Gamma_i)\end{aligned} \tag{A.2}$$

A list of the equilibrium distribution function f^{eq} for the commonly used ensembles is given in Table A.1.

There are many theoretical discussions of the derivation of the above expressions from various hypotheses which involve different levels of rigor and sophistication.[1–5] These derivations will not concern us here except to say that an extremely simple one is to assume that[5]

Table A.1

Characteristic equilibrium probabilities $f^{eq} = e^{-\beta E}/\int e^{-\beta E}$

Variables held constant	Ensemble	Characteristic energy (E)	States integrated over	β
(N,V,U)	Micocanonical	U	All states of total energy U	U
(N,V,T)	Canonical	$E(N,V)$	All states with constant N and V	$k_B T$
(N,p,T)	Constant pressure	$E(N,V) + pV$	All states with constant N	$k_B T$
(μ,V,T)	Grand canonical	$E(N,V) - N\mu$	All states with constant V and N and all values of $N \geq 0$	$k_B T$

Here μ is chemical potential of the system and k_B is the Boltzmann constant.

1. The probability only depends on the Characteristic Energy, E, of the states of the ensemble;
2. Each of these energies E is made up of contributions from different energy types which are independent.

A derivation of eq. (1.3) is then very straightforward for the ensembles covered in Table A.1 and the method of derivation works equally as well in Quantum Mechanics and Classical Mechanics provided definitions of the quantities involved are carefully made. For example one has to use a sum over discrete states and an integral over continuous states and define what is meant by the state of the system.

However, the derivation of the results for the less often used so-called Adiabtic Ensembles needs a slightly different approach.[6,7]

The usual method of carrying out this approximate procedure (eq. (A.2)) on a computer is to use a Monte Carlo (or random sampling) processes as is discussed in Appendix K.

A.2 EXPRESSIONS FOR LINEAR TRANSPORT COEFFICIENTS

The linear transport coefficients may be evaluated either as integrals over time correlation functions or in terms of the Einstein expressions. Some of these relationships are shown in Table A.2 below. The correlation function formulae may be derived in a variety of ways.[8–13] An excellent derivation of these latter relationships is given by *E.* Helfand,[12] however, it should be noted that the use of these Einstein expressions in periodic boundary conditions needs some care.[14]

Table A.2

Time correlation function relations for transport coefficients

Transport coefficient	Time correlation function relationship	Einstein relationship (Ref. 12)
Self-diffusion coefficient (D_s)	$D_s = 1/3\int_0^\infty \langle \underline{v}_i(0)\cdot \underline{v}_i(s)\rangle \mathrm{d}s$	$D_s = \lim_{\tau\to\infty}(1/6t)\langle(\underline{r}_i(0)-\underline{r}_i(0))^2\rangle$
Coefficient of viscosity (η)	$\eta = 1/(Vk_\mathrm{B}T)\int_0^\infty \langle P_{yx}(0)P_{yx}(s)\rangle \mathrm{d}s$	$\eta = 1/(2Vk_\mathrm{B}Tt)\left\langle\left[\sum_i\{x_i(0)p_{yi}(0)-x_i(t)p_{yi}(t)\}\right]^2\right\rangle$
Coefficient of bulk viscosity (ψ)	$\psi = (1/Vk_\mathrm{B}T)\int_0^\infty \langle P_{xx}(0)P_{xx}(s)\rangle \mathrm{d}s$	$\psi = (1/2Vk_\mathrm{B}Tt)\left\langle\left[\sum_i\{x_i(0)p_{xi}(0)-x_i(t)p_{xi}(t)\}\right]^2\right\rangle$
Coefficient of thermal conductivity (κ)	$\kappa = (1/Vk_\mathrm{B}T^2)\int_0^\infty \langle J_{xx}^{\kappa}(0)J_{xx}^{\kappa}(\mathrm{s})\rangle \mathrm{d}s$	$\kappa = (1/Vk_\mathrm{B}T^2t)\left\langle\left[\sum_i\{x_i(0)\bar{E}_i(0)-x_i(\mathrm{t})\bar{E}_i(\mathrm{t})\}\right]^2\right\rangle$

Here $\overline{E}_i(t) = E_i - \langle E_i\rangle$, $E_i = \left[(p_i^2/m_i) + \frac{1}{2}\sum_{i=1}^{N} V_{ij}\right]$ and $\mathbf{J}_{ij}^{\kappa} = \mathrm{d}/\mathrm{d}t\sum_{i=1}^{N}\underline{r}_i\overline{E}_i$.

A.3 SCATTERING FUNCTIONS

A.3.1 Static structure[15,16]

The time-averaged intensity of scattered radiation, $\langle I(\underline{q})\rangle$ in the first Born approximation (i.e. no multiple scattering) is described by

$$\langle I(q)\rangle = \sum_{i=1}^{N}\sum_{j=1}^{N}\langle b_i(q)b_j(q)\exp[i\underline{q}\cdot(\underline{r}_i - \underline{r}_j)]\rangle \tag{A.3}$$

where $b_i(q)$ is the scattering amplitude for the ith particle and $\underline{q}$ is the scattering vector, $|q| = 4\pi/\lambda \sin(\theta/2)$ for radiation of wavelength λ in the medium, θ is the scattering angle.

If the particles are all identical, i.e. we have a single component or mono-dispersed particulate system then $b_i(q) = b(q)$ and

$$\langle I(\underline{q})\rangle = N\,P(\underline{q})S(\underline{q}) \tag{A.4}$$

where

$$\begin{aligned} S(\underline{q}) &= \sum_{i=1}^{N}\sum_{j=1}^{N}\langle\exp[i\underline{q}\cdot(\underline{r}_i - \underline{r}_j)]\rangle \\ &= \int g(\underline{R})\exp(-i\underline{q}\cdot\underline{r})\,\mathrm{d}\underline{r} \end{aligned} \tag{A.5}$$

where

$$g(\underline{r}) = \int f_{\mathrm{eq}}^{(N)}(\underline{r}^{(N)})\,\mathrm{d}\underline{r}^{(N-2)} \quad \text{is the radial distribution function} \tag{A.6}$$

and the single particle form factor $P(q)$ is given by $P(q) = [b(q)/b(0)]$.[16]

In this case the static structure factor $S(\underline{q})$ is the Fourier transform of the radial distribution function $g(\underline{r})$.

A.3.2 Dynamic scattering[15–20]

The theoretically interesting function here is the field autocorrelation function $g^{(1)}(\underline{q},t)$, where $E(\underline{q},t)$ is the scattered field measured at the scattering vector $\underline{q}$ and time t,

$$\begin{aligned} g^{(1)}(\underline{q},t) &= \langle\underline{E}^{*}(\underline{q},0)\cdot\underline{E}(\underline{q},t)\rangle/\langle\underline{E}^{*}(\underline{q},0)\cdot\underline{E}(\underline{q},0)\rangle \\ &= F^{M}(q,t)/I(q) \end{aligned} \tag{A.7}$$

where

$$F^{M}(q,t) = [N\langle b^2(q)\rangle]^{-1} = \sum_{i=1}^{N}\sum_{j=1}^{N}\langle b_i(q)b_j(q)\exp[i\underline{q}\cdot(\underline{r}_i(0) - \underline{r}_j(t))]\rangle \tag{A.8}$$

where $\langle(b^2(q)\rangle = 1/N \sum_{i=1}^{N} b_i^2(q)$

For a single component or mono-dispersed system of particles we may directly relate this to the van Hove space–time correlation function $G(\underline{r},t)$ which gives a description of the time-dependent structure of the system, [15–20]

$$g^{(1)}(\underline{q},t) = (1/N)\sum_{i,j}^{N}\langle \exp[\mathrm{i}\underline{q}\cdot(\underline{r}_i(0)-\underline{r}_j(\mathrm{t}))]\rangle/S(\underline{q}) \quad \text{(A.9)}$$

$$= F(\underline{q},t)/S(\underline{q}) \quad \text{(A.10)}$$

$$= \int G(\underline{r},t)\exp(-\mathrm{i}\underline{q}\cdot\underline{r})\mathrm{d}\underline{r}/S(q) \quad \text{(A.11)}$$

This may be expressed as the sum of self and distinct components as follows:

$$F(\underline{q},t) = F_s(\underline{q},t) + F_d(\underline{q},t) \quad \text{(A.12)]}$$

and

$$G(\underline{r},t) = G_s(\underline{r},t) + G_d(\underline{r},t)$$

where

$$F_s(\underline{q},t) = \int G_s(\underline{r},t)\exp(-\mathrm{i}\underline{q}\cdot\underline{r})\mathrm{d}\underline{r}$$

and

$$F_d(\underline{q},t) = \int G_d(\underline{r},t)\exp(-\mathrm{i}\underline{q}\cdot\underline{r})\mathrm{d}\underline{r}$$

Computational convenient formula are[16,18]

$$G_s(\underline{r},t) = \langle N[\underline{r}_i(t) - \underline{r}_i(0) - \underline{r}]\rangle/(4\pi r^2\,\mathrm{d}r) \quad \text{(A.13)}$$

and

$$G_d(\underline{R},t) = \langle N[\underline{r}_i(t) - \underline{r}_j(0) - \underline{r}]\rangle_{i\neq j}/(4\pi r^2\,\mathrm{d}r)$$

where $\langle N[\underline{r}_i(t) - \underline{r}_j(0) - \underline{r}]\rangle$ is the average number of pairs of particles whose separation $[\underline{r}_i(t) - \underline{r}_j(0)]$ in the time interval t is equal to $\underline{r}$.

Spin echo neutron scattering[20] (SENS) measures $g^{(1)}(\underline{q},t)$ directly but in dynamic light scattering (DLS) one usually measures the intensity autocorrelation function $g^{(2)}(\underline{q},t)$ which is related to $g^{(1)}(\underline{q},t)$ (assuming Gaussian statistics) *via* the Siegert relationship,[15]

$$\begin{aligned} g^{(2)}(\underline{q},t) &= \langle I(\underline{q},0)I(\underline{q},t)\rangle/\langle I(\underline{q},0)I(\underline{q},0)\rangle \\ &= c^{1/2}\{1 + g^{(1)}(\underline{q},t)\}^{1/2} \end{aligned} \quad \text{(A.14)}$$

where c is a constant dependent on the experimental set-up.

Often it is assumed that the system is isotropic or that the experiment obtains the spherical average, then $\underline{r}$ is replaced by r and $\underline{q}$ by q.

A useful analysis of the self-intermediate scattering function, which can be measured in tracer systems, may be made as follows which allows the determination of both $\langle \Delta \underline{r}^2(t) \rangle$ and of the non-Gaussian measures $\alpha_n(t)$, [16,18]

$$\begin{aligned} F_s(q,t) &= 1/N \sum_{j=1}^{N} \langle \exp([i\underline{q}\cdot\Delta\underline{r}_j(t)]\rangle \\ &= \exp[-q^2 w(t)]\{1+\alpha_2(t)[q^2 w(t)]^2/2! + (\alpha_3(t)-\alpha_2(t))[q^2 w(\mathrm{t})]^3/3! + .\} \end{aligned} \tag{A.15}$$

where

$$w(t) = 1/6\langle \Delta \underline{r}^2(t)\rangle$$
$$\alpha_n(t) = \langle \Delta \underline{r}^{2n}(t)\rangle / (c_n \langle \Delta \underline{r}^2(t)\rangle^n) - 1$$

and

$$c_n = 1\times 3\times 5\times\cdots\times(2n+1)/3^n$$

Often rather than measuring $F(q,t)$ and $F_s(q,t)$ their Fourier transform, i.e. $S(q,\omega)$ and $S_s(q,\omega)$ are measured.[16–19]

REFERENCES

1. R.C. Tolman, *The Principles of Statistical Mechanics*, Dover, New York, 1979.
2. T.L. Hill, *Statistical Thermodynamics*, Dover, New York, 1986.
3. T.L. Hill, *Statistical Mechanics*, Dover, New York, 1987.
4. G. Wannier, *Statistical Physics*, Dover, New York, 1987.
5. T.M. Reed and K.E. Gubins, *Applied Statistical Mechanics*, McGraw-Hill, New York, 1973.
6. H.W. Graben and J.R. Ray, *Phys. Rev. A*, **43**, 4100 (1991).
7. J.R. Ray and R.J. Wolf, Springer Proceedings in Physics, Vol. 76, *Computer Simulation Studies in Condensed-Matter Physics VI* (Eds. D.P. Landau, K.K. Mon and H.-B. Scuttler), Springer-Verlag, Berlin, 1993, p. 183.
8. M.S. Green, *J. Chem. Phys.*, **20**, 1281 (1952); **22**, 398 (1954).
9. R. Kubo, M Yokota, and S Nakagima, *J. Phys. Soc. Jpn.*, **12**, 1203 (1957).
10. H. Mori, *Phys. Rev.*, **112**, 1829 (1958).
11. J.A. McLennan, *Phys. Rev. A*, **115**, 1405 (1959).
12. E. Helfand, *Phys. Rev.*, **119**, 1 (1960).
13. R.M. Mazo, *Statistical Mechanical Theories of Transport Processes*, Pergamon Press, Oxford, 1968.
14. J.J. Erpenbeck, *Phys. Rev. E*, **51**, 4296 (1995).
15. P.N. Pusey, Colloidal suspensions, Course 10. In: *Liquids, Freezing and the Glass Transition*, (Eds. J.P. Hansen, D. Levesque and J. Zinn-Justin), Elsevier, B.V., Amsterdam, 1991, pp. 764–942.
16. S.-H. Chen, Structure of Liquids, In: *Physical Chemistry, An Advanced Treatise, Vol. VIIIB, Liquid State* (Ed. D. Henderson), Academic Press, New York, 1971, pp. 108–133.

17. B.J. Berne, Time-Dependent Properties of Condensed Media, In: *Physical Chemistry, An Advanced Treatise, Vol. VIIIB, Liquid State* (Ed. D. Henderson), Academic Press, New York, 1971, pp. 539–715.
18. B.J. Berne and G.D. Harp, On the Calculation of Time Correlation Functions, In: *Advances in Chemical Physics,* Vol. XVII (Eds. I. Prigogine and S.A. Rice), Wiley, New York, 1970, pp. 63–227.
19. J.-P. Hansen, Correlation Functions and Their Relationship with Experiment, In: *Microscopic Structure and Dynamics of Liquids* (Eds. J. Dupuy and A.J. Dianoux), Plenum, New York, 1977, pp. 3–68.
20. J.B. Hayter, In: *Scattering Techniques Applied to Supramolecular and Non-Equilibrium Systems* (Eds. S.-H. Chen and R. Nossal), Plenum Press, New York, 1981, p 49.

Appendix B

Some Basic Results About Operators

$$
\begin{aligned}
\mathbf{iL}\,\underline{v}_i(t) &= \left\{\sum_{j=1}^{N}\left[\underline{v}_j \cdot \partial/\partial\underline{r}_j + \underline{F}_j/m_j \cdot \partial/\partial\underline{v}_j\right]\right\}\underline{v}_i(t) \\
&= 0 + \underline{F}_i(t)/m_i \qquad (\partial\underline{v}_i(t)/\partial\underline{v}_j = \delta_{ij} \quad \text{and} \quad \partial\underline{v}_i(t)/\partial\underline{r}_j = 0) \\
&= \underline{a}_i(t) \\
&= \{\mathrm{d}/\mathrm{d}t\}\,\underline{v}_i(t) \qquad \text{(Newton's Second Law)}
\end{aligned} \tag{B.1}
$$

So in operator language,

$$\underline{a}_i(t) = \{\mathrm{d}/\mathrm{d}t\}\,\underline{v}_i(t) = \mathbf{iL}\,\underline{v}_i(t) \tag{B.2}$$

Integrating,

$$
\begin{aligned}
\underline{v}_i(t) &= \mathrm{e}^{\mathbf{iL}t}\,\underline{v}_i(0) \\
&= \mathbf{G}(t)\,\underline{v}_i(0) \quad \text{where } \mathbf{G}(t) \text{ is the propagator } \mathrm{e}^{\mathbf{iL}t}
\end{aligned} \tag{B.3}
$$

Similarly,

$$
\begin{aligned}
\mathbf{iL}\,\underline{r}_i(t) &= \left\{\sum_{j=1}^{N}\left[\underline{v}_j \cdot \partial/\partial\underline{r}_j + \underline{F}_j/m_j \cdot \partial/\partial\underline{v}_j\right]\right\}\underline{r}_i(t) \\
&= \underline{v}_i(t) + 0 \qquad (\partial\underline{r}_i(t)/\partial\underline{r}_j = \boldsymbol{\delta}_{ij} \quad \text{and} \quad \partial\underline{r}_i(t)/\partial\underline{v}_j = 0) \\
&= \{\mathrm{d}/\mathrm{d}t\}\,\underline{r}_i(t)
\end{aligned} \tag{B.4}
$$

So in operator language,

$$\underline{v}_i(t) = \{\mathrm{d}/\mathrm{d}t\}\,\underline{r}_i(t) = \mathbf{iL}\,\underline{r}_i(t) \tag{B.5}$$

$$
\begin{aligned}
\underline{r}_i(t) &= \mathrm{e}^{\mathbf{iL}t}\,\underline{r}_i(0) \\
&= \mathbf{G}(t)\,\underline{r}_i(0)
\end{aligned} \tag{B.6}
$$

Similarly for the normalised velocity we derive from eq. (B.2),

$$\begin{aligned}\{\mathbf{iL}\underline{v}_i(t)\}/\langle\underline{v}_i\underline{v}_i\rangle^{1/2} &= \mathbf{iL}\underline{u}_i(t)\\ &= \{\mathrm{d}/\mathrm{d}t\}\{\underline{v}_i(t)/\langle\underline{v}_i\underline{v}_i\rangle^{1/2}\}\\ &= \{\mathrm{d}/\mathrm{d}t\}\underline{u}_i(t)\\ &= \underline{b}_i(t)\end{aligned} \tag{B.7}$$

In the above we have used,

$$\begin{aligned}&\{\mathrm{d}/\mathrm{d}t\}\underline{v}_i(t) = \mathbf{iL}\,\underline{v}_i(t)\\ &[\underline{v}_i(t)]^{-1}\{\mathrm{d}\underline{v}_i(t)/\mathrm{d}t\} = \mathbf{iL}\\ &[\underline{v}_i(t)]^{-1}\mathrm{d}\underline{v}_i(t) = \mathbf{iL}\,\mathrm{d}t\\ &\mathrm{d}\ln[\underline{v}_i(t)] = \mathbf{iL}\,\mathrm{d}t\\ &\int \mathrm{d}\ln[\underline{v}_i(t)] = \mathbf{iL}\int \mathrm{d}t + C\\ &\ln[\underline{v}_i(t)] = \mathbf{iL}\,t + \ln[\underline{v}_i(0)] \qquad (C = \ln[\underline{v}_i(0)] \quad \text{since at} \quad t=0 \quad \underline{v}_i(t) = \underline{v}_i(0))\\ &\quad \text{i.e. } \underline{v}_i(t)/\underline{v}_i(0) = \exp(\mathbf{iL}\,t)\end{aligned}$$

$$\text{or } \underline{v}_i(t) = \exp(\mathbf{iL}\,t)\,\underline{v}_i(0) \tag{B.8}$$

and similarly for $\underline{r}_i(t)$ or any function of t, i.e. $A(t)$ so,

$$A(t) = \exp(\mathbf{iL}\,t)A(0) = \mathbf{G}(t)\,A(0) \tag{B.9}$$

e.g.,

$$\underline{u}_i(t) = \exp(\mathbf{iL}\,t)\underline{u}_i(0) = \mathbf{G}(t)\underline{u}_i(0) \tag{B.10}$$

By definition,

$$G(t) = \mathrm{e}^{\mathbf{iL}t} = 1 + \sum_{k=1}^{\infty}(1/k!)(\mathbf{iL}t)^k$$

so

$$
\begin{aligned}
\mathbf{iL}\,G(t) &= \mathbf{iL}\,\mathrm{e}^{\mathbf{iL}t} \\
&= \mathbf{iL}\left\{1+\sum_{k=1}^{\infty}(1/k!)(\mathbf{iL}\,t)^{k}\right\} \\
&= \left\{\mathbf{iL}+\sum_{k=1}^{\infty}(1/k!)(\mathbf{iL}t)^{k+1}\right\} \\
&= \left\{1+\sum_{k=1}^{\infty}(1/k!)(\mathbf{iL}\,t)^{k}\right\}\mathbf{iL} \\
&= \mathrm{e}^{\mathbf{iL}t}\,\mathbf{iL} \\
&= G(t)\,\mathbf{iL}
\end{aligned}
\tag{B.11}
$$

The *Dyson Decomposition* for the Propagator, assuming **L** is time independent, then,

$$\mathbf{G}(t) = \mathrm{e}^{\mathbf{iL}t}$$

So the resolvent,

$$\tilde{\mathbf{G}}(s) = (s-\mathbf{iL})^{-1} \qquad \text{(Laplace transform)} \tag{B.12}$$

$$= [s-\mathrm{i}(\mathbf{Q}+\mathbf{P})\mathbf{L}]^{-1} \qquad (\mathbf{P}+\mathbf{Q}=\mathbf{1}) \tag{B.13}$$

Then define,

$$\mathbf{A} = s-\mathbf{iL} = s-\mathrm{i}(\mathbf{Q}+\mathbf{P})\mathbf{L}$$

and $\mathbf{B} = s-\mathbf{iQL}$

So $\mathbf{B}-\mathbf{A} = \mathrm{i}\mathbf{PL}$

Using the operator identity,

$$\mathbf{A}^{-1} - \mathbf{B}^{-1} = \mathbf{A}^{-1}(\mathbf{B} - \mathbf{A})\mathbf{B}^{-1} = (\mathbf{A}^{-1} - \mathbf{B}^{-1}) \tag{B.14}$$

$$\begin{aligned} \tilde{\mathbf{G}}(s) &= \mathbf{A}^{-1} \\ &= \mathbf{B}^{-1} + (\mathbf{A}^{-1} - \mathbf{B}^{-1}) \\ &= (s - \mathrm{i}\mathbf{QL})^{-1} + (s - \mathrm{i}\mathbf{L})^{-1}\,\mathrm{i}\mathbf{PL}(s - \mathrm{i}\mathbf{QL})^{-1} \end{aligned} \tag{B.15}$$

Taking the inverse Laplace transform,

$$\mathbf{G}(t) = \mathrm{e}^{\mathrm{i}\mathbf{L}t} = \mathrm{e}^{\mathrm{i}\mathbf{QL}t} + \int_0^t \mathrm{e}^{\mathrm{iL(t-\tau)}}\,\mathrm{i}\mathbf{PL}\,\mathrm{e}^{\mathrm{iQL}\tau}\,\mathrm{d}\tau \tag{B.16}$$

, this is the Dyson decomposition of $\mathbf{G}(t)$

Appendix C

Proofs Required for the GLE for a Selected Particle

1.

$$
\begin{aligned}
&(\mathrm{d}/\mathrm{d}t)\underline{u}_i(t) \\
&= \mathrm{i}\mathbf{L}\underline{u}_i(t) && \text{(from (B.7))} \\
&= \mathrm{i}\mathbf{L}\mathbf{G}(t)\underline{u}_i(0) \\
&= \mathrm{i}\mathbf{L}\,\mathrm{e}^{\mathrm{i}\mathbf{L}t}\underline{u}_i(0) && \text{(from (B.10))} \\
&= \mathrm{e}^{\mathrm{i}\mathbf{L}t}\,\mathrm{i}\mathbf{L}\underline{u}_i(0) && \text{(from (B.11))} \\
&= \mathrm{e}^{\mathrm{i}\mathbf{L}t}(\mathbf{P}+\mathbf{Q})\mathrm{i}\mathbf{L}\underline{u}_i(0) && (\mathbf{P}+\mathbf{Q}=\mathbf{1}) \\
&= \mathrm{e}^{\mathrm{i}\mathbf{L}t}\,\mathbf{P}(\mathrm{i}\mathbf{L}\underline{u}_i(0)) + \mathrm{e}^{\mathrm{i}\mathbf{L}t}(\mathbf{Q}\,\mathrm{i}\mathbf{L}\underline{u}_i(0)) \\
&= \mathrm{e}^{\mathrm{i}\mathbf{L}t}\,\mathbf{P}((\mathrm{d}/\mathrm{d}t)\underline{u}_i(0)) + \mathrm{e}^{\mathrm{i}\mathbf{L}t}(\mathbf{Q}\,\mathrm{i}\mathbf{L}\underline{u}_i(0)) && \text{(from (B.7))} \\
&= \mathrm{e}^{\mathrm{i}\mathbf{L}t}\,\underline{u}_i(0)\langle\underline{u}_i(0)(\mathrm{d}/\mathrm{d}t)\underline{u}_i(0)\rangle + \mathrm{e}^{\mathrm{i}\mathbf{L}t}(\mathbf{Q}\,\mathrm{i}\mathbf{L}\underline{u}_i(0)) && \text{(Definition of }\mathbf{P}) \\
&= 0 + \{\mathrm{e}^{\mathrm{i}\mathbf{L}t}\}(\mathbf{Q}\,\mathrm{i}\mathbf{L}\underline{u}_i(0)) && (\text{as}\langle\underline{u}_i(0)(\mathrm{d}/\mathrm{d}t)\underline{u}_i(0)\rangle = 0) \\
&= \left\{\mathrm{e}^{\mathrm{i}\mathbf{QL}t} + \int_0^t \mathrm{e}^{\mathrm{i}\mathbf{L}(t-\tau)}\,\mathrm{i}\mathbf{PL}\mathrm{e}^{\mathrm{i}\mathbf{QL}\tau}\,\mathrm{d}\tau\right\}(\mathbf{Q}\,\mathrm{i}\mathbf{L}\underline{u}_i(0)) && (\text{Using (A.16) for } \mathrm{e}^{\mathrm{i}\mathbf{L}t}) \\
&= \{\mathrm{e}^{\mathrm{i}\mathbf{QL}t}(\mathbf{Q}\,\mathrm{i}\mathbf{L}\underline{u}_i(0))\} + \int_0^t \mathrm{e}^{\mathrm{i}\mathbf{L}(t-\tau)}\,\mathrm{i}\mathbf{PL}\mathrm{e}^{\mathrm{i}\mathbf{QL}\tau}(\mathbf{Q}\,\mathrm{i}\mathbf{L}\underline{u}_i(0))\,\mathrm{d}\tau \\
&= \{\mathrm{e}^{\mathrm{i}\mathbf{QL}t}\,\mathbf{1}\mathrm{i}\mathbf{L}\underline{u}_i(0) - \mathrm{e}^{\mathrm{i}\mathbf{QL}t}\,\mathbf{P}\mathrm{i}\mathbf{L}\underline{u}_i(0)\} + \int_0^t \mathrm{e}^{\mathrm{i}\mathbf{L}(t-\tau)}\,\mathrm{i}\mathbf{PL}\mathrm{e}^{\mathrm{i}\mathbf{QL}\tau}(\mathbf{Q}\,\mathrm{i}\mathbf{L}\underline{u}_i(0))\,\mathrm{d}\tau && (\mathbf{P}+\mathbf{Q}=\mathbf{1}) \\
&= \{\mathrm{e}^{\mathrm{i}\mathbf{QL}t}\,\mathrm{i}\mathbf{L}\underline{u}_i(0) - \mathrm{e}^{\mathrm{i}\mathbf{QL}t}\,\underline{u}_i(0)\langle\underline{u}_i(0)(\mathrm{d}/\mathrm{d}t)\underline{u}_i(0)\rangle\} && \text{(Definition of }\mathbf{P}) \\
&\quad + \int_0^t \mathrm{e}^{\mathrm{i}\mathbf{L}(t-\tau)}\mathrm{i}\mathbf{PL}\mathrm{e}^{\mathrm{i}\mathbf{QL}\tau}(\mathbf{Q}\,\mathrm{i}\mathbf{L}\underline{u}_i(0))\mathrm{d}\tau \\
&= \mathrm{e}^{\mathrm{i}\mathbf{QL}t}\mathrm{i}\mathbf{L}\underline{u}_i(0) - 0 + \int_0^t \mathrm{e}^{\mathrm{i}\mathbf{L}(t-\tau)}\,\mathrm{i}\mathbf{PL}\mathrm{e}^{\mathrm{i}\mathbf{QL}\tau}(\mathbf{Q}\,\mathrm{i}\mathbf{L}\underline{u}_i(0))\,\mathrm{d}\tau && (\langle\underline{u}_i(0)\{\mathrm{d}/\mathrm{d}t\}\underline{u}_i(0)\rangle = 0)
\end{aligned}
$$

so,

$$
(\mathrm{d}/\mathrm{d}t)\underline{u}_i(t) = \mathrm{e}^{\mathrm{i}\mathbf{QL}t}\mathrm{i}\mathbf{L}\underline{u}_i(0) + \int_0^t \mathrm{e}^{\mathrm{i}\mathbf{L}(t-\tau)}\,\mathrm{i}\mathbf{PL}\mathrm{e}^{\mathrm{i}\mathbf{QL}\tau}(\mathbf{Q}\,\mathrm{i}\mathbf{L}\underline{u}_i(0))\,\mathrm{d}\tau \tag{C.1}
$$

2.
$$\int_0^t e^{iL(t-\tau)}\, \mathbf{iPL}e^{iQL\tau}(\mathbf{Q}\,iL\underline{u}_i(0))\,d\tau = \int_0^t e^{iL(t-\tau)}\underline{u}_i(0)\langle \underline{u}_i(0)\,\mathbf{L}e^{iQL\tau}\,\mathbf{1}\,iL\underline{u}_i(0)\rangle\,d\tau$$
$$-\int_0^t e^{iL(t-\tau)}\,\mathbf{iP}e^{iQL\tau}\,\underline{u}_i(0)\langle \underline{u}_i(0)\,iL\underline{u}_i(0)\rangle\,d\tau \qquad (\mathbf{P}+\mathbf{Q}=\mathbf{1})$$
$$=\int_0^t e^{iL(t-\tau)} i\underline{u}_i(0)\langle \underline{u}_i(0)\,\mathbf{L}e^{iQL\tau}\,\mathbf{1}iL\underline{u}_i(0)\rangle d\tau - 0\left(\langle \underline{u}_i(0) iL\underline{u}_i(0)\rangle = \langle \underline{u}_i(0)\underline{b}_i(0)\rangle = 0\right)$$
$$=\int_0^t (e^{iL(t-\tau)}\,\underline{u}_i(0))\langle i\underline{u}_i(0)\,\mathbf{L}e^{iQL\tau}\,iL\underline{u}_i(0)\rangle\,d\tau \qquad (1A = A)$$
$$=\int_0^t (\underline{u}_i(t-\tau))\langle i\underline{u}_i(0)\,\mathbf{L}e^{iQL\tau}\,iL\underline{u}_i(0)\rangle\,d\tau \qquad \left(e^{iL(t-\tau)}\,\underline{u}_i(0) = \underline{u}_i(t-\tau)\right)$$
$$=-\int_0^t \underline{u}_i(t-\tau)\langle iL\underline{u}_i(0)\,e^{iQL\tau}\,iL\underline{u}_i(0)\rangle\,d\tau \qquad \text{(Hermitian)}$$
$$=-\int_0^t K_u(\tau)\underline{u}_i(t-\tau)\,d\tau \qquad (C.2)$$

where the term,

$$K_u(\tau) = \langle iL\underline{u}_i(0)e^{iQL\tau}\,iL\underline{u}_i(0)\rangle \qquad (C.3)$$

is usually called the **Memory Function**.

3.
$$\langle \underline{b}_i^R(0)\cdot\underline{b}_i^R(0)\rangle = \langle e^{iQL\times 0}\,iL\underline{u}_i(0)\cdot e^{iQL\times 0}\,iL\underline{u}_i(0)\rangle \qquad (\text{Definition of } \underline{b}_i^R(0))$$
$$= \langle iL\underline{u}_i(0)\cdot iL\underline{u}_i(0)\rangle \qquad (\exp(0)=1)$$
$$= -\langle \mathbf{L}\underline{u}_i(0)\cdot \mathbf{L}\underline{u}_i(0)\rangle$$
$$= \langle \underline{b}_i(0)\cdot\underline{b}_i(0)\rangle \qquad (\text{from (B.7)})$$
$$= \langle \underline{a}_i(0)\cdot\underline{a}_i(0)\rangle/\langle \underline{v}_i(0)\underline{v}_i(0)\rangle$$
$$(C.4)$$

4.
$$\mathbf{Q}iL\underline{u}_i(t)$$
$$=(\mathbf{1}-\mathbf{P})iL\underline{u}_i(t)$$
$$=(\mathbf{1}-\mathbf{P})(iL\underline{u}_i(t))$$
$$=(\mathbf{1}-\mathbf{P})\left((d/dt)\{\underline{u}_i(t)\}\right) \qquad (iL\underline{u}_i(t) = (d/dt)\{\underline{u}_i(t)\})$$
$$=\mathbf{1}(d/dt)(\underline{u}_i(t)) - \mathbf{P}(d/dt)(\underline{u}_i(t)) \qquad (\mathbf{P}+\mathbf{Q}=\mathbf{1})$$
$$=(d/dt)(\underline{u}_i(t)) - \underline{u}_i(0)\langle \underline{u}_i(0)(d/dt)\underline{u}_i(t)\rangle \qquad (\text{Definition of } \mathbf{P})$$
$$=(d/dt)(\underline{u}_i(t)) + 0 \qquad (\text{Since} \langle \underline{u}_i(0)\underline{b}_i(t)\rangle = 0)$$
$$=(d/dt)(\underline{u}_i(t)) = iL\underline{u}_i(t) \qquad (C.5)$$

5. $$\begin{aligned}\langle \underline{b}_i^{\mathrm{R}}(0)\rangle &= \langle \mathrm{e}^{\mathbf{iQL}\times 0}\, \mathbf{iL}\underline{u}_i(0)\rangle \\ &= \langle \mathbf{iL}\underline{u}_i(0)\rangle \\ &= \langle (\mathrm{d}/\mathrm{d}t)(\underline{u}_i(0))\rangle \\ &= \langle (\mathrm{d}/\mathrm{d}t)(\underline{v}_i(0))\rangle / \langle \underline{v}_i(0)\underline{v}_i(0)\rangle^{1/2} \\ &= \langle \underline{F}_i(0)\rangle / \{m_i \langle \underline{v}_i(0)\underline{v}_i(0)\rangle^{1/2}\} \\ &= 0 \qquad \text{(Since } \langle \underline{F}_i(0)\rangle = 0\text{)} \end{aligned} \tag{C.6}$$

6. $$\begin{aligned}\langle \underline{u}_i(0)\cdot \underline{\mathrm{b}}_i^{\mathrm{R}}(t)\rangle &= \langle \underline{u}_i(0)\cdot \mathrm{e}^{\mathbf{iQL}t}(\mathbf{iL}\underline{u}_i(0))\rangle \\ &= \langle \underline{u}_i(0)\cdot \mathrm{e}^{\mathbf{iQL}t}\, \underline{b}_i(0)\rangle \\ &= \langle \underline{u}_i(0)\cdot \mathrm{e}^{\mathrm{i}(\mathbf{1-P})\mathbf{L}t}\, \underline{b}_i(0)\rangle \\ &= 0 \qquad \text{(by time reversal symmetry)} \end{aligned} \tag{C.7}$$

7. $$K_u(t) = \langle \mathbf{iL}\underline{u}_i(0)\mathrm{e}^{\mathbf{iQL}t}\, \mathbf{iL}\underline{u}_i(0)\rangle$$

and

$$\begin{aligned}&\langle \underline{b}_i^{\mathrm{R}}(0)\cdot \underline{b}_i^{\mathrm{R}}(t)\rangle \\ &= \langle \mathrm{e}^{\mathbf{iQL}0}\, \mathbf{iL}\underline{u}_i(0)\mathrm{e}^{\mathbf{iQL}t}\, \mathbf{iL}\underline{u}_i(0)\rangle \\ &= \langle \mathbf{iL}\underline{u}_i(0)\mathrm{e}^{\mathbf{iQL}t}\, \mathbf{iL}\underline{u}_i(0)\rangle \\ &= K_u(t)\end{aligned}$$

also $\mathbf{iL}\underline{u}_i(0) = \underline{b}_i(0)$ (from (B.7)

so,

$$\begin{aligned}K_u(t) &= \langle \underline{b}_i^{\mathrm{R}}(0)\cdot \underline{b}_i^{\mathrm{R}}(t)\rangle \\ &= \langle \mathbf{iL}\underline{u}_i(0)\mathrm{e}^{\mathbf{iQL}t}\, \mathbf{iL}\underline{u}_i(0)\rangle \\ &= \langle \underline{b}_i(0)\mathrm{e}^{\mathbf{iQL}t}\, \underline{b}_i(0)\rangle \end{aligned} \tag{C.8}$$

Appendix D

The Langevin Equation from the Mori–Zwanzig Approach

Now in the Brownian limit when $\lambda = (m/M_B)^{1/2} \to 0$, i.e. $M_B >> m$ it is reasonable to assume that (see Section 3.3)

$$K_v(\mathrm{t}) = \lambda_1\, \delta(t)$$

i.e.
$$\tilde{K}_v(s) = \lambda_1 = \zeta/M_B = \gamma \tag{D.1}$$

Thus, we use the simplest truncation of the Mori continued fraction then, writing the velocity of the chosen particle $\underline{v}_i$ as $\underline{V}_B$ and its mass as M_B,

$$\begin{aligned} M_B\,(\mathrm{d}/dt)\,\underline{V}_B(t) &= -M_B \int_0^t \{K_u(\tau)\}\,\underline{V}_B(t-\tau)\,\mathrm{d}\tau + M_B\, \mathrm{e}^{\mathrm{iQL}t}\, \mathrm{i}\mathbf{L}\underline{v}_B(0) \\ &= -M_B \int_0^t \{\gamma\delta(t-\tau)\}\,\underline{V}_B(t-\tau)\,\mathrm{d}\tau + \underline{F}_B^{\mathrm{R}}(t) \end{aligned}$$

$$M_B(\mathrm{d}/\mathrm{d}t)\,\underline{V}_B(t) = -\zeta\,\underline{V}_B(t) + \underline{F}_i^{\mathrm{R}}(t) \tag{D.2}$$

where,

$$\underline{F}_B^{\mathrm{R}}(t) = M_B\left[\mathrm{e}^{\mathrm{iQL}t}\,\mathrm{i}\mathbf{L}\underline{V}_B(0)\right] = M_B\langle \underline{V}_B \underline{V}_B\rangle^{1/2}\left[\mathrm{e}^{\mathrm{iQL}t}\,\mathrm{i}\mathbf{L}\underline{u}_B(0)\right] \tag{D.3}$$

$$\langle \underline{F}_B^{\mathrm{R}}(0)\rangle = 0 \tag{D.4}$$

$$\begin{aligned} \langle \underline{F}_B^{\mathrm{R}}(0)\cdot\underline{F}_B^{\mathrm{R}}(t)\rangle &= (3k_B T\, M_B)\, K_v(t) \\ &= (3k_B T M_B)\lambda_1\, \delta(t) \\ &= 3\gamma\, M_B\, k_B T\, \delta(t) \\ &= 3\zeta\, k_B T\, \delta(t) \end{aligned} \tag{D.5}$$

From eq. (D.5),

$$\gamma = \frac{\zeta}{M_{\mathrm{B}}} = \frac{1}{(3M_{\mathrm{B}}\,k_{\mathrm{B}}T)}\int_0^\infty \langle \underline{F}_{\mathrm{B}}^{\mathrm{R}}(0)\cdot\underline{F}_{\mathrm{B}}^{\mathrm{R}}(t)\rangle\,\mathrm{d}t$$

$$\begin{aligned}\underline{F}_{\mathrm{B}}^{\mathrm{R}}(t) &= M_{\mathrm{B}}\,\underline{a}_{\mathrm{B}}^{\mathrm{R}}(t)\\ &= M_{\mathrm{B}}\{\mathrm{e}^{\mathrm{iQL}t}\,\mathbf{iL}[\underline{v}_{\mathrm{B}}(0)]\}\\ &= M_{\mathrm{B}}\{\mathrm{e}^{\mathrm{iQL}t}\,\underline{a}_{\mathrm{B}}(0)\}\\ &= \mathrm{e}^{\mathrm{iQL}t}\,\underline{F}_{\mathrm{B}}(0)\end{aligned} \qquad \text{(D.6)}$$

So,

$$\gamma = \frac{\zeta}{M_{\mathrm{B}}} = \frac{1}{(3M_{\mathrm{B}}\,k_{\mathrm{B}}T)}\int_0^\infty \langle \mathrm{e}^{\mathrm{i}\mathbf{QL}_0}\,\underline{F}_{\mathrm{B}}(0)\cdot\mathrm{e}^{\mathrm{i}\mathbf{QL}t}\,\underline{F}_{\mathrm{B}}(0)\rangle\,\mathrm{d}t$$

$$= \frac{1}{(3M_{\mathrm{B}}\,k_{\mathrm{B}}T)}\int_0^\infty \langle \underline{F}_{\mathrm{B}}(0)\cdot\mathrm{e}^{\mathrm{i}\mathbf{QL}t}\,\underline{F}_{\mathrm{B}}(0)\rangle\,\mathrm{d}t \qquad \text{(D.7a)}$$

$$= \frac{1}{(3M_{\mathrm{B}}\,k_{\mathrm{B}}T)}\int_0^\infty f_{\mathrm{eq}}\,\underline{F}_{\mathrm{B}}(0)\cdot\mathrm{e}^{\mathrm{i}\mathbf{QL}t}\,\underline{F}_{\mathrm{B}}(0)\,\mathrm{d}t \qquad \text{(D.7b)}$$

This is still exact but if we make the following approximations where the terms below are defined in Section 5.1:

$$\begin{aligned}\mathrm{e}^{\mathrm{i}\mathbf{QL}} &= \mathrm{e}^{(\mathrm{i}\mathbf{QL}_0+\mathrm{i}\mathbf{QL}_B)}\\ &\approx \mathrm{e}^{\mathrm{i}\mathbf{L}_0}\end{aligned}$$

$$H = H_0 + H_{\mathrm{B}}$$

$$\begin{aligned}f_{\mathrm{eq}} &= \mathrm{e}^{-H(\Gamma)/k_{\mathrm{B}}T}\Big/\int \mathrm{e}^{-H(\Gamma)/k_{\mathrm{B}}T}\,\mathrm{d}\Gamma\\ &\approx \mathrm{e}^{-H0/k_{\mathrm{B}}T}\Big/\int \mathrm{e}^{-H0/k_{\mathrm{B}}T}\,\mathrm{d}\Gamma_0\\ &= f_{\mathrm{eq}}^{\ 0}\end{aligned}$$

Then eq. (D.7) becomes

$$\begin{aligned}\gamma_B &= \frac{1}{(3M_B\,k_BT)}\int_0^\infty f_{eq}{}^0\,\underline{F}_B(0)\cdot e^{i\mathbf{L}_0 t}\,\underline{F}_B(0)\,dt \\ &= \frac{1}{(3M_B\,k_BT)}\int_0^\infty \langle\underline{F}_B(0)\cdot\underline{F}_B(t)\rangle_0\,dt\end{aligned} \tag{D.8}$$

Appendix E

The Friction Coefficient and Friction Factor

If the friction coefficient γ and the friction factor ζ are defined in general by

$$D_s = k_B T/(m_i \gamma) = kT/\zeta \tag{E.1}$$

where D_s is the self-diffusion coefficient given by

$$D_s = 1/3 \int_0^\infty \langle \underline{v}_i(0) \cdot \underline{v}_i(t) \rangle \, dt \tag{E.2}$$

It should be noted that only if the Stokes–Einstein relation holds do we have $\zeta = 6\pi\eta a_i$ and $\gamma = 6\pi m_i \eta a_i$ where a_i is the particle radius and η is the coefficient of Newtonian viscosity of the background fluid.

So,

$$\begin{aligned} \gamma &= \zeta/m_i \\ &= (k_B T)/(m_i D_s) \\ &= (k_B T/m_i) \bigg/ \left\{ 1/3 \int_0^\infty \langle \underline{v}_i(0) \cdot \underline{v}_i(t) \rangle \, dt \right\} \end{aligned} \tag{E.3}$$

But from eq. (3.15) putting $n = 1$,

$$\tilde{C}(s) = C(0)/(s + \tilde{K}_v(s))$$

and putting $s = 0$,

$$\tilde{C}(0) = C(0)/\tilde{K}_v(0)$$

But

$$C(0) = \langle \underline{v}_i(0) \cdot \underline{v}_i(0) \rangle = (3k_B T/m_i)$$

then

$$\tilde{C}(0) = (3k_B T/m_i)[\tilde{K}_v(0)]^{-1}$$

However,

$$\begin{aligned}\tilde{C}(0) &= \lim_{s\to 0}\int_0^\infty e^{ist}\langle \underline{v}_i(0)\cdot\underline{v}_i(t)\rangle \mathrm{d}t\\ &= \int_0^\infty \langle \underline{v}_i(0)\cdot\underline{v}_i(t)\rangle \mathrm{d}t\\ &= 3D_s\\ &= 3\{k_B T/(m_i\gamma)\}\\ &= 3(k_B T/\zeta)\end{aligned}$$

So,

$$\zeta/m_i = \gamma = \tilde{K}_v(0) \tag{E.4}$$

So the friction coefficient is given by

$$\gamma = \tilde{K}_v(0)$$

and friction factor is given by

$$\zeta = m_i\,\tilde{K}_v(0)$$

Now,

$$\begin{aligned}\tilde{K}_v(0) &= \lim_{s\to 0}\int_0^\infty e^{-st}\langle \underline{b}_i^R(0)\cdot\underline{b}_i^R(t)\rangle \mathrm{d}t\\ &= \int_0^\infty \langle \underline{b}_i^R(0)\cdot\underline{b}_i^R(t)\rangle \mathrm{d}t\\ &= \int_0^\infty \langle \underline{a}_i^R(0)\cdot\underline{a}_i^R(t)\rangle \mathrm{d}t/\langle \underline{v}_i\underline{v}_i\rangle\\ &= \int_0^\infty \langle \underline{F}_i^R(0)\cdot\underline{F}_i^R(t)\rangle \mathrm{d}t/(\langle \underline{v}_i\underline{v}_i\rangle m_i^2)\\ &= 1/(3k_B T m_i)\int_0^\infty \langle \underline{F}_i^R(0)\cdot\underline{F}_i^R(t)\rangle \mathrm{d}t\end{aligned}$$

and so from eq. (E.4),

$$\gamma = \zeta/m_i = 1/(3m_i k_B T)\int_0^\infty \langle \underline{F}_i^R(0)\cdot\underline{F}_i^R(t)\rangle \mathrm{d}t \tag{E.5}$$

Appendix F

Mori Coefficients for a Two-Component System

The Mori coefficients may be used to construct approximate expressions for correlation functions (see Section 3.3) and upon which generalised Brownian dynamics algorithms may be based, and expressions for many of these coefficients may be found in references of Chapter 3. Often, simple theories use only the first Mori coefficient (see Section 5.5) for their use in studying the approach to the Brownian limit and (see Chapter 6) for their application in developing schemes to solve GLEs. Thus, we give here an analysis of this coefficient for the case of velocity.

F.1 BASICS

We will consider a set of N particles of which N_b are classified as bath particles of mass m_b and N_s are classified as suspended particles of mass m_s, interacting according to central pair-wise additive conservative forces,

$$\begin{aligned} \underline{F}_i(t) &= -\sum_{k\neq i}^{Nb} (\mathrm{d}/\mathrm{d}r_{ik})\phi_{ik}(r_{ik})(\underline{r}_{ik}/r_{ik}) - \sum_{l\neq i}^{Ns} (\mathrm{d}/\mathrm{d}r_{il})\phi_{il}(r_{il})(\underline{r}_{il}/r_{il}) \\ &= \underline{F}_{ib}(t) + \underline{F}_{is}(t) \end{aligned} \tag{F.1}$$

where k labels the bath species and l labels the suspended species, $\underline{F}_{ib}(t)$ is the force on particle i due to the bath particles and $\underline{F}_{is}(t)$ is the force on particle i due to the suspended particles.

If i is a bath particle then,

$$\underline{F}_b(t) = \underline{F}_{bb}(t) + \underline{F}_{bs}(t) \tag{F.2}$$

and if i is a suspended particle then,

$$\underline{F}_s(t) = \underline{F}_{bs}(t) + \underline{F}_{ss}(t) \tag{F.3}$$

F.2 SHORT TIME EXPANSIONS

$$\begin{aligned} c_b(t) &= \langle \underline{v}_b(0)\cdot \underline{v}_b(t)\rangle / \langle \underline{v}_b(0)\cdot \underline{v}_b(0)\rangle \\ &= 1-\{1/(6k_\mathrm{B}T/m_b)\}\langle \underline{a}_b(0)\cdot \underline{a}_b(0)\rangle t^2 + \cdots \\ &= 1 - A_b t^2 + \cdots \end{aligned} \tag{F.4}$$

and,

$$\begin{aligned} K_{vb}(0) &= \langle \underline{a}_b(0)\cdot \underline{a}_b(0)\rangle / \langle \underline{v}_b(0)\cdot \underline{v}_b(0)\rangle \\ &= \langle \underline{F}_b{}^2(0)\rangle/(3k_\mathrm{B}Tm_b) \\ &= \langle (\nabla\Phi_b)^2\rangle/(3k_\mathrm{B}Tm_b) \\ &= \langle (\nabla^2\Phi_b)\rangle/(3m_b) \\ &= \left\langle \sum_{k\neq i}^{Nb} (\mathrm{d}^2/\mathrm{d}r_{ik}{}^2)\phi_{ik}(r_{ik}) + \sum_{l\neq i}^{Ns} (\mathrm{d}^2/\mathrm{d}r_{il}^2)\phi_{il}(r_{il})\rangle/(3m_b)\right\rangle \\ &= N_b/(3m_b)\langle (\mathrm{d}^2/\mathrm{d}r^2)\phi_{bb}(r)\rangle/3m_b + N_s/(3m_b)\langle (\mathrm{d}^2/\mathrm{d}r^2)\phi_{bs}(r)\rangle \\ &= (1/3m_b)\{\rho_b \int (\mathrm{d}^2/\mathrm{d}r^2)\phi_{bb}(r)g_{bb}(r)\mathrm{d}\underline{r} + \rho_s \int (\mathrm{d}^2/\mathrm{d}r^2)\phi_{bs}(r)g_{bs}(r)\mathrm{d}\underline{r}\}\} \end{aligned} \tag{F.5}$$

Similarly,

$$K_{vs}(0) = (1/3m_s)\left\{\rho_b \int (\mathrm{d}^2/\mathrm{d}r^2)\phi_{bs}(r)g_{bs}(r)\mathrm{d}\underline{r} + \rho_s \int (\mathrm{d}^2/\mathrm{d}r^2)\phi_{ss}(r)g_{ss}(r)\mathrm{d}\underline{r}\right\} \tag{F.6}$$

F.3 RELATIVE INITIAL BEHAVIOUR OF $c(t)$

This will be given by the ratio A_b/A_s and so if the b and s particles only differ in mass then

$$\rho = \rho_s = \rho_b \text{ and } \int (\mathrm{d}^2/\mathrm{d}r^2)\phi_{bs}(r)g_{bs}(r)\mathrm{d}\underline{r} = \int (\mathrm{d}^2/\mathrm{d}r^2)\phi_{ss}(r)g_{ss}(r)\mathrm{d}\underline{r} \text{ so,}$$

$$A_s/A_b = m_b/m_s \tag{F.7}$$

And, thus, the normalised velocity autocorrelation function for the suspended particles will initially decay much more slowly than that for the bath particles, if $m_s > m_b$, as then $A_s/A_b < 1$ because $A_s < A_b$.

Appendix G

Time-Reversal Symmetry of Non-Equilibrium Correlation Functions

Consider the correlation function defined by

$$\langle A(t+\tau)B^*(t)\rangle = \int d\Gamma\, f(\Gamma)e^{iLt}B^*(\Gamma)e^{iLt}A(\Gamma) \tag{G.1}$$

where the propagator e^{iLt} acts on all phase variables to its right. We would like to find an expression similar to the expression for the time-reversal symmetry of an equilibrium correlation function,

$$\langle A(t+\tau)B^*(t)\rangle = p_{AB}\langle A(t-\tau)B^*(t)\rangle \tag{G.2}$$

where p_{AB} is the parity of the product AB under the time-reversal operation. If we consider the case where $B = 1$ and $t = 0$, we see that the average of A must equal zero if A has odd time-reversal parity. Therefore it is convenient to define all A variables to be the fluctuations of the relevant phase variables from their average values at time $t = 0$.

The frequency relevant to our discussion of eq. (4.20) in Chapter 4 is obtained by substituting $B=\dot{A}$ and $\tau = 0$. In this case, $p_{AB}=-1$ and the frequency must vanish. This has been proven for the equilibrium case[1–3] and we wish to extend this proof to the non-equilibrium steady state.

For this illustration, we will consider a specific case that is of particular interest in this work, where the external field corresponds to a shear strain rate, with the streaming velocity in the x-direction and the gradient in the y-direction. We begin by defining a change of coordinates (also called the Kawasaki mapping[4]) in the ensemble average (eq. (G.1)) where the new phase space coordinates are given by

$$\Gamma^K{}_i = (x_i, -y_i; z_i, -p_{xi}, p_{yi}, -p_{zi}) \tag{G.3}$$

which corresponds to the usual time-reversal transformation (momentum reversal), followed by reflection in the x–z plane (field reversal). This change of coordinates has the following properties:

$$\mathrm{i}L^K = -\mathrm{i}L \tag{G.4}$$

$$\alpha^K = -\alpha \tag{G.5}$$

$$\int \mathrm{d}\Gamma^K = \int \mathrm{d}\Gamma \tag{G.6}$$

$$A^K(\Gamma) = p_A^K A(\Gamma) \tag{G.7}$$

Application of this change of variables to the ensemble average in eq. (G.1) gives

$$\langle A(t+\tau)B^*(t)\rangle = p^K{}_{AB}\langle A(-t-\tau)B^*(-t)\rangle \tag{G.8}$$

The right-hand side of eq. (G.8) can be written as

$$\begin{aligned}\langle A(-t-\tau)B^*(-t)\rangle &= \int \mathrm{d}\Gamma\, f(\Gamma)\mathrm{e}^{-\mathrm{i}Lt} B^*(\Gamma)\mathrm{e}^{-\mathrm{i}L\tau} A(\Gamma)\\ &= \int \mathrm{d}\Gamma\, f(\Gamma,-t)B^*(\Gamma)\mathrm{e}^{-\mathrm{i}L\tau} A(\Gamma)\\ &= \int \mathrm{d}\Gamma\, f(\Gamma,t)B^*(\Gamma)\mathrm{e}^{-\mathrm{i}L\tau} A(\Gamma)\\ &\quad +\beta F_{\mathrm{e}}\int_{-t}^{t} \mathrm{d}s\langle J(0)B^*(s)A(s-\tau)\rangle\end{aligned} \tag{G.9}$$

where we have used $f(\Gamma,t) = f(\Gamma,-t) + \int_{-t}^{t} \mathrm{d}s\, \partial f/\partial s$ with eq. (1.24) in Chapter 1. This may be further simplified by using the relationship $A(s) = A(s-\tau) + \int_{s-\tau}^{s} \dot{A}\mathrm{d}u$, giving,

$$\begin{aligned}\langle A(-t-\tau)B^*(-t)\rangle &= \langle A(t-\tau)B^*(t)\rangle + \beta F_{\mathrm{e}}\int_{-t}^{t}\langle J(0)B^*(s)A(s)\rangle \mathrm{d}s\\ &\quad +\beta F_{\mathrm{e}}\int_{-t}^{t}\mathrm{d}s\int_{s}^{s-\tau}\langle J(0)B^*(s)\dot{A}(u)\rangle \mathrm{d}u\end{aligned} \tag{G.10}$$

Equation (G.10), together with eq. (G.8) gives the general result for time-reversal symmetry of a non-equilibrium correlation function about the time origin of the correlation function, $\tau = 0$.

This can be used to evaluate the frequency, defined by eq. (4.20) in Chapter 4 by substituting $B = \dot{A}$ and $\tau = 0$ in eq. (G.10). Using the facts that the field F_e is constant, the dissipative flux (in the case of the shear stress) is odd with respect to the Kawasaki transformation and the product $A\dot{A}$ must be odd, we obtain

$$\int_{-t}^{t} \mathrm{d}s \langle J(0)\,\dot{A}(s)A(s)\rangle = 2\int_{0}^{t} \mathrm{d}s \langle J(0)\,\dot{A}(s)A(s)\rangle \tag{G.11}$$

in agreement with eq. (1.31) of Chapter 1 where B $= A$ and $\tau = 0$. This expression can be directly integrated giving the following expression:

$$2\int_{0}^{t} \mathrm{d}s \langle J(0)\,\dot{A}(s)\,A(s)\rangle = \langle J(0)[A^2(t) - A^2(0)]\rangle = \langle J(0)\,A^2(t)\rangle \tag{G.12}$$

The last equality results from the fact that $\langle J(0)A^2(t)\rangle$ is odd and $\langle J(0)\rangle = 0$, giving $\langle J(0)A^2(0)\rangle = 0$. In the long time limit, the final correlation function factorises and becomes equal to zero. Thus, we recover the result that the frequency tends to zero in the steady state for variables of the same time-reversal symmetry in the presence of the external field. A discussion of non-equilibrium time-reversal symmetry in the context of generalised Onsager symmetry has been given by Dufty and Rubi.[5]

REFERENCES

1. B.J. Berne, Projection Operator Techniques in the Theory of Fluctuations in Statistical Mechanics, In: *Modern Theoretical Chemistry 6. Statistical Mechanics, Part B: Time-Dependent Processes* (Ed. B.J. Berne), Plenum Press, New York, 1977, Chapter 5, pp. 233–257.
2. B.J. Berne and R. Pecora, *Dynamic Light Scattering*, Wiley, New York, 1976.
3. J.-P. Hansen and I.R. McDonald, *Theory of Simple Liquids,* Academic Press, London, 1986.
4. D.J. Evans and G.P. Morriss, *Statistical Mechanics of Nonequilibrium Liquids,* Academic Press, London, 1990.
5. J.W. Dufty and J.M. Rubi, *Phys. Rev. A*, 36, 222 (1987).

Appendix H

Some Proofs Needed for the Albers, Deutch and Oppenheim Treatment

Now first,

$$\begin{aligned}
\mathbf{G}(t)[(\mathbf{1}-\mathbf{P})\mathrm{i}\mathbf{L}\underline{v}_j(0)] &= \mathrm{e}^{\mathrm{i}\mathbf{L}t}[(\mathbf{1}-\mathbf{P})\mathrm{i}\mathbf{L}\underline{V}_{\mathrm{B}}(0)] \\
&= \mathrm{e}^{\mathrm{i}\mathbf{L}t}[\mathbf{1}\mathrm{i}\mathbf{L}\underline{V}_{\mathrm{B}}(0)] - \mathrm{e}^{\mathrm{i}\mathbf{L}t}[\mathbf{P}\mathrm{i}\mathbf{L}\underline{V}_{\mathrm{B}}(0)] \\
&= \mathrm{e}^{\mathrm{i}\mathbf{L}t}[(\mathrm{i}\mathbf{L}_0\underline{V}_{\mathrm{B}}(0)] + \mathrm{e}^{\mathrm{i}\mathbf{L}t}[(\mathrm{i}\mathbf{L}_{\mathrm{B}}\underline{V}_{\mathrm{B}}(0)] \\
&\quad -\mathrm{e}^{\mathrm{i}\mathbf{L}t}[\mathbf{P}\mathrm{i}\mathbf{L}_0\underline{V}_{\mathrm{B}}(0)] - \mathrm{e}^{\mathrm{i}\mathbf{L}t}[\mathbf{P}\mathrm{i}\mathbf{L}_{\mathrm{B}}\,\underline{V}_{\mathrm{B}}(0)] \\
&= 0 + \mathrm{e}^{\mathrm{i}\mathbf{L}t}\,\mathrm{d}\,\underline{V}_{\mathrm{B}}(0)/\mathrm{d}t - 0 - 0 \\
&= \mathrm{d}\,\underline{V}_{\mathrm{B}}(t)/\mathrm{d}t
\end{aligned} \tag{H.1}$$

and

$$\begin{aligned}
\mathrm{e}^{\mathrm{i}(\mathbf{1}-\mathbf{P})\mathbf{L}t}[(\mathbf{1}-\mathbf{P})\mathrm{i}\mathbf{L}\underline{V}_{\mathrm{B}}(0)] &= \mathrm{e}^{\mathrm{i}(\mathbf{1}-\mathbf{P})\mathbf{L}t}[(\mathbf{1})\mathrm{i}\mathbf{L}\underline{V}_{\mathrm{B}}(0)] - \mathrm{e}^{\mathrm{i}(\mathbf{1}-\mathbf{P})\mathbf{L}t}[(\mathbf{P})\mathrm{i}\mathbf{L}\underline{V}_{\mathrm{B}}(0)] \\
&= \mathrm{e}^{\mathrm{i}(\mathbf{1}-\mathbf{P})\mathbf{L}t}[\mathrm{i}(\mathbf{L}_0 + \mathbf{L}_{\mathrm{B}})\underline{V}_{\mathrm{B}}(0)] + 0 \\
&= \mathrm{e}^{\mathrm{i}(\mathbf{1}-\mathbf{P})\mathbf{L}t}[\mathrm{i}\mathbf{L}_{\mathrm{B}}\underline{V}_{\mathrm{B}}(0)] \\
&= \mathrm{e}^{\mathrm{i}(\mathbf{1}-\mathbf{P})\mathbf{L}t}\underline{F}_{\mathrm{B}}(0)/M_{\mathrm{B}} \\
&= K^{+}(t)
\end{aligned} \tag{H.2}$$

Also,

$$\begin{aligned}
&\int_0^t \mathrm{e}^{\mathrm{i}\mathbf{L}(t-\tau)}\mathrm{i}\mathbf{P}\mathbf{L}\,\mathrm{e}^{\mathrm{i}(\mathbf{1}-\mathbf{P})\mathbf{L}\tau}\,\mathrm{d}\tau[(\mathbf{1}-\mathbf{P})\mathrm{i}\mathbf{L}\underline{V}_{\mathrm{B}}(0)] \\
&= \int_0^t \mathrm{e}^{\mathrm{i}\mathbf{L}(t-\tau)}\,\mathrm{i}\mathbf{P}\mathbf{L}\{\mathrm{e}^{\mathrm{i}(\mathbf{1}-\mathbf{P})\mathbf{L}\tau}[(\mathbf{1}-\mathbf{P})\mathrm{i}\mathbf{L}\underline{V}_{\mathrm{B}}(0)]\}\,\mathrm{d}\tau \\
&= \int_0^t \mathrm{e}^{\mathrm{i}\mathbf{L}(t-\tau)}\,\mathrm{i}\mathbf{P}\mathbf{L}\,K^{+}(\tau)\,\mathrm{d}\tau
\end{aligned} \tag{H.3}$$

$$= \int_0^t e^{i\mathbf{L}(t-\tau)} \mathbf{P}(i\mathbf{L}_0 + i\mathbf{L}_B) K^+(\tau)\, d\tau$$
$$= \int_0^t e^{i\mathbf{L}(t-\tau)} \mathbf{P}(i\mathbf{L}_B) K^+(\tau)\, d\tau$$
$$= \int_0^t e^{i\mathbf{L}(t-\tau)} \langle i\,\mathbf{L}_B\, K^+(\tau)\rangle_0\, d\tau$$

Next,

$$\begin{aligned}
\langle K^+(t)\rangle_0 &= \langle K^+(0)\rangle_0 \\
&= \langle e^{i(\mathbf{1}-\mathbf{P})\mathbf{L}_0}[(\mathbf{1}-\mathbf{P})i\mathbf{L}\underline{V}_B(0)]\rangle_0 \\
&= \langle[(\mathbf{1}-\mathbf{P})i\mathbf{L}\underline{V}_B(0)]\rangle_0 \\
&= \langle[i\mathbf{L}_B\underline{V}_B(0)]\rangle_0 \\
&= \langle \underline{F}_B(0)/M_B\rangle_0 \\
&= 0
\end{aligned} \tag{H.4}$$

Thus,

$$\begin{aligned}
\langle i\mathbf{L}_B K^+(t)\rangle_0 &= \langle(\underline{P}_B/M_B\cdot\underline{\nabla}_R + \underline{F}_B\cdot\underline{\nabla}_P)K^+(t)\rangle_0 \\
&= \langle \underline{P}_B/M_B\cdot\underline{\nabla}_R K^+(t)\rangle + \langle \underline{F}_B\cdot\underline{\nabla}_P K^+(t)\rangle_0 \\
&= \underline{P}_B/M_B\cdot\langle\underline{\nabla}_R K^+(t)\rangle + \underline{\nabla}_P\cdot\langle\underline{F}_B K^+(t)\rangle_0
\end{aligned} \tag{H.5}$$

So from $\langle K^+(t)\rangle_0 = 0$, it follows that

$$\begin{aligned}
&\underline{\nabla}_R \langle K^+(t)\rangle_0 = 0 \\
&\int \underline{\nabla}_R\, f_{eq}^0\, K^+(t)\, d\underline{r}^{(N)}\, d\underline{p}^{(N)} = 0 \\
&\int f_{eq}^0\, \underline{\nabla}_R\, K^+(t)\, d\underline{r}^{(N)}\, d\underline{p}^{(N)} + \int K^+(t)\, \underline{\nabla}_R\, f_{eq}^0\, d\underline{r}^{(N)}\, d\underline{p}^{(N)} = 0 \\
&\langle \underline{\nabla}_R K^+(t)\rangle_0 \\
&= -\int\left(K^+(t)\, \underline{\nabla}_R\, e^{-H_0/k_BT} d\underline{r}^{(N)} d\underline{p}^{(N)}\right) / \int e^{-H_0/k_BT} d\underline{r}^{(N)} d\underline{p}^{(N)} \\
&= -\int\left(K^+(t)\underline{\nabla}_R\{-H_0/k_BT\} e^{-H_0/k_BT}\, d\underline{r}^{(N)} d\underline{p}^{(N)}\right) / \left(\int e^{-H_0/k_BT} d\underline{r}^{(N)} d\underline{p}^{(N)}\right) \\
&= -(1/k_BT)\int\left(K^+(t)\{\underline{F}_B\} e^{-H_0/k_BT}\, d\underline{r}^{(N)} d\underline{p}^{(N)}\right) / \left(\int e^{-H_0/k_BT} d\underline{r}^{(N)} d\underline{p}^{(N)}\right) \\
&= -(1/k_BT)\int \underline{F}_B K^+(t) f_{eq}^0\left(\underline{p}^{(N)}, \underline{r}^{(N)}\right) d\underline{p}^{(N)} d\underline{r}^{(N)} \\
&= -(1/k_BT)\langle \underline{F}_B K^+(t)\rangle
\end{aligned} \tag{H.6}$$

Finally, for the integrals needed in the final limiting process,

$$
\begin{aligned}
&\int_0^t e^{i\mathbf{L}(t-\tau)}[\underline{\nabla}_P - \underline{V}_B/k_BT]\cdot\langle \underline{F}_B\, e^{i(\mathbf{1-P})\mathbf{L}t}\, \underline{F}_B(0)\rangle_0\, d\tau \\
&= \int_0^t e^{i\mathbf{L}(t-\tau)}[\underline{\nabla}_P - \underline{V}_B/k_BT]\cdot[\langle \underline{F}_B e^{i(\mathbf{1-P})\mathbf{L}t}\, \underline{F}_B(0)\rangle_0 + \langle \underline{F}_B e^{i\mathbf{L}_0 t}\, \underline{F}_B(0)\rangle_0 - \langle \underline{F}_B e^{i\mathbf{L}_0 t}\, \underline{F}_B(0)\rangle_0]\, d\tau \\
&= \int_0^t e^{i\mathbf{L}(t-\tau)}[\underline{\nabla}_P - \underline{V}_B/k_BT]\cdot\langle \underline{F}_B\, e^{i(\mathbf{1-P})\mathbf{L}t}\, \underline{F}_B(0)\rangle_0\, d\tau \\
&\quad + \int_0^t e^{i\mathbf{L}(t-\tau)}[\underline{\nabla}_P - \underline{V}_B/k_BT]\cdot[\langle \underline{F}_B e^{i(\mathbf{1-P})\mathbf{L}t}\, \underline{F}_B(0)\rangle_0 - \langle \underline{F}_B e^{i\mathbf{L}_0 t}\, \underline{F}_B(0)\rangle_0]\, d\tau \\
&= \int_0^t [e^{i\mathbf{L}t} - e^{i\mathbf{L}t} + e^{i\mathbf{L}(t-\tau)}][\underline{\nabla}_P - \underline{V}_B/k_BT]\cdot\langle \underline{F}_B\, e^{i\mathbf{L}_0 t}\, \underline{F}_B(0)\rangle_0\, d\tau \\
&\quad + \int_0^t e^{i\mathbf{L}(t-\tau)}[\underline{\nabla}_P - \underline{V}_B/k_BT]\cdot[\langle \underline{F}_B e^{i(\mathbf{1-P})\mathbf{L}t}\, \underline{F}_B(0)\rangle_0 - \langle \underline{F}_B e^{i\mathbf{L}_0 t}\, \underline{F}_B(0)\rangle_0]\, d\tau \\
&= \int_0^t e^{i\mathbf{L}t}[\underline{\nabla}_P - \underline{V}_B/k_BT]\cdot\langle \underline{F}_B e^{i\mathbf{L}_0 t}\, \underline{F}_B(0)\rangle_0\, d\tau \\
&\quad + \int_0^t [e^{i\mathbf{L}(t-\tau)} - e^{i\mathbf{L}t}][\underline{\nabla}_P - \underline{V}_B/k_BT]\cdot\langle \underline{F}_B e^{i\mathbf{L}_0 t}\, \underline{F}_B(0)\rangle_0\, d\tau \\
&\quad + \int_0^t e^{i\mathbf{L}(t-\tau)}[\underline{\nabla}_P - \underline{V}_B/k_BT]\cdot[\langle \underline{F}_B e^{i(\mathbf{1-P})\mathbf{L}t}\, \underline{F}_B(0)\rangle_0 - \langle \underline{F}_B e^{i\mathbf{L}_0 t}\, \underline{F}_B(0)\rangle_0]\, d\tau \qquad \text{(H.7)} \\
&= e^{i\mathbf{L}t}\int_0^t [\underline{\nabla}_P - \underline{V}_B/k_BT]\cdot\langle \underline{F}_B e^{i\mathbf{L}_0 t}\, \underline{F}_B(0)\rangle_0\, d\tau \\
&\quad + \int_0^t e^{i\mathbf{L}(t-\tau)}[\underline{\nabla}_P - \underline{V}_B/k_BT]\cdot[\langle \underline{F}_B e^{i(\mathbf{1-P})\mathbf{L}t}\, \underline{F}_B(0)\rangle_0 - \langle \underline{F}_B e^{i\mathbf{L}_0 t}\, \underline{F}_B(0)\rangle_0]\, d\tau \\
&\quad + \int_0^t [e^{i\mathbf{L}(t-\tau)} - e^{i\mathbf{L}t}][\underline{\nabla}_P - \underline{V}_B/k_BT]\cdot\langle \underline{F}_B e^{i\mathbf{L}_0 t}\, \underline{F}_B(0)\rangle_0\, d\tau \\
&= MI_0(t) + MI_1(t) + MI_2(t)
\end{aligned}
$$

Appendix I

A Proof Needed for the Deutch and Oppenheim Treatment

Here,

$$\sum_{\mu}\sum_{\nu}\int_0^t e^{i\mathbf{L}(t-\tau)}\left\{\left[\underline{\nabla}_{P\mu}-1/(Mk_BT)\underline{P}_{\mu}\right]\cdot\left\langle \underline{E}_{B\mu}(0)e^{i(1-\mathbf{P})\mathbf{L}\tau}\,\underline{E}_{B\nu}\right\rangle_0\cdot\underline{\nabla}_{P\nu}\,\underline{v}_B(0)\right\}d\tau$$

$$=\sum_{\mu}\sum_{\nu}\int_0^t e^{i\mathbf{L}(t-\tau)}\left\{\left[\underline{\nabla}_{P\mu}-1/(Mk_BT)\underline{P}_{\mu}\right]\right.$$

$$\left.\cdot\left(\left\langle \underline{E}_{B\mu}(0)e^{i\mathbf{L}_0\tau}\,\underline{E}_{B\nu}\right\rangle_0+\left\langle \underline{E}_{B\mu}(0)e^{i(1-\mathbf{P})\mathbf{L}\tau}\,\underline{E}_{B\nu}\right\rangle_0-\left\langle \underline{E}_{B\mu}(0)e^{i\mathbf{L}_0\tau}\,\underline{E}_{B\nu}\right\rangle_0\right)\cdot\underline{\nabla}_{P\nu}\,\underline{v}_B(0)\right\}d\tau$$

$$=\sum_{\mu}\sum_{\nu}\int_0^t e^{i\mathbf{L}(t-\tau)}\left\{\left[\underline{\nabla}_{P\mu}-1/(Mk_BT)\underline{P}_{\mu}\right]\cdot\left(\left\langle \underline{E}_{B\mu}(0)e^{i\mathbf{L}_0\tau}\,\underline{E}_{B\nu}\right\rangle_0\right)\cdot\underline{\nabla}_{P\nu}\,\underline{v}_B(0)\right\}d\tau$$

$$+\sum_{\mu}\sum_{\nu}\int_0^t e^{i\mathbf{L}(t-\tau)}\left\{\left[\underline{\nabla}_{P\mu}-1/(Mk_BT)\underline{P}_{\mu}\right]\cdot\left(\left\langle \underline{E}_{B\mu}(0)e^{i(1-\mathbf{P})\mathbf{L}\tau}\,\underline{E}_{B\nu}\right\rangle_0-\left\langle \underline{E}_{B\mu}(0)e^{i\mathbf{L}_0\tau}\,\underline{E}_{B\nu}\right\rangle_0\right)\right.$$

$$\left.\cdot\underline{\nabla}_{P\nu}\,\underline{v}_B(0)\right\}d\tau$$

$$=\sum_{\mu}\sum_{\nu}\int_0^t \left(e^{i\mathbf{L}t}+e^{i\mathbf{L}(t-\tau)}-e^{i\mathbf{L}t}\right)\left\{\left[\underline{\nabla}_{P\mu}-1/(Mk_BT)\underline{P}_{\mu}\right]\cdot\left(\left\langle \underline{E}_{B\mu}(0)e^{i\mathbf{L}_0\tau}\,\underline{E}_{B\nu}\right\rangle_0\right)\right.$$

$$\left.\cdot\underline{\nabla}_{P\nu}\,\underline{v}_B(0)\right\}d\tau$$

$$+\sum_{\mu}\sum_{\nu}\int_0^t e^{i\mathbf{L}(t-\tau)}\left\{\left[\underline{\nabla}_{P\mu}-1/(Mk_BT)\underline{P}_{\mu}\right]\cdot\left(\left\langle \underline{E}_{B\mu}(0)e^{i(1-\mathbf{P})\mathbf{L}\tau}\,\underline{E}_{B\nu}\right\rangle_0-\left\langle \underline{E}_{B\pi}(0)e^{i\mathbf{L}_0\tau}\,\underline{E}_{B\nu}\right\rangle_0\right)\right.$$

$$\left.\cdot\underline{\nabla}_{P\nu}\,\underline{v}_B(0)\right\}d\tau$$

$$
\begin{aligned}
&= \sum_{\mu}\sum_{\nu}\int_0^t \mathrm{e}^{\mathrm{i}\mathbf{L}t}\left\{\left[\underline{\nabla}_{\mathrm{P}\mu} - 1/(Mk_{\mathrm{B}}T)\underline{P}_{\mu}\right]\cdot\left(\left\langle \underline{E}_{\mathrm{B}\mu}(0)\mathrm{e}^{\mathrm{i}\mathbf{L}_0\tau}\,\underline{E}_{\mathrm{B}\nu}\right\rangle_0\right)\cdot\underline{\nabla}_{\mathrm{P}\nu}\,\underline{v}_{\mathrm{B}}(0)\right\}\mathrm{d}\tau \\
&+\sum_{\mu}\sum_{\nu}\int_0^t \mathrm{e}^{\mathrm{i}\mathbf{L}(t-\tau)}\left\{\left[\underline{\nabla}_{\mathrm{P}\mu} - 1/(Mk_{\mathrm{B}}T)\underline{P}_{\mu}\right]\cdot\left(\left\langle \underline{E}_{\mathrm{B}\mu}(0)\mathrm{e}^{\mathrm{i}(\mathbf{1}-\mathbf{P})\mathbf{L}\tau}\,\underline{E}_{\mathrm{B}\nu}\right\rangle_0 - \left\langle \underline{E}_{\mathrm{B}\mu}(0)\mathrm{e}^{\mathrm{i}\mathbf{L}_0\tau}\,\underline{E}_{\mathrm{B}\nu}\right\rangle_0\right)\right. \\
&\left.\cdot\underline{\nabla}_{\mathrm{P}\nu}\,\underline{v}_{\mathrm{B}}(0)\right\}\mathrm{d}\tau \\
&+\sum_{\mu}\sum_{\nu}\int_0^t \left(\mathrm{e}^{\mathrm{i}\mathbf{L}(t-\tau)} - \mathrm{e}^{\mathrm{i}\mathbf{L}t}\right)\left\{\left[\underline{\nabla}_{\mathrm{P}\mu} - 1/(Mk_{\mathrm{B}}T)\underline{P}_{\mu}\right]\cdot\left(\left\langle \underline{E}_{\mathrm{B}\mu}(0)\mathrm{e}^{\mathrm{i}\mathbf{L}_0\tau}\,\underline{E}_{\mathrm{B}\nu}\right\rangle_0\right)\cdot\underline{\nabla}_{\mathrm{P}\nu}\,\underline{v}_{\mathrm{B}}(0)\right\}\mathrm{d}\tau \\
&= I_0(t) + I_1(t) + I_2(t) \qquad \text{(I.1)}
\end{aligned}
$$

Appendix J

The Calculation of the Bulk Properties of Colloids and Polymers

If we can generate the phase space trajectory of the N colloidal particles and polymer solutions interacting via the combined effect of the solvent averaged forces and the hydrodynamic interactions we may calculate the properties from this information provided that we can write the properties in terms of Γ_1 only (see the introduction to Chapter 5).

If the diffusive limit is used then only the time dependant positions $(\underline{R}^{(N)}(t))$ of the large particles are generated which further restricts the dynamic properties which can be calculated or rather restricts the formulae which may be used to calculate them.

J.1 EQUILIBRIUM PROPERTIES

Typically, one is only interested in the excess thermodynamic properties of the suspension or solution, e.g. the osmotic pressure.

Usually equilibrium properties are the sum of separate functions of momenta only and of positions only and as the momentum dependent part (see Appendices A and K) is the sum of one body terms then the only remaining problems involved are

1. To express the excess equilibrium properties in terms of $(\underline{R}^{(N)})$.
2. To calculate the solvent averaged potential of mean force, defined in eq. (5.56) but it should be noted that allowance must be made for the fact that this force is a function of the physical conditions imposed on the system, e.g. the temperature and density.
3. To perform ensemble averages.

J.2 STATIC STRUCTURE

The expressions in Appendix A are directly applicable to polymers and colloidal suspensions.

J.3 TIME CORRELATION FUNCTIONS

As outlined in Appendix A the linear transport coefficients and scattering properties of a system of $N + n$ particles may be related to time correlation functions of appropriate dynamical variables $A(\Gamma(t))$ and $B(\Gamma(t))$, i.e.

$$\langle A(0)\,B(t)\rangle = \int f_{\text{eq}}\left(p^{(N+n)}, r^{(N+n)}\right) A(0)\,B(t)\mathrm{d}\underline{p}^{(N+n)}\,\mathrm{d}\underline{r}^{(N+n)} \tag{J.1}$$

However, in the Brownian limit we only calculate the phase space trajectory of the large particles, Γ_1 and not that for all the particles, Γ. Thus, one should derive time correlation function expressions for the dynamic properties in terms of the appropriate variables, $A(\Gamma_1(t))$ and $B(\Gamma_1(t))$.

In the diffusive limit only the variable $(\underline{R}^{(N)}(t))$ is generated and one must allow for this. Although this has been done for scattering and for diffusion the general problem does not seem to have been considered. Some of the appropriate results which may be readily written down are collected below.

J.3.1 Self-diffusion

Here we use the result that

$$\langle \Delta \underline{R}^2(t)\rangle = \langle (\underline{R}_i(0) - \underline{R}_i(t))^2\rangle = \int_0^t (1 - s/t)\langle \underline{V}_i(0)\cdot \underline{V}_i(s)\rangle\,\mathrm{d}s \tag{J.2}$$

So as the self or tracer diffusion coefficient D_{T} is given by,

$$D_{\mathrm{T}} = 1/3\int_0^\infty \langle \underline{V}_i(0)\cdot\underline{V}_i(t)\rangle\,\mathrm{d}t \tag{J.3}$$

$$= \operatorname*{Slope}_{t\to\infty}\,[\langle (\underline{R}_i(0) - \underline{R}_i(t))^2\rangle/(6t)] \tag{J.4}$$

this latter result being particularly useful in the diffusive limit.

Experimentally D_{T} may be measured from a variety of scattering techniques, e.g. forced Rayleigh Scattering, Tracer Scattering and from radioactive tracer measurements.

J.3.2 Time-dependent scattering

The expressions given in Appendix A are still applicable.

J.3.3 Bulk stress

In order to calculate the rheological properties in the diffusive limit we need to be able to evaluate the bulk stress $\langle\Sigma\rangle$. Brady and Bossis[1] show that this may be written as (ignoring isotropic terms which do not contribute to the rheological properties),

$$\langle\Sigma\rangle = 2\eta_0\,\mathbf{E}^{\infty} + N/V\{\langle\mathbf{S}^{\mathrm{H}}\rangle + \langle\mathbf{S}^{\mathrm{P}}\rangle + \langle\mathbf{S}^{\mathrm{B}}\rangle\} \tag{J.5}$$

where,

1. $\langle S^H\rangle$ is the mechanical or contact stress transmitted by the fluid due to bulk shear flow.
2. $\langle S^P\rangle$ is the "elastic" stress due to the interparticle forces.
   and
3. $\langle S^P\rangle$ is the stress contribution due to the Brownian force.

These may be written as,

$$\langle\mathbf{S}^{\mathrm{H}}\rangle = -\langle\mathbf{R}_{\mathbf{SU}}\cdot(\mathbf{U}-\mathbf{U}^{\infty}) - \mathbf{R}_{\mathbf{SE}} : \mathbf{E}^{\infty}\rangle \tag{J.6}$$

$$\langle\mathbf{S}^{\mathbf{P}}\rangle = -\langle\mathbf{x}\mathbf{F}^{\mathbf{P}}\rangle$$

and

$$\langle\mathbf{S}^{\mathbf{P}}\rangle = -k_{\mathrm{B}}T\langle\underline{\nabla}\cdot(\mathbf{R}_{\mathbf{SU}}\cdot\mathbf{R}_{\mathbf{FU}}{}^{-1})\rangle$$

where $\mathbf{R_{SU}}$ and $\mathbf{R_{SE}}$ are configuration-dependent resistance matrices, similar to $\mathbf{R_{FU}}$ and $\mathbf{R_{FE}}$, both relating the particle "stresslet" $\mathbf{S}$ to the particle's velocities in $\mathbf{R_{SU}}$ and the imposed strain rate $\mathbf{R_{SE}}$. The velocities U in eq. (J.6) are those arising from the deterministic contributions to the particle displacements Δx given by,

$$\mathbf{U}-\mathbf{U}^{\infty} = \mathbf{R}_{\mathbf{FU}}{}^{-1}\cdot[\mathbf{R}_{\mathbf{FE}} : \mathbf{E}^{\infty} + \gamma^{*-1}\,\mathbf{F}^{\mathbf{P}}] \tag{J.7}$$

and the divergence $\langle\nabla\cdot(\mathbf{R_{SU}}\cdot\mathbf{R_{FU}}^{-1})\rangle$ in $\langle\mathbf{S}^{\mathbf{P}}\rangle$ is with respect to the last index of $\mathbf{R_{FU}}^{-1}$.

J.3.4 Zero time (high frequency) results in the diffusive limit

In the diffusive limit at zero time (infinite frequency) we may express the q dependent diffusion constant $D(q)$,[2–7], the local self diffusion constant D_{S}, the collective (or mutual) diffusion constant $D_{\mathrm{c},}$[8–12] the permeability[13] K and the Newtonian shear viscosity[13] η_0 as equilibrium ensemble averages over the particle positions alone as follows,

$$\partial F(q,t)/\partial t\,|_{t=0} = -S(q)q^2D(q) \tag{J.8}$$

i.e.

$$D(q) = (k_B T/q^2\, S(q)/N) \sum_{i,j}^{N} \langle \underline{q} \cdot \mu_{ij} \cdot \underline{q} \exp([i\underline{q} \cdot \{\underline{R}_i(0) - \underline{R}_j(t)\}]) \rangle$$
$$= H(q)/S(q) \tag{J.9}$$

$$D_S = \lim_{q\to\infty} D(q) = (k_B T/3N)\, tr\left\langle \sum_{i=1}^{N} \mu_{ii} \right\rangle \tag{J.10}$$

$$D_c = \lim_{q\to\infty} D(q) = (k_B T/3N)\, tr\left\langle \sum_{i,j}^{N} \mu_{ij} \right\rangle / S(0) \tag{J.11}$$

$$K^{-1} = (3\eta_f V)^{-1}\, tr\left\langle \sum_{i,j}^{N} \zeta_{ij} \right\rangle \tag{J.12}$$

$$\eta_0/\eta_f = 1 + 5/2\phi\langle\gamma\rangle \tag{J.13}$$

where γ is a component of the normalised dipole–dipole friction tensor,[13] η_f is the Newtonian viscosity of the background fluid and $H(q)$ is the so-called hydrodynamic factor.[11]

We may similarly write the sedimentation velocity[12] s and the collective mobility μ as,[13]

$$s = D_c\, S(0) = D_c\, [(1/\rho)(\partial\rho/\partial p)_T] \tag{J.14}$$

and

$$\mu/\mu_0 = s/s_0\ .$$

where $\mu_0 = 6\pi\eta_f a = k_B T/D_0$, D_0 is the free-particle diffusion constant and η_f is again the solvent viscosity.

While K, η and D_c may be measured by conventional, direct methods,[13] e.g. D_c by means of boundary relaxation techniques[14] the quantities D(q), D_S and D_c are conveniently measured by means of DLS[11] or by SENS. [7].
Both the zero-time (infinite-frequency) shear viscosity η_0 and the long time shear viscosity may be measured using rheometers.

There have been a large number of calculations of many of the above properties using a wide variety of techniques and approximations.[1,10,15–19,21–25] This is principally because these properties may be evaluated by means of equilibrium methods, e.g. the Monte

Carlo, equilibrium Molecular Dynamics simulation methods or approximate equilibrium theories.

Van Megen, Snook and Pusey[22] showed that directly using the cluster series result for μ_{ij} from the MIF cluster method (see Section 7.2) does not necessarily give good results for D_c or D_S since the convergence of the resultant series is extremely non-uniform. However, Beenakker and Mazur[15–19] were able to get good agreement with experiment for a variety of zero time properties by essentially selectively resumming these series.

Of particular note is the extensive work of Ladd[13] who calculated μ/μ_0, D_S/D_0, η/η_0 and K for a hard sphere system as a function of particle volume fraction using the MIF method. He found that convergence is enhanced by the use of lubrication theory and the convergence as a function of the number of moments used is then reasonably rapid. However, the N dependence of the results is quite strong and extrapolation to $N = \infty$ is necessary to get accurate results. Similar results have been obtained by Phillips, Brady and Bossis[26] and Wajnryb and Dahler[27] have calculated the long time values of η.

REFERENCES

1. J. Brady and G. Bossis, *Ann. Rev. Fluid Mech.*, **20**, 111 (1988); *J. Chem. Phys.* **87**, 5437 (1987).
2. P.N. Pusey and R.J.A. Tough, *Adv. Coll. Interf. Sci.*, **16**, 143 (1982).
3. P.N. Pusey and R.J.A. Tough, *J. Phys. A: Math. Gen.*, **15**, 1291 (1982).
4. P.N. Pusey and R.J.A. Tough, *Faraday Disc. Chem. Soc.*, **76**, 123 (1983).
5. P.N. Pusey and R.J.A. Tough, *Particle interactions*. In: *Dynamic Light Scattering* (Ed. R. Pecora), Plenum, New York, 1985.
6. R.B. Jones and P.N. Pusey, Dynamics of suspended colloidal spheres, *Ann. Rev. Phys. Chem.*, **42**, 137 (1991).
7. J.B. Hayter, *Scattering Techniques Applied to Supramolecular and Non-Equilibrium Systems* (Eds. S-H. Chen and R. Nossal), Plenum Press, New York, p. 49, 1981.
8. P.N. Pusey and R.J.A. Tough, *J. Phys. A: Math. Gen.*, **15**, 1291 (1982).
9. P.N. Pusey and R.J.A. Tough, *Faraday Disc. Chem. Soc.*, **76**, 123 (1983).
10. C.W. J. Beenakker and P. Mazur, *Physica*, **126A**, 349 (1984).
11. W. van Megen, R.H. Ottewill, S.M. Owens and P.N. Pusey, *J. Chem. Phys.*, **82**, 508 (1985).
12. I. Snook and W. van Megen, *J. Coll. Interf. Sci.*, **100**, 194 (1984).
13. A. Ladd, *J. Chem. Phys.*, **93**, 3484 (1990).
14. B.N. Preston, W.D. Comper, A.E. Hughes, I. Snook and W. van Megen, *J. Chem. Soc., Faraday Trans. I*, **78**, 1209 (1982).
15. C.W.J. Beenakker and P. Mazur, *Phys. Lett.*, **91**, 290 (1982).
16. C.W.J. Beenakker and P. Mazur, *Phys. Lett.*, **98A,** 22 (1983).
17. C.W.J. Beenakker and P. Mazur, *Physica*, **120A**, 388 (1983).
18. C.W.J. Beenakker and P. Mazur, *Physica*, **126A**, 349 (1984).
19. C.W.J. Beenakker, *Physica A*, **128**, 48 (1984).
20. W. van Megen, S. Underwood, R.H. Ottewill, N. St. J. Williams and P.N. Pusey, *Faraday Disc. Chem. Soc.*, **83**, 47 (1987).
21. I. Snook and W. van Megen, *J. Coll. Interf. Sci.*, **100**, 194 (1984).
22. W. van Megen, I. Snook and P.N. Pusey, *J. Chem. Phys.*, **78**, 931 (1983).
23. B. Cichocki and B.U. Felderhoff, *J. Chem. Phys.*, **89**, 1049, 3705 (1988).
24. M. Tokuyama and I. Oppenheim, *Phys. Rev. E*, **50**, R16 (1994).

25. D.M.E. Theis-Weesie, A.P.Philipse, G. Nagele, B. Mandl and R. Klein, *J. Coll. Interf. Sci.*, **176**, 43 (1995).
26. R.J. Phillips, J.F. Brady and G. Bossis, *Phys. Fluids*, **31**, 3462 (1988); *ibid.* **31**, 3473 (1973).
27. E. Wajnryb and J.S. Dahler, Viscosity of moderately dense suspensions. In: Adv. Chem. Phys., Vol. 102 (Eds I. Prigogine and S.A. Rice), Wiley, New York, 1997 and *Physica*, **253**, 77 (1998).

Appendix K

Monte Carlo Methods

In order to implement some of the theoretical approaches described in this book on a computer one needs to evaluate ensemble averages (multi-dimensional integrals) or to generate states of a system by means of a stochastic approach. Both these procedures may be carried out by means of a class of techniques termed Monte Carlo (MC) methods.[1–8] In order to enable the reader to understand how these methods work we will outline one of these techniques.

K.1 METROPOLIS MONTE CARLO TECHNIQUE

In order to evaluate equilibrium ensemble averages one uses the formulae given in Appendix A, that is,

$$\begin{aligned}\langle B\rangle &= \int B(\Gamma) f^{\mathrm{eq}}(\Gamma)\mathrm{d}\Gamma \\ &\approx \sum_{i=1}^{N_c} B(\Gamma_i) f^{\mathrm{eq}}(\Gamma_i)\end{aligned} \tag{K.1}$$

In Classical equilibrium statistical mechanics we may usually write the expression for properties as the sum of a function of the particle momenta ($\underline{p}^{(N)}$) only and a function of particle co-ordinates ($\underline{r}^{(N)}$) only, that is,

$$B = B_p\left(\underline{p}^{(N)}\right) + B_r\left(\underline{r}^{(N)}\right) \tag{K.2}$$

For example, the Hamiltonian H, that is, the total mechanical energy E is the sum of the kinetic energy K, the potential energy Φ, that is,

$$H = E = K\left(\underline{p}^{(N)}\right) + \Phi\left(\underline{r}^{(N)}\right) \tag{K.3}$$

This combined with the fact that $f^{\text{eq}}(\Gamma_i)$ in the commonly used ensembles is an exponential (see Table A.1, Appendix A) which means that we may express the ensemble average expression for $\langle B \rangle$ into the sum of two terms, one an integral over $(\underline{p}^{(N)})$ and the other an integral over $(\underline{r}^{(N)})$. Furthermore, because the functions of $(\underline{p}^{(N)})$ are only functions of sums of one particle properties the former integrals can usually be performed analytically and, thus, we only need to evaluate the integrals over $(\underline{r}^{(N)})$ numerically, that is,

$$\langle B \rangle = \int B_p\left(\underline{p}^{(N)}\right) f^{\text{eq}}{}_p\left(\underline{p}^{(N)}\right) \mathrm{d}\underline{p}^{(N)} + \int B_r\left(\underline{r}^{(N)}\right) f^{\text{eq}}{}_r\left(\underline{r}^{(N)}\right) \mathrm{d}\underline{r}^{(N)} \tag{K.4}$$

$$\begin{aligned} \langle B \rangle_r &= \int B_r\left(\underline{r}^{(N)}\right) f^{\text{eq}}{}_r\left(\underline{r}^{(N)}\right) \mathrm{d}\underline{r}^{(N)} \\ &\approx \sum_{i=1}^{N_c} B_r\left(\underline{r}^{(N)}{}_i\right) f^{\text{eq}}\left(\underline{r}^{(N)}{}_i\right) \end{aligned} \tag{K.5}$$

Hence, we need to compute the sum eq. (K.5) for some set of configurations $\{\underline{r}^{(N)}{}_i;\ i = 1, N_c\}$. Now the choice of these configurations could be done in a variety of ways, for example, using points chosen by product formulae based on one dimensional quadrature methods or by uniformly generating random points in $3N$ dimensional space. The use of product formulae proves to be very inefficient for integrals of dimension, d greater than about 8 as the number of points required for acceptable accuracy increases as $N_1{}^d$ where N_1 is the number of points used in each dimension but the error scales only as $N_1{}^{-4/d}$ [8]. Merely randomly sampling points in $3N$ dimensional co-ordinate space from a uniform distribution can be more efficient than simple quadrature and the method gives errors scale as $N_c{}^{-1/2}$ irrespective of the value of $3N$, however, this method also fails for integrals of the type give by eq. (K.5) because the factor $f^{\text{eq}}(\underline{r}^{(N)}{}_i)$ is usually reasonably strongly peaked in co-ordinate space meaning that $B_r(\underline{r}^{(N)})\, f^{\text{eq}}{}_r(\underline{r}^{(N)})$ is also peaked and not uniform in space.

Thus, a method must be sought which combines the advantages of $N_c{}^{-1/2}$ error but which biases the sampling to regions of co-ordinate space where $B_r(\underline{r}^{(N)})\, f^{\text{eq}}{}_r(\underline{r}^{(N)})$, that is $f^{\text{eq}}{}_r(\underline{r}^{(N)})$ is large. In essence we wish to replace $\Sigma B_r(\underline{r}^{(N)}{}_i)\, f^{\text{eq}}(\underline{r}^{(N)}{}_i)$ by $\Sigma B_r(\underline{r}^{(N)}{}_j)$ where the set of points $\{\underline{r}^{(N)}{}_j, j = 1, N_c\}$ are sampled from the distribution f^{eq}. A method for doing this was developed by Metropolis, Rosenbluth, Rosenbluth, Teller and Teller[1] which is usually referred to as the Metropolis Monte Carlo method.

In order to explain how the method works we will follow the clever (but not entirely rigorous) argument presented by Foulkes, Mitas, Needs and Rajagopal.[8] The rules for generating configuration are those for a random walk and involve the following sequence of steps:

1. Choose a starting configuration in $3N$ dimensional space, $\underline{r}^{(N)}$.
2. Make a move, called a trial move, to a new position $\underline{r}^{(N)\prime}$ chosen from some probability density function $T(\underline{r}^{(N)\prime} \leftarrow \underline{r}^{(N)})$.
3. Accept the trial move with probability,
   $A(\underline{r}^{(N)\prime} \leftarrow \underline{r}^{(N)}) = \mathrm{Min}[1, T(\underline{r}^{(N)} \leftarrow \underline{r}^{(N)\prime})\, f^{\text{eq}}(\underline{r}^{(N)\prime}) / T(\underline{r}^{(N)\prime} \leftarrow \underline{r}^{(N)})\, f^{\text{eq}}(\underline{r}^{(N)})]$
4. If the trial move is accepted then the point $\underline{r}^{(N)\prime}$ in co-ordinate space replaces the point $\underline{r}^{(N)}$ otherwise use the point $\underline{r}^{(N)}$ again.

5. Accumulate the sum $B_r(\underline{r}^{(N)})$
6. Repeat steps 2–5 as many times as necessary.

Let us assume that the system reaches a steady state in which the number of configurations in a volume element $\mathrm{d}\underline{r}^{(N)}$ is $n(\underline{r}^{(N)})\,\mathrm{d}\underline{r}^{(N)}$. Now since the probability that the next move from $\underline{r}^{(N)}$ to $\underline{r}^{(N)}+\mathrm{d}\underline{r}^{(N)}$ is $\mathrm{d}\underline{r}^{(N)'}\,A(\underline{r}^{(N)'}\leftarrow\underline{r}^{(N)})T(\underline{r}^{(N)'}\leftarrow\underline{r}^{(N)'})$ the average number moving from $\mathrm{d}\underline{r}^{(N)}$ to $\mathrm{d}\underline{r}^{(N)'}$ during a single move is $\mathrm{d}\underline{r}^{(N)'}\,A(\underline{r}^{(N)'}\leftarrow\underline{r}^{(N)})T(\underline{r}^{(N)'}\leftarrow\underline{r}^{(N)'})\,n(\underline{r}^{(N)})\,\mathrm{d}\underline{r}^{(N)}$.

At steady state this must be balanced by the number moving from $\mathrm{d}\underline{r}^{(N)'}$ to $\mathrm{d}\underline{r}^{(N)}$ i.e.

$$A\left(\underline{r}^{(N)}\leftarrow\underline{r}^{(N)'}\right)T\left(\underline{r}^{(N)}\leftarrow\underline{r}^{(N)'}\right)\times n\left(\underline{r}^{(N)}\right)\mathrm{d}\underline{r}^{(N)'}$$

So

$$\begin{aligned}&A\left(\underline{r}^{(N)'}\leftarrow\underline{r}^{(N)}\right)T\left(\underline{r}^{(N)'}\leftarrow\underline{r}^{(N)}\right)\times n\left(\underline{r}^{(N)}\right)\mathrm{d}\underline{r}^{(N)}\,\mathrm{d}\underline{r}^{(N)'}\\&=A\left(\underline{r}^{(N)}\leftarrow\underline{r}^{(N)'}\right)T\left(\underline{r}^{(N)}\leftarrow\underline{r}^{(N)'}\right)n\left(\underline{r}^{(N)}\right)\mathrm{d}\underline{r}^{(N)'}\end{aligned}$$

i.e.

$$n\left(\underline{r}^{(N)}\right)/n\left(\underline{r}^{(N)'}\right)=A\left(\underline{r}^{(N)}\leftarrow\underline{r}^{(N)'}\right)T\left(\underline{r}^{(N)}\leftarrow\underline{r}^{(N)'}\right)/A\left(\underline{r}^{(N)'}\leftarrow\underline{r}^{(N)}\right)T\left(\underline{r}^{(N)'}\leftarrow\underline{r}^{(N)}\right)$$

But the ratio of Metropolis algorithm acceptance probabilities is

$$A\left(\underline{r}^{(N)}\leftarrow\underline{r}^{(N)'}\right)/A\left(\underline{r}^{(N)'}\leftarrow\underline{r}^{(N)}\right)=T\left(\underline{r}^{(N)'}\leftarrow\underline{r}^{(N)}\right)f^{\mathrm{eq}}\left(\underline{r}^{(N)}\right)/T\left(\underline{r}^{(N)}\leftarrow\underline{r}^{(N)'}\right)f^{\mathrm{eq}}\left(\underline{r}^{(N)'}\right)$$

Then,

$$n\left(\underline{r}^{(N)}\right)/n\left(\underline{r}^{(N)'}\right)=f^{\mathrm{eq}}\left(\underline{r}^{(N)}\right)/f^{\mathrm{eq}}\left(\underline{r}^{(N)'}\right)\tag{K.6}$$

So this procedure generates a set of configurations with probability of occurrence proportional to $f^{\mathrm{eq}}(\underline{r}^{(N)})$.
More rigorous approaches to this proof may be found in the literature[2,3] than the above argument used by Foulkes *et al.*.[8]

K.2 AN MC ROUTINE

Listed below is a simple Fortran 77 routine to carry out MC calculations in an (NVT) ensemble. In order to use it other routines must be supplied, for example, to calculate energies and set up starting lattices etc., for example see Program MEANBD in Appendix O. Note this is a very basic routine.

```
SUBROUTINE MOVE

C
C SUBROUTINE WHICH CHOOSES A PARTICLE AT RANDOM AND GIVES IT A
RANDOM MOVE
C THE RESULTING CHANGE IN ENERGY IS THEN COMPUTED AND THE MOVE
ACCEPTED OR
C REJECTED ACCORDING TO THE NORMAL MONTE CARLO PROCEDURE FOR
AN (N V T)
C ENSEMBLE, TOTAL PRESSURE AND N(R) ARE ALSO CALCULATED
C
      COMMON/A/NSW,NMOL,NDIV,NN(10,200),DEL,DELTA,XL,YL,ZL,MSW
      COMMON/B/UTOTAL,PTOTAL,R2MAX,XL2,YL2,ZL2,RCORE2,RCORE
      COMMON/C/XC(256),YC(256),ZC(256)
      COMMON/DELR/DELRES
C
      NSW=1
C
C    CHOOSE A PARTICLE AT RANDOM
C
      X=RAND()
      I=NMOL*X
      I=I+1
      IF(I.GT.NMOL)  I=NMOL
      IF(I.LT.0)  I=1
C
C    SET UP PREVIOUS POSITION OF PARTICLE I
C
      XIO=XC(I)
      YIO=YC(I)
      ZIO=ZC(I)
C
C    GIVE IT A RANDOM DISPLACEMENT OF MAXIMUM SIZE DEL
C
      XI=(RAND()-0.5)*DEL
      YI=(RAND()-0.5)*DEL
      ZI=(RAND()-0.5)*DEL
C
```

```
C   OBSERVE PERIODIC BOUNDARY CONDITIONS FOR NEW POSITIONS
C   FOR A CUBIC BOX OF SIDE XL
C
      XIN=AMOD(XIO +XI+XL,XL)
      YIN=AMOD(YIO +YI+YL,YL)
      ZIN=AMOD(ZIO +ZI+ZL,ZL)
C
      UOLD=0.0
      UNEW=0.0
      PROL=0.0
      PREW=0.0
C
C   FORM THE ENERGY SUMMATION SUM OVER J .NE . I U(RIJ) J= 1 UP TO
    NMOL
C   AND THE CORRESPONDING VIRIAL TERM FOR THE OLD AND NEW POSI-
    TION OF PARTICLE I
C
      DO 1 J=1,NMOL
      IF(J.EQ.I)  GO TO 1
      XCJ=XC(J)
      YCJ=YC(J)
      ZCJ=ZC(J)
C
C   MINIMIUM IMAGE DISTANCES FOR NEW POSITIONS
C
      XWN=XCJ-XIN
      IF(XWN.GT.XL2)  XWN=XWN-XL
      IF(XWN.LT. -XL2)  XWN=XWN+XL
      YWN=YCJ-YIN
      IF(YWN.GT.YL2)  YWN=YWN-YL
      IF(YWN.LT. -YL2)  YWN=YWN+YL
      ZWN=ZCJ-ZIN
      IF(ZWN.GT.ZL2)  ZWN=ZWN-ZL
      IF(ZWN.LT.-ZL2)  ZWN=ZWN+ZL
      SJN=XWN*XWN+YWN*YWN+ZWN*ZWN
      RIJNEW(J)=4000.0
      SJNEW(J)=SJN
C   CHECK TO SEE IF RIJ < RCORE ANDIF SO REJECT THE CONFIGURATION
    I.E. USE THE HARD
C   SPHERE CRITERION IF TWO SPHERES OVERLAP REJECT I.E. IF R< RCORE
    U(RIJ) IS
C   ENORMOUS AND THE MOVE IS HIGHLY UNLIKELY
```

```
C
      IF(SJN.LE.RCORE2) GO TO 2
C      CHECK TO SEE RIJ > RMAXIF SO GO TO THE NEXT J
C       I.E. SPHERICAL TRUNCATION
C
      IF(SJN.GE.R2MAX)   GO TO 3
C       FORM U(RIJ) AND RIJ*V(RIJ)
C
      RNEW=SQRT(SJN)
      RIJNEW(J)=RNEW
C
C   SUBROUTNE RETURNS THE VALUE OF THE PAIR POTENTIAL POTNEW
C   AND PAIR VIRIAL VIRNEW FOR PARTICLES I AND J SEPARATED BY
C   RIJ = RNEW
C
      CALL POTFN(SJN,POTNEW,VIRNEW,I,J)
      UNEW=UNEW+POTNEW
      PREW=PREW+VIRNEW
3     CONTINUE
C
C       FOR THE OLD POSITION
C
      XWO=XCJ-XIO
      IF(XWO.GT.XL2)  XWO=XWO-XL
      IF(XWO.LT. -XL2)  XWO=XWO+XL
      YWO=YCJ-YIO
      IF(YWO.GT.YL2)  YWO=YWO-YL
      IF(YWO.LT. -YL2)  YWO=YWO+YL
      ZWO=ZCJ-ZIO
      IF(ZWO.GT.ZL2)  ZWO=ZWO-ZL
      IF(ZWO.LT. -ZL2)  ZWO=ZWO+ZL
      SJO=XWO*XWO+YWO*YWO+ZWO*ZWO
      RIJOLD(J)=4000.0
C
C       SPHERICAL TRUNCATION IGNORE INTERACTION IF
C       RIJ**2 > OR = R2MAX WHERE R2MAX  = RMAX*RMAX
C
      IF(SJO.GE.R2MAX)   GO TO 1
      ROLD=SQRT(SJO)
      RIJOLD(J)=ROLD
      CALL POTFN(SJO,POTOLD,VIROLD,I,J)
```

```
      UOLD=UOLD+POTOLD
      PROL=PROL+VIROLD
      UOLDP(KK)=UOLDP(KK)+POTOLD
      PROLP(KK)=PROLP(KK)+VIROLD
    1 CONTINUE
C
C     CHECK TO SEE IF U FOR NEW CONFIGURATION  IS > ,= OR < U FOR THE
      PREVIOUS ONE
C     AND  IF U NEW < U OR = U OLD  ACCEPT OTHERWISE ONLY ACCEPT
      WITH A PROBABILITY
C     PROPORTIONAL TO  EXP(- (UNEW-UOLD)/KT).
C
      IF(UNEW.LE.UOLD)   GO TO 8
      DELTU=UOLD-UNEW
      CEX=EXP(UOLD-UNEW)
       DEX=RAND()
       IF(CEX.LT.DEX)   GO TO 2
    8   NSW=-1
      MSW=MSW+1
C
C     CALCULATE THE TOTAL ENERGY AND PRESSURE OF THE CONFIGURA-
      TION
C
       UO=UOLD
       UN=UNEW
       UTOTAL=UTOTAL+UN-UO
       PO=PROL
       PN=PREW
       PTOTAL=PTOTAL+PN-PO
C
C     UPDATE PARTICLE COORDINATES
C
       XC(I)=XIN
       YC(I)=YIN
       ZC(I)=ZIN
C
C     ACCUMULATE THE HISTOGRAM N(R) FOR G(R)
C
      DO 5 J=1,NMOL
      IF(I.EQ.J) GO TO 5
      RIJO=RIJOLD(J)
```

```
      RIJN=RIJNEW(J)
      N1=(RIJO)*DELRES+1
10    CONTINUE
      IF(N1.LE.NDIV) NN(N1)=NN(N1) -1
      N1=(RIJN)*DELRES+1
      IF(N1.LE.NDIV) NN(N1)+NN(N1)+1
5     CONTINUE
2     CONTINUE
      RETURN
      END
```

REFERENCES

1. N. Metropolis, A.W. Rosenbluth, M.N. Rosenbluth, A.H. Teller and E. Teller, *J. Chem. Phys.*, **21**, 1087 (1953).
2. W. Feller, *An Introduction to Probability Theory and its Applications*, Wiley, New York, 1968.
3. G. Bhanot, *The Metropolis Algorithm, Rep. Prog. Phys.*, **51**, 429–457 (1988).
4. W.W. Wood, Monte Carlo studies of simple liquid models. In *Physics of Simple Liquids* (Eds. H.N.V. Temperley, J.S. Rowlinson and G.S. Rushbrooke), North-Holland, Amsterdam, 1968, pp. 115–230.
5. J.P. Valleau and S.G. Whittington, A guide to Monte Carlo for statistical mechanics: 1. Highways. In *Statistical Mechanics, Part A: Equilibrium Techniques* (Ed. B.J. Berne), Plenum Press, New York, 1977, pp. 137–168.
6. J.P. Valleau and G.M. Torrie, A guide to Monte Carlo for statistical mechanics: 1. Byways. In *Statistical Mechanics, Part A: Equilibrium Techniques* (Ed. B.J. Berne), Plenum Press, New York, 1977, pp. 169–194.
7. M.P. Allen and D.J. Tildesley, *Computer Simulation of Liquids*, Oxford University Press, Oxford, UK, 1994.
8. W.M.C. Foulkes, L. Mitas, R.J. Needs and G. Rajagopal, Quantum Monte Carlo simulations of solids, *Rev. Mod. Phys.*, **73**, 33–83 (2001).

Appendix L

The Generation of Random Numbers

Most numerical applications involving GLE and LE methods require the generation of random numbers and this is a much discussed and researched topic. The generation of random numbers means finding a procedure to calculate numbers with no strong statistical correlations. There are many procedures to test whether a given routine has statistical correlation which we shall not discuss here. Suffice to say that not all random number generators are equally good and some produce numbers with strong serial correlation. Thus, each routine should be tested before using in a major program especially if many such numbers are to be generated. The key reference in this area is clearly Knuth,[1] however, the various versions of Numerical Recipes[2] have a very good collection of routines to generate random numbers and an excellent discussion of practical methods for their generation is given by Hammond, Lester and Reynolds.[3]

L.1 GENERATION OF RANDOM DEVIATES FOR BD SIMULATIONS

As the random displacements, X_i must be generated by calculating coupled random deviates their calculation is one of the most time consuming operations and so efficient methods for their generation are vital. This may be done by use of a multivariate normal deviate generator or by the following scheme[4].

First one generates normal un-coupled random deviates,

$$\left[\{x_i\};\langle x_i\rangle = 0, \langle x_i x_j\rangle = 2\delta_{ij}\,\Delta t\right]$$

and the coupled random displacements are given by a weighted sum of these,

$$X_i(\Delta t) = \sum_{j=1}^{i} \sigma_{ij}\, x_j \tag{L.1}$$

where the weighting factors are given by

$$\sigma_{ii} = \left(D_{ii} - \sum_{k=1}^{i-1} \sigma_{ik}^{2} \right)^{1/2} \tag{L.2}$$

$$\sigma_{ij} = \left(D_{ij} - \sum_{k=1}^{j-1} \sigma_{ik}\sigma_{jk} \right) \bigg/ \sigma_{jj}, \qquad i > j \tag{L.3}$$

This method also shows that there are limitations on the form of the D_{ij} used as they must form a positive definite matrix or else eq. (L.2) will not yield real displacements.

REFERENCES

1. D.E. Knuth, *The Art of Computer Programming, Vol. 2, Semi-numerical Algorithms*, 2nd edn., Chapter 3, Addison-Wesley, Reading, MA, 1981.
2. See for example, W.H. Press, B.P. Flannery, S.A. Teukolsky and W.T. Vetterling, *Numerical Recipes in Fortran: The Art of Scientific Computing*, Chapter 7, Cambridge University Press, Cambridge, UK, 1981.
3. B.L. Hammond, W.A. Lester Jr. and P.J. Reynolds, *Monte Carlo Methods in Ab Initio Quantum Chemistry*, World Scientific, Singapore, 1994.
4. D.L. Ermak and J.A. McCammon, J. Chem. Phys., **69**, 1352 (1978).

Appendix M

Hydrodynamic Interaction Tensors

One of the key elements in performing BD simulations on concentrated polymers and colloidal suspensions is to evaluate the hydrodynamic interaction tensors, both friction tensors and mobility tensors, which represent the indirect, solvent-mediated interaction between the particles. As detailed in Chapter 7, there are many methods available to calculate these tensors and references where these methods and results of their application may be found are given in that chapter.

However, there are some forms of the mobility tensors for translational effects, which are commonly used, and we have thus included formulae for these particular cases below.

In all these formulae, $\mu_0 = 1/(6\pi\eta a_i)$, a_i is the radius of particle i and **rr** is a dyadic.

M.1 THE OSEEN TENSOR FOR TWO BODIES[1]

$$\begin{aligned}\boldsymbol{\mu}_{ii}/\mu_0 &= \mathbf{1} \\ \boldsymbol{\mu}_{ij}/\mu_0 &= \mathbf{r}_{ij}^{-1}(3a_i/4)\left[\mathbf{1}+\mathbf{r}_{ij}\mathbf{r}_{ij}/\mathbf{r}_{ij}^2\right]\end{aligned} \tag{M.1}$$

M.2 THE ROTNE–PRAGER TENSOR FOR TWO BODIES[1]

$$\begin{aligned}\boldsymbol{\mu}_{ii}/\mu_0 &= \mathbf{1} \\ \boldsymbol{\mu}_{ij}/\mu_0 &= \left(3a_i/4r_{ij}^{-1}\right)\left[\mathbf{1}+\mathbf{r}_{ij}\mathbf{r}_{ij}/r_{ij}^2\right]-r_{ij}^{-3}(3a_i/4)(a_i^2+a_j^2)\left[\mathbf{r}_{ij}\mathbf{r}_{ij}/r_{ij}^2-1/3\mathbf{1}\right]\end{aligned} \tag{M.2}$$

M.3 THE SERIES RESULT OF JONES AND BURFIELD FOR TWO BODIES[2]

Exact results are available for $\boldsymbol{\mu}_{ij}$ for two spheres with stick boundary conditions.[3] However, a very convenient expansion in terms of (a/r) has been made by Jones and Burfield,[2]

$$\boldsymbol{\mu}_{ij} = \alpha_{ij}(r)\mathbf{r}_{ij}\mathbf{r}_{ij}/r_{ij}^2+\beta_{ij}(r)\left[\mathbf{1}-\left(\mathbf{r}_{ij}\mathbf{r}_{ij}/r_{ij}^2\right)\right] \tag{M.3}$$

where $\alpha_{ij}(r)$ and $\beta_{ij}(r)$ are functions of the function (a/r) given by

$$\begin{aligned}\alpha_{ij}(r) &= (4\pi\eta a)^{-1}\sum_{n=0} a_n^{ij}(a/r)^{2n+|i-j|}\\ \beta_{ij}(r) &= (4\pi\eta a)^{-1}\sum_{n=0} b_n^{ij}(a/r)^{2n+|i-j|}\end{aligned} \tag{M.4}$$

where the expansion coefficients are tabulated for slip and stick boundary conditions.

Many two-body tensors have been expressed in a similar manner by Jones and co-workers, for example for permeable spheres and droplets.[4]

M.4 MAZUR AND VAN SAARLOOS RESULTS FOR THREE BODIES[5]

$$\begin{aligned}\boldsymbol{\mu}_{ij}/\mu_0 = {} & \delta_{ij}\mathbf{1} + r_{ij}^{-1}(3a_i/4r_{ij})\left[\mathbf{1}+\mathbf{r}_{ij}\mathbf{r}_{ij}/r_{ij}^2\right] - r_{ij}^{-3}(3a_i/4)\left(a_i^2+a_j^2\right)\left[\mathbf{r}_{ij}\mathbf{r}_{ij}/r_{ij}^2 - 1/3\mathbf{1}\right]\\ & + \sum_{k\neq i,j}\Big\{r_{ik}^{-2}r_{kj}^{-2}(-15/8)a_i a_k^3\,(a_i a_k^3)\left[(1-3(\underline{r}_{ik}\cdot\underline{r}_{kj}/r_{ij}^2)^2)(\mathbf{r}_{ik}\mathbf{r}_{kj}/r_{ik}r_{kj})\right]\\ & + r_{ik}^{-2}r_{kj}^{-4}(3/8)(a_i a_k^3)(5a_j^2+3a_k^2)\Big[\left[(1-5(\underline{r}_{ik}\cdot\underline{r}_{kj}/r_{ik}r_{kj})^2)(\mathbf{r}_{ik}\mathbf{r}_{kj}/r_{ik}r_{kj})\right]\\ & + 2(\underline{r}_{ik}\cdot\underline{r}_{kj}/r_{ik}r_{kj})(\mathbf{r}_{ik}\mathbf{r}_{kj}/r_{ik}r_{ik})\Big]\\ & + r_{ik}^{-4}r_{kj}^{-2}(3/8)\left(a_i a_k^3\right)(5a_i^2+3a_k^2)\Big[\left[(1-5(\underline{r}_{ik}\cdot\underline{r}_{kj}/r_{ik}r_{kj})^2)(\mathbf{r}_{ik}\mathbf{r}_{kj}/r_{ik}r_{kj})\right]\\ & + 2(\underline{r}_{ik}\cdot\underline{r}_{jk}/r_{ik}r_{kj})(\mathbf{r}_{kj}\mathbf{r}_{kj}/r_{kj}r_{kj})\Big]\\ & + r_{ik}^{-3}r_{kj}^{-3}(1/64)a_i a_k^5\Big[(49-117(\underline{r}_{ik}\cdot\underline{r}_{kj}/r_{ik}r_{kj})^2)\mathbf{1}\\ & - (93-315(\underline{r}_{ik}\cdot\underline{r}_{kj}/r_{ik}r_{kj})^2)(\mathbf{r}_{ki}\mathbf{r}_{ik}/r_{ik}r_{ik} + \mathbf{r}_{kj}\mathbf{r}_{kj}/r_{kj}r_{kj}) + 54(\underline{r}_{ik}\cdot\underline{r}_{kj}/r_{ik}r_{kj})\,\mathbf{r}_{ik}\mathbf{r}_{kj}/r_{kj}r_{ik}\\ & + (729-1575(\underline{r}_{ik}\cdot\underline{r}_{kj}/r_{ik}r_{kj})^2)(\underline{r}_{ik}\cdot\underline{r}_{kj}/r_{ik}r_{kj})\mathbf{r}_{ik}\mathbf{r}_{kj}/r_{ik}r_{kj}\Big]\Big\}\end{aligned} \tag{M.5}$$

Results for four-body translation, three-body translation−rotation and rotation−rotation effects are also available in ref. 5.

M.5 RESULTS OF LUBRICATION THEORY[6]

Ladd[6] gives the form of the friction tensor for two particles using lubrication theory as,

$$\boldsymbol{\zeta}_{ij} = X_{ij}^A(r_{ij})\mathbf{r}_{ij}\mathbf{r}_{ij}/r_{ij}^{\,2} + Y_{ij}^A(r_{ij})\left[\mathbf{1}-\mathbf{r}_{ij}\mathbf{r}_{ij}/r_{ij}^{\,2}\right] \tag{M.6}$$

where details of the functions $X_{ij}^A(r_{ij})$ and $Y_{ij}^A(r_{ij})$ are available in the literature[7].

M.6 THE ROTNE–PRAGER TENSOR IN PERIODIC BOUNDARY CONDITIONS[8]

As outlined in Chapter 7, the Rotne–Prager tensor may be summed in PBC's and the result put in a convenient form by Graham *et al.*,[8]

$$\mu_{ij} = \left[1-(6/\sqrt{\pi})\alpha r + 40/(3\sqrt{\pi})\alpha^3 r^3\right] I + \sum_{\underline{n}\in Z^3}{}' \boldsymbol{M}^{(1)}(\underline{r}_{ij\underline{n}}) + (1/V)\sum_{\underline{k}\neq 0}{}' \boldsymbol{M}^{(2)}(\underline{k})\cos(\underline{k}\cdot\underline{r}_{ij\underline{n}}) \quad \text{(M.7)}$$

where $\underline{r}_{ij\underline{n}} = \underline{r}_i - \underline{r}_j + \underline{n}L$, L is the length of a side of the cubic box, α a parameter, the first summation is over all lattice points $\underline{n} = (n_x, n_y, n_z)$ with n_x, n_y, n_z integers but excluding the $\underline{n} = \underline{0} = (0,0,0)$ and the second sum is over reciprocal lattice vectors $\underline{k} = 2\pi\underline{n}/L$.

The explicit expressions for $\boldsymbol{M}^{(1)}$ and $\boldsymbol{M}^{(2)}$ are,

$$\boldsymbol{M}^{(1)}(\underline{r}) = \{C_1 \,\mathrm{erfc}(\alpha r) + C_2 \exp(-\alpha^2 r^2)/(\sqrt{\pi})\}\mathbf{1} + \{C_3 \,\mathrm{erfc}(\alpha r) + C_4 \exp(-\alpha^2 r^2)/(\sqrt{\pi})\}\boldsymbol{rr}/r^2 \quad \text{(M.8)}$$

and

$$\boldsymbol{M}^{(2)}(\underline{k}) = (r - 1/3 r^3 k^2)(1 + 1/4\alpha^{-2}k^2 + 1/8\alpha^{-4}k^4)6\pi/k^2 \exp(-1/4\alpha^{-2}k^2)[\mathbf{1} - \boldsymbol{kk}/k^2]$$

where

$$C_1 = (3/4r + 1/2r^3 r^{-3})$$
$$C_2 = (4\alpha^7 r^3 r^4 + 3\alpha^3 rr^2 - 20\alpha^5 r^3 r^2 - 9/2\alpha r + 14\alpha^3 r^3 + \alpha r^3 r^{-2})$$
$$C_3 = (3/4rr^{-1} - 3/2r^3 r^{-3})$$

and

$$C_4 = (-4\alpha^7 r^3 r^4 - 3\alpha^3 rr^2 + 16\alpha^5 r^3 r^2 + 3/2\alpha r - 2\alpha^3 r^3 - 3\alpha r^3 r^{-2})$$

REFERENCES

1. J. Happel and H. Brenner, *Low Reynolds Number Hydrodynamics*, Noorhoff, Lyden, 1991.
2. R.B. Jones and G.S. Burfield, *Physica*, **133A**, 152 (1985).

3. D.J. Jeffrey and Y. Onishi, *J. Fluid Mech.*, **139**, 261 (1984); R. Schmidtz and B.U. Felderhof, *Physica*, **113A**, 90, 103 (1982); *Physica*, **116A**, 163 (1982).
4. R.B. Jones and R. Schmitz, *Physics*, **149A**, 373 (1988).
5. P. Mazur and W. van Saarloos, *Physica*, **115A**, 21 (1982).
6. A.J.C. Ladd, *J. Chem. Phys.*, **93**, 3484 (1990).
7. L. Durlofsky, J.F. Brady and G. Bossis, *J. Fluid Mech.*, **180**, 21 (1987); D.J. Jeffrey and Y. Onishi, J. Fluid Mech., **139**, 261 (1984); S. Kim and R.T. Miflin, *Phys. Fluids*, **28**, 2033 (1985); D.J. Jeffrey and R.M. Corless, *Physico Chem. Hydrodyn.*, **10**, 461 (1988).
8. C. Stolz, J.J. de Pablo and M. D. Graham, *J. Rheol.*, **50**, 137 (2006).

Appendix N

Calculation of Hydrodynamic Interaction Tensors

For convenience, let us rewrite eq. (7.9) in the form,[1]

$$\begin{aligned} \underline{V}_i &= \sum_{j=1}^{n} \boldsymbol{\mu}(j,k;0,0)\cdot\underline{F}_j + \sum_{j=1}^{n} \boldsymbol{\mu}(j,k;0,1)\cdot\underline{T}_j \\ \underline{\Omega}_i &= \sum_{j=1}^{n} \boldsymbol{\mu}(j,k;1,0)\cdot\underline{F}_j + \sum_{j=1}^{n} \boldsymbol{\mu}(j,k;1,1)\cdot\underline{T}_j \end{aligned} \tag{N.1}$$

that is, we denote the mobility matrix by $\boldsymbol{\mu}(j,k;\sigma,\sigma')$.

Let us assume that the velocity $\underline{\upsilon}(\underline{r})$ obeys boundary condition x, for example either free boundaries or PBC, then

$$\underline{\upsilon}(\underline{r}) = \int \mathbf{T}_x(\underline{r},\underline{r}')\cdot\mathbf{F}(\underline{r}')\,\mathrm{d}\underline{r} \tag{N.2}$$

where $\mathbf{F}(\underline{r}')$ is the induced hydrodynamic force density in the fluid, the integral is over R^3 for free boundary conditions and over the periodic cell of side L for PBC and the Green function is given by,

$$\mathbf{T}_{\mathrm{Free}}(\underline{r},\underline{r}') = (2\pi)^{-3}\int_{R^3} (1/\eta_f)(\mathbf{1}-\mathbf{kk}/\mathrm{kk})k^{-2}\exp[\mathrm{i}\underline{k}\cdot(\underline{r}-\underline{r}')]\mathrm{d}\underline{k} \tag{N.3}$$

and

$$\mathbf{T}_{\mathrm{PBC}}(\underline{r},\underline{r}') = \sum_{\underline{k}\neq\underline{0}} (4\pi^2\eta_f L)^{-1}(\mathbf{1}-\mathbf{kk}/\mathrm{kk})k^{-2}\exp[2\pi\mathrm{i}\underline{k}\cdot(\underline{r}-\underline{r}')/L]$$

and if we write the induced force density as,

$$\underline{F}(\underline{r}) = \sum_{k=1}^{n} \underline{f}_k(\underline{r}-\underline{r}_k)\delta(|\underline{r}-\underline{r}_k|-a) \tag{N.4}$$

where $\underline{f}_k(\underline{r})$ is the traction on the surface of sphere k.

In order to find a series solution, we introduce the spherical harmonics, $Y_{lm}(\underline{r}/r)$ and the vector harmonics A_{lm} ,B_{lm} and C_{lm} by,[2]

$$A_{lm}(\underline{r}/r) = r^{-l+1}\underline{\nabla} r^{l} Y_{lm}(\underline{r}/r)$$
$$B_{lm}(\underline{r}/r) = r^{l+2}\underline{\nabla} r^{-l-1} Y_{lm}(\underline{r}/r) \tag{N.5}$$

and

$$C_{lm}(\underline{r}/r) = r^{-l}\,\underline{\nabla}\times[\underline{r}r^{l} Y_{lm}(\underline{r}/r)]$$

and this leads to two families of vector functions for $l \geq 1$,

$$\underline{v}_{lm0}(\underline{r}) = r^{l-1} A_{lm}(\underline{r}/r)$$
$$\underline{v}_{lm1}(\underline{r}) = \mathrm{i}\, r^{l} C_{lm}(\underline{r}/r) \tag{N.6}$$

and

$$\underline{v}_{1\mathrm{mk}}(\underline{r}) = r^{l+1}[(1/2l)\,(l+1)\,(2l+1)\,A_{lm}(\underline{r}/r) + B_{lm}(\underline{r}/r)]$$

which are solutions of the linearised Navier−Stokes equations and the adjoint set is

$$\underline{w}_{lm0}(\underline{r}) = 1/\{l(2l+1)\}r^{-1}[A_{lm}(\underline{r}/r) - \tfrac{1}{2}(2l+3)\,B_{lm}(\underline{r}/r)]$$
$$\underline{w}_{lm1}(\underline{r}) = i/\{l(l+1)\}r^{-l-1}C_{lm}(\underline{r}/r)$$

and

$$\underline{w}_{lm2}(\underline{r}) = 1/\{(l+1)(2l+1)\}r^{-l-2}B_{lm}(\underline{r}/r)$$

so,

$$\int_{R^3} (1/a)\delta(r-a)\,\underline{w}_{lm\sigma}(\underline{r})^{*}\cdot\underline{v}_{l'm'\sigma}(\underline{r})\,\mathrm{d}^3\underline{r} = \delta_{ll'}\,\delta_{mm'}\,\delta_{\sigma\sigma'} \tag{N.8}$$

Introducing the orthonormal set $\underline{e}_{-1}$ = $(1,-i,0)/\surd 2$, $\underline{e}_0$ = $(0,0,1)$ and $\underline{e}_1$ = $-(1,i,0)/\surd 2$ then

$$A_{lm}(\underline{r}/r) = (3/4\pi)^{1/2}\,\underline{e}_m \text{ and } C_{lm}(\underline{r}/r) = (3/4\pi)^{1/2}\,\underline{e}_m \times \underline{r}$$

Now, we may expand the fluid velocity $\underline{\upsilon}$ about particle j in terms of the functions as,

$$\underline{\upsilon}(\underline{r}_j + \underline{r}) = \sum_{lm\sigma} V_{lm\sigma}(j)\,\underline{v}_{lm\sigma}(\underline{r}) \tag{N.9}$$

and the induced traction as,

$$\underline{f}_k(a\underline{r}'/r') = \sum_{l'm'\sigma'} F_{l'm'\sigma'}(k)\underline{w}_{l'm'\sigma'}(a\underline{r}'/r') \tag{N.10}$$

which leads to

$$V_{lm\sigma}(j) = \sum_{k=1}^{n}\sum_{l'm'\sigma'} \mathrm{T}_x(j, lm\sigma; l'm'\sigma')F_{l'm'\sigma'}(k) \tag{N.11}$$

where

$$\mathbf{T}_{\text{Free}}(j, lm\sigma;\ k, l'm'\sigma') = a^4(8\pi^3\eta_f)^{-3}\int_{\partial S_1} \mathrm{d}^2\underline{r}\int_{\partial S_2} \mathrm{d}^2r'\int_{R^3} \underline{w}_{lm\sigma}(a\underline{r})\cdot(\mathbf{1} - \mathbf{kk}/\mathrm{kk})k^{-2}\cdot\underline{w}_{l'm'\sigma}(a\underline{r}')$$
$$\times\exp[\mathrm{i}\underline{k}\cdot(\underline{r}_j + a\underline{r} - \underline{r}_k - a\underline{r}')]\mathrm{d}\underline{k} \tag{N12}$$

and ∂S_1 is a unit sphere with its centre at the origin and,

$$\mathbf{T}_{\text{PBC}}(j, lm\sigma;\ k, l'm'\sigma') = a^4(3\pi^2\eta_f L)^{-1}\int_{\partial S_1} \mathrm{d}^2\underline{r}\int_{\partial S_2} \mathrm{d}^2\underline{r'}\sum_{k\neq 0} \underline{w}_{lm\sigma}(a\underline{r})\cdot(\mathbf{1} - \mathbf{kk}/\mathrm{kk})k^{-2}\cdot\underline{w}_{l'm'\sigma}(a\underline{r}')$$
$$\times\exp[2\pi\mathrm{i}\underline{k}\cdot(\underline{r}_j + a\underline{r} - \underline{r}_k - a\underline{r}')/L]$$

The stick boundary conditions at the particle surfaces means that $\underline{v}(\underline{r}_j + a\underline{r}) = \underline{V}_j + \underline{\Omega}_j \times \underline{r}$ and thus,

$$\begin{aligned} V_{1m0}(j) &= (4\pi/3)^{1/2}\underline{V}_j\cdot\underline{e}_m^* \\ V_{1m1}(j) &= -\mathrm{i}(4\pi/3)^{1/2}\,\underline{\Omega}_j\cdot\underline{e}_m^* \\ V_{1m2}(j) &= 0 \\ V_{lm\sigma}(j) &= 0 \text{ if } l \geq 2 \end{aligned} \tag{N.13}$$

$$F_{1m0}(j) = (3/4\pi)^{1/2}\underline{F}_j\cdot\underline{\mathrm{e}}_m^*/a \tag{N.14}$$

and

$$F_{1m1}(j) = -\mathrm{i}(3/4\pi)^{1/2}\,\underline{T}_j\cdot\underline{e}_m^*/a$$

Equation (N.11) may be truncated at various levels to give approximate solutions and we will call this the Lth approximation if $1 \le l$ and $l' \le L$ which gives,

$$V^L_{lm\sigma}(j) = \sum_{k=1}^{n} \sum_{l'=1}^{L} \sum_{m'\sigma'} T^{LL}_x(j,lm\sigma;\ k,l'm'\sigma')F^L_{l'm'\sigma'}(k) \tag{N.15}$$

for $1 \le l \le L$.

Equation (N.15) may then be used to construct approximations to $\mu(j,k;\sigma,\sigma')$, which may be done by regarding the objects in eq. (N.15) as elements in a vector space with components indexed by (j,l,m,σ), and the components of the matrix T_x^{LL} are indexed in this way. This vector space may be partitioned into a subspace A composed of the $(j,1,m,\sigma)$ components with $\sigma = 0,1$ and a subspace B consisting of the other components, and then, $V^L = V^L_A + V^L_B$, $F^L = F^L_A + F^L_B$ and $T^{LL}_x = T^{LL}_{x\ \mathrm{AA}} + T^{LL}_{x\ \mathrm{AB}} + T^{LL}_{x\ \mathrm{BA}} + T^{LL}_{x\ \mathrm{BB}}$, where $T^{LL}_{x\ \mathrm{CD}}$ maps a vector from subspace D to subspace C.

Then,

$$\begin{aligned} V^L_A &= T^{LL}_{x\ AA} F^L_A + T^{LL}_{x\ AB} F^L_B \\ V^L_B &= T^{LL}_{x\ BA} F^L_A + T^{LL}_{x\ BB} F^L_B \end{aligned} \tag{N.16}$$

and from eq. (N.13) $V^L_{\mathrm{B}} = 0$ and so,
$F^L_B = -\,{T^{LL}_{x\ BB}}^{-1} T^{LL}_{x\ BA} F^L_A$ and so eq. (N.16) yields,

$$\begin{aligned} V^L_A &= (T^{LL}_{x\ AA} - T^{LL}_{x\ AB} {T^{L}_{x\ BB}}^{-1} T^{LL}_{x\ BA}) F^L_A \\ &= M^{LL}_x F^L_A \end{aligned} \tag{N.17}$$

Then we may construct $\underline{V}_j$ and $\underline{\Omega}_j$ (or $\underline{F}_k$ and $\underline{T}_k$) and thus from eq. (N.1) obtain,

$$\mu^{LL}_x(j,k;\sigma,\sigma') = (3/4\pi a) i^{\sigma-\sigma'} \sum_{m=-1}^{1} \sum_{m'=-1}^{1} M^{LL}_x(j,1m\sigma;k,1m\sigma')\underline{e}_m\, \underline{e}_{m'} \tag{N.18}$$

It may be shown that[1]

1. For both free and PBC μ^{LL}_x from eq. (N.18) is positive definite at all levels of truncation;
2. Incompressibility holds for all L;

3. At the lowest level of truncation if (eq. (N.18)) $L = 1$ then for free-boundary conditions,

$$\mu_{ij}^{11}(j,k;0,0)/\mu_0 = \delta_{jk}\mathbf{1} + (1-\delta_{jk})[r_{jk}^{-1}(3a/4)[\mathbf{1} + \mathbf{r_{jk}r_{jk}}/r_{jk}^2] - r_{jk}^{-3}(a^3/4)[3\mathbf{r_{jk}r_{jk}}/r_{jk}^2 - \mathbf{1}]$$
$$-(1/20)\sum_{J=1}^{n}(1-\delta_{jJ})(1-\delta_{kJ})r_{jJ}^{-3}r_{kJ}^{-3}a^6[(\mathbf{1} - 3\mathbf{r_{jJ}}\,\mathbf{r_{jJ}}/r_{jJ}^2$$
$$-3\mathbf{r_{kJ}r_{kJ}}/r_{kJ}^2 + 9(\underline{r}_{jJ}/r_{jJ}\cdot\underline{r}_{Jk}/r_{Jk})\mathbf{r_{jJ}}\,\mathbf{r_{Jk}}/r_{jJ}\,r_{Jk})]$$

(N.19)

and the other components of $\mu_{ij}^{11}(j,k;\sigma,\sigma')/\mu_0$ are the standard Rotne–Prager ones (see Appendix M, Section M.2).

REFERENCES

1. K. Briggs, E.R. Smith, I.K. Snook and W. Van Megen, *Phys. Lett. A*, **154**, 149 (1991).
2. I.S. Gradshteyn and I.M. Ryzhik, *Table of Integrals, Series and Products*, Academic Press, New York, 1980.

Appendix O

Some Fortran Programs

Some may have noted the paucity of diagrams in this book. This is quite deliberate as I believe that it is preferable for the reader to go to the original literature and also to generate data for themselves. I also believe that the reader will profit greatly from generating their own data for various parameter values, plotting this data for themselves and finding out what effect changing various parameters has on the calculated values of the quantity of interest.

To this end I have included below three FORTRAN programs, these are rather old programs and are not intended to be research tools but are only included as aids to understanding.

These programs are as follows:

1. VOLSOL which computes a time correlation given an assumed memory function.
2. MEANBD which is a BD program based on the simple Ermak BD algorithm (eqs. (6.45) and (7.27)) in the diffusive limit. It will calculate the pressure, energy, radial distribution function and mean-square displacement.
3. SPHERE3 which is a program to calculate the positions for three hydrodynamically interacting spheres falling in a gravitational field. It uses the mobility tensor of Mazur and van Saaloos, *Physica*, **115A**, 21–57 (1982) and was written by Dr. Keith Briggs.

Two excellent introductions to algorithms suitable for implementing many of the numerical schemes discussed in this book are as follows:

1. S.E. Koonin and D.C. Meredith, *Computational Physics, Fortran Version*, Addison-Wesley, CA, USA, 1990, this book includes Fortran programs in particular some to carry out Monte Carlo calculations including basic Quantum Monte Carlo ones.
2. F.J. Vesely, *Computational Physics, An Introduction,* Kluwer Academic/Plenum Publishers, New York, 2001.

Other programs may be found in the literature as a search on the World Wide Web will reveal. Some of these programs are as follows:

1. BROWND which performed BD simulations in the diffusive limit using the Ermak-McCammon algorithm with either Oseen or Rotne-Prager two-body hydrodynamic

interactions with a simple cutoff. This program may be found at the time of writing at http://www.mpikg-golm.mpg.de/th/physik/allen_tildesley/allen_tildesley.33

2. T.J. Walls, *Kinetic Monte Carlo Simulations of Perovskite Crystal Growth with Long Range Coulomb Interactions*, Thesis for Bachelor of Science with Honors in Physics, College of William and Mary, VA, USA, 1999, which contains a KMC program.
3. P. Stoltze, *Microkinetic Simulation of Catalytic Reactions*, Prog. Surf. Sci., **65**, 65–150 (2000), which contains a KMC program.
4. Routines to construct a Quantum Monte Carlo code are available in: B.L. Hammond, W.A. Lester Jr. and P.J. Reynolds, *Monte Carlo Methods in Ab Initio Quantum Chemistry*, World Scientific, Singapore, 1994.
5. The Quantum Monte Carlo code CASINO is available from http://www.tcm.phys.cam.ac.uk/~mdt26/casino2.html but registration is necessary.

Various full MD codes are also currently available, some of which are free and some which require the paying of fees, details of many of these program may be found on the World Wide Web.

Subroutines are also available which may be used to construct an MD program, for example:

1. M.P. Allen and D.J. Tildesley, *Computer Simulation of Liquids,* Oxford University Press, 1994.
2. D. Fincham and D.M. Heyes, *Recent Advances in Molecular-Dynamics Computer Simulation*, M.W. Evans (ed.), Adv. Chem. Phys., LXIII, Wiley, 1985.
3. Frenkel and Smit have, at the time of writing, much useful information on the World Wide Web, e.g. see http://molsim.chem.uva.nl/frenkel_smit/README.html
4. At the time of writing Allen and Tildesley also have very useful information available at http://www.mpikg-golm.mpg.de/th/physik/allen_tildesley/allen_tildesley.33
5. Also at the time of writing, the following web site is worth visiting http://www.CCP5

```
PROGRAM VOLSOL
C
C
C     SOLVES THE VOLTERRA EQUATION FOR A TIME CORRELATION FUNC-
TION, Y(T), GIVEN THE
C    CORRESPONDING  MEMORY FUNCTION , FUNKY(T)
C     B.J. Berne and G.D. Harp, On the Calculation of Time Correlation Functions, in
Advances in Chemical
C     Physics, XVII, Editors I. Prigogine and S.A. Rice, Wiley, New York, 1970, p. 217,
218, 221-222.
C    TO CALCULATE THE MEMORY FUNCTION FROM A CORRELATION FUNC-
TION IS MORE
C    DIFFICULT, SEE THE ABOVE REFERENCE, pp. 217-221
C
      DIMENSION TIME(700),FUNKA(700),Y(700)
```

```
      COMMON/PARAMS/A,B
      NP=700
      TIMAX=100.0
      DT=TIMAX/NP
      DT2=DT*DT
      T0=0.0
      Y0=1.0
      A=1.0
      B=-1.0/2.00
      FUNKYO=FUNKY(T0)
      PRINT 1
      PRINT 2,NP,TIMAX,DT
      PRINT 3,T0,Y0,FUNKYO
 1    FORMAT(///////,40X,* MEMORY FUNCTION PROGRAM *,//////)
 2    FORMAT(* NP = *,I5,2X,* MAXIMIUM TIME =*,E16.9,
     12X,* DT = *,E16.9,//)
 3    FORMAT(*  INITIAL TIME = *,E16.9,5X,* YO = *,E16.9,5X
     1,* MO = *,E16.9,///)
      T=T0+DT
       TIME(1)=T
      Y(1)=1.0-0.5*FUNKYO*T*T
      T=T+DT
      TIME(2)=T
      Y(2)=1-DT2*(FUNKY(TIME(1))+FUNKYO*Y(1))
      T=T0
      DO 10 I=1,NP
      T=T+DT
      TIME(I)=T
      FUNKA(I)=FUNKY(T)
10    CONTINUE
      DO 4 I=3,NP
      IM1=I-1
      IM2=I-2
      Y(I)=Y(IM2)-DT2*(FUNKA(IM1)+FUNKYO*Y(IM1))
      SUM=0.0
      DO 5 J=2,IM1
      IMJ=I-J
      SUM=SUM+FUNKA(J)*Y(IMJ)
 5    CONTINUE
      Y(I)=Y(I)-2.0*DT2*SUM
 4    CONTINUE
      DO 6 I=1,NP
      PRINT 7,I,TIME(I),Y(I),FUNKA(I)
```

```
7   FORMAT(*I = *,I5,2X,* TIME = *,E16.9,2X,* Y(I)= *,E16.9,
   12X,* M(I) = *,E15.9)
6   CONTINUE
    STOP
    END
C##########################################################################
################
C
    FUNCTION FUNKY(T)
C
C     MEMORY FUNCTION
C
    COMMON/PARAMS/A,B
    FUNKY=A*EXP(B*T)
    RETURN
    END
```

```
PROGRAM MEANBD
C
C      BROWNIAN DYNAMICS PROGRAM FOR A SPHERICALLY SYMMETRIC
PAIR POTENTIAL
C    THIS VERSION FOR A HYDRODYNAMICALLY DILUTE DISPERSION WITH
DIRECT
C     COMPUTATION OF MEAN SQUARE DISPLACEMENTS
C    IT CAN USE TWO DIFFERENT TIME STEPS, ONE DTP FOR EQUILIBRATION
C     TWO DT  FOR RUN PROPER
     COMMON/COREL/X2R(15,256),Y2R(15,256),Z2R(15,256),
    1X2S(15,256),Y2S(15,256),Z2S(15,256),
    2SUMX2(256),SUMY2(256),SUMZ2(256),IACCUM
     COMMON/MAID/DIAM
     COMMON/CORDS/XH(256),YH(256),ZH(256),X(256),Y(256),Z(256)
    1,NN(200)
     COMMON/INFO/NMOL,XL,XL2,XL5,RCUT,RCORE,RCORE2,NTYPE,IBCC,
    1NPRE,NTSTEPS,ITTOT,NPRINT,NWRITE,DT,DKTDT,DKTDTP,SQRVAR,
    SQRVARP
    2,YL,YL2,ZL,ZL2
     COMMON/FORCOR/INDIV,NCHAN,MULT,KMAX,TPIL
     DIMENSION CXA(256),CYA(256),CZA(256),CRA(256)
C
C     IF NTYPE=1 RUN STARTS FROM FCC LATTICE
C     IF NTYPE=2 CONTINUES PREVIOUS RUN
C     IF NTYPE=3 RUN RE-ZEROS VARIABLES AND STARTS FROM END
```

```
C     OF PREVIOUS RUN
C
C    SET UP DATA TO START A RUN
C    AND IF NTYPE NE 2 DO NPRE EQUILIBRATION STEPS
      CALL STARTER
      NNMOL=NMOL
C
C     CORRELATOR ROUTINE
C
C     NCHAN= NO. OF CHANNELS IN CORRELATION FUNCTION
C     EACH CHANNEL BEING SEPARATED BY A TIME TOR = MULT * H
C
      TORT=MULT*DT
      TORMAX=NCHAN*TORT
C     NBIG=TIME STEPS EQUIVALENT TO TORMAX
      NBIG=MULT*NCHAN
      WRITE(9,134)TORT,TORMAX,NBIG
 134  FORMAT(//,'TOR = ',E16.9,' TORMAX = ',E16.9,
     1' NUMBER OF STEPS EQUIVALENT TO TORMAX = ',I5,//)
      ISTART=ITTOT
      KOUNT=KOUNT+1
C
      IF(NTYPE.EQ.2) GO TO 2345
      ICHAN=NCHAN+1
C
C     INITIALIZE THE NCHAN REGISTERS
C
      ITSTEPS=0
 6    CONTINUE
      ITSTEPS=ITSTEPS+1
      IF(ICHAN.LE.1)GOTO 6000
      CALL MOVE(ITSTEPS,1)
      IF(MOD(ITSTEPS,MULT).NE.0) GO TO 60
      ICHAN=ICHAN-1
      DO 61 IM=1,NMOL
      X2R(ICHAN,IM)=XH(IM)
      Y2R(ICHAN,IM)=YH(IM)
      Z2R(ICHAN,IM)=ZH(IM)
 61   CONTINUE
60    CONTINUE
      GO TO 6
 6000 CONTINUE
C
```

```
C     ZERO STORES
C
      IACCUM=0
      ITTOT=0
      ISTART=1
      DO 899 ICHAN=1,NCHAN
      DO 899 IMOL=1,NMOL
      X2S(ICHAN,IMOL)=0.0
      Y2S(ICHAN,IMOL)=0.0
      Z2S(ICHAN,IMOL)=0.0
 899  CONTINUE
      DO 888 IMOL=1,NMOL
      SUMX2(IMOL)=0.0
      SUMY2(IMOL)=0.0
      SUMZ2(IMOL)=0.0
 888  CONTINUE
 2345 CONTINUE
      IT=ISTART-1
 7    CONTINUE
      IT=IT+1
      CALL MOVE(IT,2)
      IF(MOD(IT,MULT).NE.0) GO TO 7000
      DO 70 IMOL=1,NMOL
C     MAKE FIRST CHANNEL = CURRENT SAMPLE
C
      X2O=XH(IMOL)
      Y2O=YH(IMOL)
      Z2O=ZH(IMOL)
C
C     ACCUMULATE SUMS FOR X*X,Y*Y AND Z*Z
C
      XOXO=X2O*X2O
      YOYO=Y2O*Y2O
      ZOZO=Z2O*Z2O
      SUMX2(IMOL)=SUMX2(IMOL)+XOXO
      SUMY2(IMOL)=SUMY2(IMOL)+YOYO
      SUMZ2(IMOL)=SUMZ2(IMOL)+ZOZO
      DO 700 KK=1,NCHAN
      K=NCHAN-KK
C
C     MULTIPLY AND ADD INTO STORES
C
      KP1=K+1
```

```
      X2RT=X2R(KP1,IMOL)
      Y2RT=Y2R(KP1,IMOL)
      Z2RT=Z2R(KP1,IMOL)
      X2S(KP1,IMOL)=X2S(KP1,IMOL)-2.0*X2O*X2RT
     1+XOXO+X2RT*X2RT
      Y2S(KP1,IMOL)=Y2S(KP1,IMOL)-2.0*Y2O*Y2RT
     1+YOYO+Y2RT*Y2RT
      Z2S(KP1,IMOL)=Z2S(KP1,IMOL)-2.0*Z2O*Z2RT
     1+ZOZO+Z2RT*Z2RT
      IF(K.EQ.0) GO TO 777
C
C     SHIFT REGISTERS
C
      X2R(KP1,IMOL)=X2R(K,IMOL)
      Y2R(KP1,IMOL)=Y2R(K,IMOL)
      Z2R(KP1,IMOL)=Z2R(K,IMOL)
 700  CONTINUE
 777  CONTINUE
      X2R(1,IMOL)=X2O
      Y2R(1,IMOL)=Y2O
      Z2R(1,IMOL)=Z2O
 70   CONTINUE
      IACCUM=IACCUM+1
 7000   CONTINUE
      IF(MOD(IT,NPRINT).NE.0) GO TO 77
      FACTOR=FLOAT(NMOL)
      FAC=FLOAT(IACCUM)
      WRITE(9,800)IT,IACCUM
 800  FORMAT('IT = ',I9,' IACCUM = ',I9,//)
C
C     CALCULATE AVERAGE VALUE OF X2,Y2 AND Z2
C
      WRITE(9,8000)
 8000 FORMAT('EQUILIBRIUM VALUES OF X2 Y2 AND Z2',//)
      CX=0.0
      CY=0.0
      CZ=0.0
      CXYZ=0.0
      DO 11 IMOL=1,NMOL
      CX1=SUMX2(IMOL)/FAC
      CY1=SUMY2(IMOL)/FAC
      CZ1=SUMZ2(IMOL)/FAC
      CXYZ1=CX1+CY1+CZ1
```

```
      CX=CX+CX1
      CY=CY+CY1
      CZ=CZ+CZ1
      CXYZ=CXYZ+CXYZ1
      IF(INDIV.NE.0)WRITE(9,12)IMOL,CX1,CY1,CZ1,CXYZ1
 12   FORMAT('I = ',I3,' X2 = ',E12.6,'Y2 = ',E12.6,
     1' Z2 = ',E12.6,' R2 = ',E12.6)
C
 11   CONTINUE
      CX=CX/FACTOR
      CY=CY/FACTOR
      CZ=CZ/FACTOR
      CXYZ=CXYZ/FACTOR
C
      WRITE(9,7776)CX,CY,CZ,CXYZ
 7776 FORMAT('EQUILIBRIUM M.S.D. S AVERAGED OVER PARTICLES',//,
     1'X2 = ',E12.6,' Y2 = ',E12.6,' Z2 = ',E12.6,' R2 = ',E12.6)
      WRITE(9,8001)
 8001 FORMAT('TIME DEPENDENT MEAN SQUARE DISPLACEMENTS',//)
      DO 555 IC=1,NCHAN
      CXA(IC)=0.0
      CYA(IC)=0.0
      CZA(IC)=0.0
      CRA(IC)=0.0
 555  CONTINUE
      DO 8 IMOL=1,NMOL
      CX=0.0
      CY=0.0
      CZ=0.0
      CXYZ=0.0
      DO 9 ICHAN=1,NCHAN
      CX1=X2S(ICHAN,IMOL)/FAC
      CY1=Y2S(ICHAN,IMOL)/FAC
      CZ1=Z2S(ICHAN,IMOL)/FAC
      CXYZ1=CX1+CY1+CZ1
      TORT=DT*ICHAN*MULT
      IF(INDIV.NE.0)WRITE(9,10)ICHAN,TORT,CX1,CY1,CZ1,CXYZ1
 10   FORMAT('CHAN NO.',I3,' TOR = ',E12.6,
     1' X2 = ',E12.6,' Y2 = ',E12.6,' Z2 = ',E12.6,' R2 = ',E12.6)
      CXA(ICHAN)=CXA(ICHAN)+CX1
      CYA(ICHAN)=CYA(ICHAN)+CY1
      CZA(ICHAN)=CZA(ICHAN)+CZ1
      CRA(ICHAN)=CRA(ICHAN)+CXYZ1
```

```
 9   CONTINUE
 8   CONTINUE
     WRITE(9,15)
 15  FORMAT('MEAN SQUARE DISPLACEMENTS AVERAGED OVER ALL PARTI-
CLES'
    1,' IN DIAMETERS SQUARED',//)
     DO 16 ICHAN=1,NCHAN
     CX=CXA(ICHAN)/FACTOR
     CY=CYA(ICHAN)/FACTOR
     CZ=CZA(ICHAN)/FACTOR
     CXYZ=CRA(ICHAN)/FACTOR
     TORT=DT*MULT*ICHAN
     WRITE(9,10)ICHAN,TORT,CX,CY,CZ,CXYZ
 16  CONTINUE
     WRITE(9,17)
 17  FORMAT('MEAN SQUARE DISPLACEMENTS AVERAGED OVER ALL
      PARTICLES'
    1, 'IN METRES SQUARED PER SEC',//)
     FACT=FACTOR/(DIAM*DIAM)
     DO 18 ICHAN=1,NCHAN
     CX=CXA(ICHAN)/FACT
     CY=CYA(ICHAN)/FACT
     CZ=CZA(ICHAN)/FACT
     CXYZ=CRA(ICHAN)/FACT
     TORT=DT*MULT*ICHAN
     WRITE(9,10)ICHAN,TORT,CX,CY,CZ,CXYZ
 18   CONTINUE
 77  CONTINUE
     ITTOT=ITTOT+1
C    DO 9822 I=1,NMOL
C    WRITE(9,5497)I,X(I),Y(I),Z(I)
C9822 CONTINUE
C    DO 8776 I=1,NMOL
C    WRITE(9,5497)I,XH(I),YH(I),ZH(I)
C8776 CONTINUE
 5497 FORMAT(' I = ',I3' X = ',E16.9,' Y = ',E16.9,
    1' Z = ',E16.9)
     IF(MOD(IT,NWRITE).NE.0)GO TO 7777
C
C     WRITE OUT RESTART INFORMATION
C
     REWIND 10
     WRITE(10)XH,YH,ZH,X,Y,Z,ITTOT
```

```
      REWIND 10
      REWIND 11
      WRITE(11)NN,X2R,Y2R,Z2R,X2S,Y2S,Z2S,SUMX2,SUMY2,SUMZ2,IACCUM
      REWIND 11
      REWIND 12
      WRITE(12) ITTOT,NN
      REWIND 12
 7777 CONTINUE
      IF(IT.LT.NTSTEPS) GO TO 7
      STOP
      END
C
C#################################################################
C
      SUBROUTINE STARTER
C
C     ROUTINE WHICH INITIALIZES A RUN
C
      COMMON/COFM/CMXO,CMYO,CMZO
      COMMON/CORDS/XH(256),YH(256),ZH(256),X(256),Y(256),Z(256)
     1,NN(200)
      COMMON/COREL/X2R(15,256),Y2R(15,256),Z2R(15,256),
     1X2S(15,256),Y2S(15,256),Z2S(15,256),
     2SUMX2(256),SUMY2(256),SUMZ2(256),IACCUM
      COMMON/INFO/NMOL,XL,XL2,XL5,RCUT,RCORE,RCORE2,NTYPE,IBCC,
      1NPRE,NTSTEPS,ITTOT,NPRINT,NWRITE,DT,DKTDT,DKTDTP,SQRVAR,SQR-
VARP
     2,YL,YL2,ZL,ZL2
      COMMON/FORCOR/INDIV,NCHAN,MULT,KMAX,TPIL
      COMMON/FORDF/DELREC,DELTA,PRODEL,NDIV
      COMMON/STORE/IGAP,IFILE,KOUNT
C    1,IXYZ(256,10)
C
C     READ INPUT INFORMATION AND SET UP RUN
C
      CALL SETUP
      IF(NTYPE.NE.1) GO TO 1
      IF(IBCC.EQ.1)GO TO 984
C
C     SET UP AN FCC LATTICE
C
        NC=(NMOL/4)**(1./3.)+0.01
CUBE=XL/FLOAT(NC)
```

```
      WRITE(9,5)
 5    FORMAT(//,'F.C.C. LATTICE  ',//)
      X(1)=0.0
      X(2)=0.5*CUBE
      X(3)=0.0
      X(4)=0.5*CUBE
      Y(1)=0.0
      Y(2)=0.5*CUBE
      Y(3)=0.5*CUBE
      Y(4)=0.0
      Z(1)=0.0
      Z(2)=0.0
      Z(3)=0.5*CUBE
      Z(4)=0.5*CUBE
      M=0
      DO 13 I=1,NC
      DO 13 J=1,NC
      DO 13 K=1,NC
      DO 14 IJ=1,4
      L=IJ+M
      X(L)=X(IJ)+CUBE*(K-1)
      Y(L)=Y(IJ)+CUBE*(J-1)
      Z(L)=Z(IJ)+CUBE*(I-1)
      XH(L)=X(L)
      YH(L)=Y(L)
      ZH(L)=Z(L)
      WRITE(9,2)L,X(L),Y(L),Z(L)
 2    FORMAT('L = ',I5,' X = ',E12.6,' Y = ',E12.6,' Z = '
     1,E12.6)
 14   CONTINUE
      M=M+4
 13   CONTINUE
      GO TO 3
 984  CONTINUE
C
C     SET UP B.C.C. LATTICE
C
      WRITE(9,6)
 6    FORMAT(//,'B.C.C LATTICE',//)
      IL=(NMOL/2)**0.333333+0.01
      IF(2*IL**3.LT.NMOL) IL=IL+1
      D=XL/IL
      D1=D/2.0
```

```
      L=0
      DO 200 I=1,IL
      DO 200 J=1,IL
      DO 200 K=1,IL
      L=L+1
      X(L)=(I-1)*D
      Y(L)=(J-1)*D
      Z(L)=(K-1)*D
      XH(L)=X(L)
      YH(L)=Y(L)
      ZH(L)=Z(L)
      L=L+1
      X(L)=(I-1)*D+D1
      Y(L)=(J-1)*D+D1
      Z(L)=(K-1)*D+D1
      XH(L)=X(L)
      YH(L)=Y(L)
      ZH(L)=Z(L)
      IF(L.GE.NMOL) GO TO 33
200   CONTINUE
33    CONTINUE
3     CONTINUE
      CMX=0.0
      CMY=0.0
      CMZ=0.0
      DO 201 I=1,NMOL
      WRITE(9,211) I,X(I),Y(I),Z(I)
      CMX=CMX+X(I)
      CMY=CMY+Y(I)
      CMZ=CMZ+Z(I)
201   CONTINUE
211   FORMAT(' I = ',I3,' X = ',E12.6,' Y = ',E12.6,' Z = ',E12.6)
      FM=FLOAT(NMOL)
      CMXO=CMX/FM
      CMYO=CMY/FM
      CMZO=CMZ/FM
      GO TO 300
1     CONTINUE
C
C     IF NOT A NTYPE = 1 READ RESTART INFORMATION
C
      REWIND 10
      READ(10)XH,YH,ZH,X,Y,Z,ITTOT
```

```
      REWIND 10
C
C     READ CORRELATOR RESTART INFORMATION
C
      REWIND 11
      READ(11)NN,X2R,Y2R,Z2R,X2S,Y2S,Z2S,SUMX2,SUMY2,SUMZ2,IACC
      REWIND 11
C
C     READ RADIAL DISTRIBUTION INFORMATION
C
      REWIND 12
      READ(12)ITTOT,NN
      REWIND 12
C
      IF(NTYPE.EQ.2) GO TO 4
 300  CONTINUE
      DO 2222 IG=1,NDIV
      NN(IG)=0
 2222 CONTINUE
C     DO EQUILIBRATION
C
      ITTOT=0
      DO 111 IT=1,NPRE
      CALL MOVE(IT,3)
      ITTOT=ITTOT+1
 111  CONTINUE
C
C     SET UP CHECKERBOARD CO-ORDINATES IN MAIN CELL
C     AFTER EQUILIBRATION
C
      DO 19 I=1,NMOL
      XH(I)=X(I)
      YH(I)=Y(I)
      ZH(I)=Z(I)
 19   CONTINUE
      DO 20 IG=1,NDIV
      NN(IG)=0
 20   CONTINUE
      ITTOT=1
 4    CONTINUE
      RETURN
      END
C
```

```
C###############################################################
C
      SUBROUTINE MOVE(IT,KL)
C
C     BROWNIAN DYNAMICS MOVE ROUTINE SEARCHES ALL PARTICLES FOR
INTERACTIONS
C
      COMMON/COFM/CMXO,CMYO,CMZO
      COMMON/CORDS/XH(256),YH(256),ZH(256),X(256),Y(256),Z(256)
     1,NN(200)
      COMMON/INFO/NMOL,XL,XL2,XL5,RCUT,RCORE,RCORE2,NTYPE,IBCC,
     1NPRE,NTSTEPS,ITTOT,NPRINT,NWRITE,DT,DKTDT,DKTDTP,SQRVAR,
     SQRVARP
     2,YL,YL2,ZL,ZL2
      COMMON/FORDF/DELREC,DELTA,PRODEL,NDIV
      DIMENSION TFX(256),TFY(256),TFZ(256)
      DELT=DKTDT
      SQRV=SQRVAR
C
C     ZERO POSITION OF CENTRE OF MASS
C
      CMX=0.0
      CMY=0.0
      CMZ=0.0
      IF(KL.EQ.3)DELT=DKTDTP
      IF(KL.EQ.3)SQRV=SQRVARP
C
C     ZERO THE TOTAL FORCE ON EACH PARTICLE
C
      DO 10 I=1,NMOL
      TFX(I)=0.0
      TFY(I)=0.0
      TFZ(I)=0.0
 10   CONTINUE
      NMUL=NMOL-1
       DO 11 I=1,NMUL
      XI=X(I)
      YI=Y(I)
      ZI=Z(I)
      K=I+1
      DO 11 J=K,NMOL
C
C     MINIMIUM IMAGE CONVENTION
```

```
C
      DXI=XI-X(J)
      IF(DXI.GT.XL2)   DXI=DXI-XL
      IF(DXI.LT.-XL2) DXI=DXI+XL
      DYI=YI-Y(J)
      IF(DYI.GT.YL2)    DYI=DYI-YL
      IF(DYI.LT.-YL2) DYI=DYI+YL
      DZI=ZI-Z(J)
      IF(DZI.GT.ZL2)    DZI=DZI-ZL
      IF(DZI.LT.-ZL2) DZI=DZI+ZL
      XNER=(DXI**2+DYI**2+DZI**2)
C
      RDIS=SQRT(XNER)
C
C     CHECK FOR SPHERICAL CUT-OFF
C
      IF(RDIS.GE.XL2) GOTO 11
      RS=RDIS
C
C     UPDATE COUNTER FOR G(R)
C
      NR=RS*DELREC+1
      IF(NR.LE.NDIV) NN(NR)=NN(NR)+1
C
C     WORK OUT PAIR FORCE AND POTENTIAL
C
      CALL POTFN(POT,FORCE,XNER)
C
C     F = -DU/DU *X/R = X COMPONENT OF F
C
      F=DXI*FORCE
      TFX(I)=TFX(I)+F
      TFX(J)=TFX(J)-F
C
C     Y COMPONENT OF F
C
      F=DYI*FORCE
      TFY(I)=TFY(I)+F
      TFY(J)=TFY(J)-F
C
C     Z COMPONENT OF F
C
      F=DZI*FORCE
```

```
      TFZ(I)=TFZ(I)+F
      TFZ(J)=TFZ(J)-F
 11   CONTINUE
C
C     WORK OUT RANDOM FORCE
C
      AM=0.0
      DO 12 I=1, NMOL
      CALL GAUSS(SQRV,AM,RFX)
      CALL GAUSS(SQRV,AM,RFY)
      CALL GAUSS(SQRV,AM,RFZ)
      TFX(I)=TFX(I)*DELT
      TFY(I)=TFY(I)*DELT
      TFZ(I)=TFZ(I)*DELT
C
C     ERMAK ALGORITHM FOR R(T+DT) - R(T)
C
      XH(I)=XH(I)+RFX+TFX(I)
      YH(I)=YH(I)+RFY+TFY(I)
      ZH(I)=ZH(I)+RFZ+TFZ(I)
C
C     ACCUMULATE CENTRE OF MASS POSITION
C
      CMX=CMX+XH(I)
      CMY=CMY+YH(I)
      CMZ=CMZ+ZH(I)
 12   CONTINUE
C
C     SUBTRACT OUT CENTRE OF MASS MOVEMENT
C
      RNMOL=FLOAT(NMOL)
      CMX=CMX/RNMOL
      CMY=CMY/RNMOL
      CMZ=CMZ/RNMOL
      DO 14 I=1,NMOL
      XH(I)=XH(I)+(CMXO-CMX)
      YH(I)=YH(I)+(CMYO-CMY)
      ZH(I)=ZH(I)+(CMZO-CMZ)
 14   CONTINUE
C
C     NOW APPLY THE PERIODIC BOUNDARY CONDITIONS
C
      DO 15 I=1,NMOL
```

```
      X(I)=AMOD(XH(I)+XL5,XL)
      Y(I)=AMOD(YH(I)+XL5,YL)
      Z(I)=AMOD(ZH(I)+XL5,ZL)
 15   CONTINUE
      IF(MOD(IT,NPRINT).NE.0) GO TO 13
C
C     WRITE OUT DATA TO OUTPUT FILE 9
C
      IF(KL.EQ.1)GO TO 13
C
C     R.D.F. PRINTING   ROUTINE
C
      YDB=(ITTOT+1)*NMOL
      WRITE(9,123)
      DO 18 II=1,NDIV
      XDB=NN(II)
      ZZ=XDB/YDB
       Z1=(II-0.5)*DELTA
      ZY=Z1
      Z1=Z1*Z1*PRODEL
      ZZ=ZZ/Z1
      WRITE(9,124)II,ZY,ZZ
 123  FORMAT(//,'RADIAL DISTRIBUTION FUNCTION  ',//)
 124  FORMAT('I = ',I4,' R = ',F12.6,' G(R) = ',F12.6)
 18   CONTINUE
 13   CONTINUE
      RETURN
      END
C
C#################################################################
C
      SUBROUTINE SETUP
C
      COMMON/INFO/NMOL,XL,XL2,XL5,RCUT,RCORE,RCORE2,NTYPE,IBCC,
      1NPRE,NTSTEPS,ITTOT,NPRINT,NWRITE,DT,DKTDT,DKTDTP,SQRVAR,SQR-
VARP
     2,YL,YL2,ZL,ZL2
      COMMON/FORCOR/INDIV,NCHAN,MULT,KMAX,TPIL
      COMMON/MAID/DIAM
      COMMON/FORPOT/CONST,TOR
      COMMON/RANF/ISEED
      COMMON/FORDF/DELREC,DELTA,PRODEL,NDIV
      DATA PI/3.141592654/,BK/1.38E-23/,ELCH/1.602E-2/,
```

```
     1PERM/8.854/,AVAG/60.225 /
      OPEN(UNIT=8,FILE='BDSC.IN',FORM='FORMATTED',STATUS='OLD')
      OPEN(UNIT=9,FILE='BDSC.OUT',FORM='FORMATTED',STATUS='OLD')
      OPEN(UNIT=10,FILE='BDSC.POS',FORM='UNFORMATTED',STATUS='OLD')
      OPEN(UNIT=11,FILE='BDSC.RES',FORM='UNFORMATTED',STATUS='OLD')
      OPEN(UNIT=12,FILE='BDSC.GR',FORM='UNFORMATTED',STATUS='OLD')
C
C     READ IN CONTROL DATA
C
      WRITE(9,9876)
 9876 FORMAT('BROWNAIN DYNAMICS PROGRAM'//)
      REWIND 8
      READ(8,1000)NMOL,CONC,TEMP,FRACT,VISC,EPS,RAD,RCORE,SURPOT,DTP
     1,DT,NTYPE,NPRE,NTSTEPS,NPRINT,NWRITE,NDIV,NCHAN,MULT,KMAX,IPOT
     2,INDIV,IBCC
 1000 FORMAT(I6,/,F10.0,/,F10.0,/,F10.0,/,F10.0,/,F10.0,/,F10.0,
     1/,F10.0,/,F10.0,/,F10.0,/,F10.0,/,I3,/,I6,/,I6,/,I6,/,I6,/,
     2I3,/,I3,/,I3,/,I3,/,I3,/,I3,/,I3)
C
      RAD=RAD*1.0E-10
      ISEED=-2000
      DIAM=RAD+RAD
      BKT=TEMP*BK
 1234 FORMAT('TOR = ',E16.9,' CONST = ',E16.9,//)
      DEBKAP=2.0*AVAG*CONC/(EPS*PERM*BKT)
 1235 FORMAT('DEBKAP = ',E16.9,' KAPPA D = ',E16.9,
     1' 1.0/DEBKAP = ',E16.9,//)
      DEBKAP=SQRT(DEBKAP)*ELCH
      DEB=1.0/DEBKAP
      RT=DEBKAP*RAD*2.0
      WRITE(9,1235)DEBKAP,RT,DEB
      CONST=2.0*PI*RAD*EPS*PERM*SURPOT*SURPOT
      CONST=CONST*(1.0E-12)
      TOR=-DEBKAP*RAD*2.0
      CONST=CONST/BKT
      CONSTT=CONST*2.0*RAD
      WRITE(9,1234)TOR,CONSTT
       IF(IPOT.EQ.1) GOTO 543
C
C     WRITE OUT PAIR POTENTIAL AND FORCE VERSES R
C
      WRITE(9,8709)
 8709 FORMAT('PAIR POTENTIAL AND PAIR FORCE',//)
```

```
      RR=RCORE
      RHO=6.0*FRACT/(100.0*PI)
      XL=(NMOL/RHO)**0.3333333
      XL2=XL/2.0
      DELR=(XL2-RCORE)/100.0
      DO 1 I=1,100
      RR=RR+DELR
      R2=RR*RR
      CALL POTFN(POT,FORCE,R2)
      WRITE(9,2)RR,POT,FORCE
2     FORMAT('R  = ',E12.6,' POT  = ',E12.6,' FORCE  = ',E12.6)
1     CONTINUE
543   CONTINUE
      RHO=6.0*FRACT/(100.0*PI)
      X=(NMOL/RHO)**0.3333333333
      WRITE(9, 3)CONC,RAD,DEB,RHO,FRACT
3     FORMAT(//,'CONC = ',F10.6,' RADIUS = ',E12.6,' 1/KAPPA = ',E12.5
     1,' RHO = ',F10.6,' FRACT = ',F10.6,//)
      XL=X
      XL2=XL/2.0
      YL=XL
      YL2=YL/2.0
      ZL=XL
      ZL2=ZL/2.0
      RCUT=XL2
      RCUT2=RCUT*RCUT
      XL5=5.0*XL
      RCORE2=RCORE*RCORE
      TPIL=2.0*PI/XL
      DELTA=(XL2)/NDIV
      PRODEL=2.0*PI*DELTA*RHO
      DELREC=1.0/DELTA
      DKTDT=DT/(6.0*PI*VISC*RAD)
      D0=DKTDT/DT
      D0=BKT*D0
      DKTDTP=DKTDT*(DTP/DT)
      WRITE(9,4390)D0
4390  FORMAT('D0 = ',E16.9,//)
      SQRVAR=SQRT(2.0*DKTDT*BK*TEMP)
      SQRVARP=SQRT(2.0*DKTDTP*BK*TEMP)
      SQRVAR=SQRVAR/DIAM
      SQRVARP=SQRVARP/DIAM
      DKTDT=DKTDT*BKT/(DIAM*DIAM)
```

```
      DKTDTP=DKTDTP*BKT/(DIAM*DIAM)
      WRITE(9,212)SQRVAR,DKTDT,SQRVARP,DKTDTP,DELTA
 212  FORMAT('SQRVAR = ',E16.9,' DKTDT = ',E16.9
     1,/,'SQRVARP = ',E16.9,' DKDTP = ',E16.9,//,
     2'DELTA FOR RDF = ',F16.9,//)
      WRITE(9,318)XL,XL2,XL5,RCORE,RCORE2
 318  FORMAT('XL = ',F16.9,' XL/2 = ',F16.9,' 5 XL = ',F16.9,
     1/,'RCORE = ',E16.9,' RCORE X RCORE = ',E16.9,//)
      TORT=MULT*DT
      WRITE(9,414)NCHAN,MULT,KMAX,TORT
 414  FORMAT('NO. OF CHANNELS = ',I3,' MULT = ',I3,
     1' KMAX = ',I3,' TOR = ',E16.9,//)
      RETURN
      END
C
C###############################################################
C
      SUBROUTINE POTFN(STAT,DERIV,S)
C     PAIR POTENTIAL AND PAIR FORCE/R
C     THIS VERSION FOR A SCREENED COULOMB POTENTIAL
C
      COMMON/FORPOT/CONST,TOR
      Y=SQRT(S)
      X=Y-1.0
C
C     PAIR POTENTIAL
C
      STAT=CONST*EXP(TOR*X)/Y
C
C     PAIR FORCE/R
C
      DERIV=-STAT*(TOR*Y-1.0)/Y
      DERIV=DERIV/Y
      RETURN
      END
C
C####################################################################
C
      SUBROUTINE GAUSS(S,AM,V)
C
C     AN IBM ROUTINE TO RETURN A GAUSSIAN RANDOM NUMBER
C     IT NEEDS A FUNCTION RAND(ISEED) WHICH GENERATES RANDOM
C     NUMBERS UNIFORMLY DISTRIBUTED ON (0,1)
```

```
C     SUCH A FUNCTION MAY BE FOUND IN,
C     "Numerical Recipes in Fortran: The Art of Scientific Computing", Chapter 7,
C     Cambridge University Press, UK.
C
      COMMON/RANF/ISEED
      A=0.0
      DO 1  I=1,12
 1    A=A+RAND(ISEED)
      V=(A-6.0)*S+AM
      RETURN
      END
C
  PROGRAM SPHERE3
C
C      VERSION OF KEITH BRIGG S N-SPHERE  STOKESIAN DYNAMICS PRO-
GRAM
C    THIS VERSION USES THE MOBILITY TENSOR OF  MAZUR AND VAN SAAR-
LOOS,
C     PHYSICA 115A,21-57,1982. AND  WAS WRITTEN ON 24/9/1990 FOR THE 3-
SPHERE CASE
C
      COMMON/DATAIN/A(3)
      COMMON/INFO/IOUT
      COMMON/FT/FORCE(3,3),TORQUE(3,3)
      DIMENSION POS(18),DPOS(18),VEL(18)
C    EXTERNAL F
C
C     SET UP FORCES ,TORQUES AND INITIAL POSITIONS  FOR THREE SEDI-
MENTING SPHERES
C
C     N.B. THE FIRST 9 LOCATIONS IN ARRAY POS ARE THE X,Y AND Z CO-
ORDINATES OF THE
C    PARTICLES AND THE LAST 9 LOCATIONS ARE THE ANGLES
C    AND THEY ARE STORED AS X1,Y1,Z1,X2,Y2,Z2,X3,Y3,Z3
C    TX1,TY1,TZ1,TX2,TY2,TZ2,TX3,TY3 AND TZ3
C
C     FIRST INITIALIZE POSITIONS, FORCES AND TORQUES TO ZERO
C
       PRINT 500
500 FORMAT(//,' 3 SPHERE SEDIMENTATION MAZUR TENSOR TO O(R-4)',//)
C
      CALL INIT(POS)
C
```

```
      PRINT 1239,POS(1),POS(2),POS(3),POS(4),POS(5),
     1POS(6),POS(7),POS(8),POS(9)
 1239 FORMAT(3(2X,E12.6))
C     SET UP SPHERE RADII
C
      A(1)=0.4
      A(2)=0.4
      A(3)=0.4
C
C     NOW FOR THE INITIAL POSITIONS
C
C     X1, X2 AND X3
C
      POS(1)=0.0
      POS(2)=0.00
      POS(3)=0.0
      POS(4)=0.0
      POS(5)=1.0
      POS(6)=1.0
      POS(7)=0.0
      POS(8)=1.0
      POS(9)=-1.0
C
C     FORCE DUE TO GRAVITY IN X - DIRECTION
C
      FORCE(1,1)=1.0
      FORCE(2,1)=1.0
      FORCE(3,1)=1.0
C
C     REPOSITION YI 'S FOR CONVENIENT DISPLAY
C
C     DO 999 I=1,3
C     IT=3*I-1
C     POS(IT)=170./0.1
C999  CONTINUE
C
C     SOLVE  VEL = MU TT . F + MUTR . T
C              OMEGA =MURT . F + MURR . T
C     FOR R AND THETA ,WHICH ARE STORED IN ARRAY POS
C     BY USE OF THE RUNGE-KUTTA O.D.E. SOLVER
C
C     N.B. DPOS(K) CONTAINS DPOS(K)/DT =LINEAR AND ANGULAR VELOCI-
TIES, WITH THE SAME
```

```
C     STORAGE NOTATION AS FOR POS(K)
C     ALSO  IN THIS CASE TORQUE,T =0.0 SO SECOND TERMS IN EACH
C     EQUATION = 0.0
C
C     INITIALIZE THE TIME,T,THE TIME STEP,DT AND THE INTEGRATION TOL-
ERANCE,EPS
C
      T=0.0
      DT=10.0
      EPS=1.0E-13
      PRINT 550,A(1),A(2),A(3),DT,EPS
 550  FORMAT(' SPHERE RADII',/,' A1 = ',E12.3,
     1' A2 = ',E12.3,' A3 = ',E12.3,/,' TIME STEP = ',E12.5,
     2' INTEGRATION TOLERANCE,EPS = ',E12.5,//)
      IOUT=2
      PRINT 555
 555  FORMAT(//,' STARTING POSITIONS ',//)
      CALL OUT(T,POS)
      PRINT 560
 560  FORMAT(' FORCES ON PARTICLES ',//)
      DO 556 J=1,3
      PRINT 570,J,FORCE(J,1),FORCE(J,2),FORCE(J,3)
 570  FORMAT(' J= ',I3,3(1X,E12.3))
 556  CONTINUE
C
C     INTEGRATE THE EQUATIONS OF MOTION FROM T = 0 TO NTSTEPS*DT
C     BUT ONLY PRINT OUT THE TRAJECTORY EVERY NPRINT TIME STEPS
C
      NTSTEPS=1000
      NPRINT=10
C
C     SET UP CONSTANTS FOR DOPRI5
C
      CALL DOPCON(1)
C
      DO 100 I=1,NTSTEPS
       TEND=T+DT
      CALL DOPRI5(T,POS,TEND,EPS,DT,DT)
      T=TEND
      IF(MOD(I,NPRINT).EQ.0)CALL OUT(T,POS)
 100  CONTINUE
      STOP
      END
```

```
C
C#################################################################
C
C    OUTPUT ROUTINE
C
      SUBROUTINE OUT(T,POS)
C
C     OUTPUT ROUTINE WHICH EITHER PRINTS
C     A) X,Y,Z CO-ORDINATES OF EACH PARTICLE
C
C     OR
C
C     B)CALCULATES THE COMBINATIONS OF X,Y AND Z
C     DISPLAYED BY  CAFLISCH ET AL:PHYS.FLUIDS,31,3175 (1988)
C     AND PRINTS THEM
      DIMENSION  POS(18)
      COMMON/INFO/IOUT
      PRINT 1,T
 1    FORMAT(' AT TIME T = ',D12.5)
C
C     DECIDE WHAT TYPE OF OUTPUT IS NEEDED
C
      IF(IOUT.NE.1) GO TO 2
C
C     CO-ORDINATE OUTPUT
C
      PRINT 1,T
      PRINT 200,POS(1),POS(2),POS(4),POS(5),POS(7),POS(8)
 200  FORMAT(6D12.3)
      GO TO 3
 2    CONTINUE
C
C     CAFLISCH ET AL CO-ORDINATES
C
C     EXPRESS CO-ORDINATES ABOUT THOSE OF PARTICLE 1
C
      Y1=POS(4)-POS(1)
      Y2=POS(5)-POS(2)
      Y3=POS(6)-POS(3)
      Z1=POS(7)-POS(1)
      Z2=POS(8)-POS(2)
      Z3=POS(9)-POS(3)
C
```

```
C     A = \Y + Z\/2 = LENGTH OF MIDLINE
C     B = \Y - Z\/2  =  LENGTH OF BASE
C     C = (Y1 + Z1)/(\Y+Z\) = COSINE OF ANGLE BETWEEN MIDLINE
C    AND VERTICAL
C     D = (Y1 - Z1)/(\Y-Z\)  =  COSINE OF ANGLE BETWEEN BASELINE
C    AND VERTICAL
C     E = COSINE OF ANGLE BETWEEN BASE AND MIDLINE
C
      A=SQRT((Y1+Z1)**2+(Y2+Z2)**2+(Y3+Z3)**2)/2.0D0
      B=SQRT((Y1-Z1)**2+(Y2-Z2)**2+(Y3-Z3)**2)
      C=(Y1+Z1)/(2.0D0*A)
      D=(Y1-Z1)/B
      PRINT 2000,T,A,B,C,D
 2000 FORMAT(' T= ',E12.5,' A= ' ,E9.3,' B= ' ,E9.3,' C= ' ,E9.3,
     1' D= ',E9.3)
 3    CONTINUE
      RETURN
      END
C
C##############################################################
C
      SUBROUTINE INIT(POS)
C
C     INITIALIZE POSITIONS, FORCES AND TORQUES TO ZERO
C
      COMMON/FT/FORCE(3,3),TORQUE(3,3)
      DIMENSION POS(18)
      DO 1 I=1,18
      POS(I)=0.0
 1    CONTINUE
      DO 2 I=1,3
      FORCE(I,1)=0.0
      FORCE(I,2)=0.0
      FORCE(I,3)=0.0
      TORQUE(I,1)=0.0
      TORQUE(I,2)=0.0
      TORQUE(I,3)=0.0
 2    CONTINUE
      RETURN
      END
C
C##################################################################
C
```

```
C     MATRIX HANDLING ROUTINES
C
      SUBROUTINE RMV(A,B,C)
C
C     FINDS THE VECTOR C WHICH IS THE DOT PRODUCT OF  MATRIX A BY
VECTOR B
C
      DIMENSION A(3,3),B(3),C(3)
      C(1)=A(1,1)*B(1)+A(1,2)*B(2)+A(1,3)*B(3)
      C(2)=A(2,1)*B(1)+A(2,2)*B(2)+A(2,3)*B(3)
      C(3)=A(3,1)*B(1)+A(3,2)*B(2)+A(3,3)*B(3)
      RETURN
      END
C
C#################################################################
C
      SUBROUTINE ADDPOS(X,A,K)
C
C     ADD A VECTOR A TO COMPONENTS OF VECTOR X WHICH REPRESENTS
C     THE POSITION OF SPHERE K
C
      DIMENSION X(18),A(3)
      KT=2*K+K-3
      DO 1 I=1,3
      KPI=KT+I
      X(KPI)=X(KPI)+A(I)
 1    CONTINUE
      RETURN
      END
C
C#################################################################
C
      SUBROUTINE ADDANG(X,A,K)
C
C     ADDS A VECTOR A TO THE COMPONENTS OF VECTOR X
C     WHICH REPRESENTS THE ANGLES OF SPHERE K
C
      DIMENSION X(18),A(3)
      KT=3*(3+K-1)
      DO 1 I=1,3
      KPI=KT+I
      X(KPI)=X(KPI)+A(I)
 1    CONTINUE
```

```
      RETURN
      END
C
C######################################################################
C
      SUBROUTINE TRANPOS(A,B)
C
C     FIND THE MATRIX B WHICH IS THE TRANSPOSE OF INPUT MATRIX A
C
      DIMENSION A(3,3),B(3,3)
      DO 1 I=1,3
      DO 2 J=1,3
      B(J,I)=A(I,J)
 2    CONTINUE
 1    CONTINUE
      RETURN
      END
C
C###################################################################
C
C     FORCE AND TORQUE ROUTINES
C
      SUBROUTINE GETFORC(J,T,FORCEJ)
C
C       THIS PLACES THE ELEMENTS OF 2-D FORCE MATRIX IN 1-D ARRAY
FORCEJ
C
      COMMON/FT/FORCE(3,3),TORQUE(3,3)
      DIMENSION FORCEJ(3)
      DO 1 I=1,3
      FORCEJ(I)=FORCE(J,I)
 1    CONTINUE
      RETURN
      END
C
C###################################################################
C
C     SIMILARLY FOR TORQUES
C
      SUBROUTINE GETTORC(J,T,TORCJ)
C
C     ANALOGOUS TO GET FORCE
C
```

```
      COMMON/FT/FORCE(3,3),TORQUE(3,3)
      DIMENSION TORCJ(3)
      DO 1 I=1,3
      TORCJ(I)=TORQUE(J,I)
 1    CONTINUE
      RETURN
      END
C
C##################################################################
C
C     FIND MU . FORCE AND MU . TORQUE
C
      SUBROUTINE F(TIME,POS,DPOS)
C
C     FINDS THE VECTORS ARISING FROM THE DOT PRODUCT
C     OF THE TENSOR MU WITH VECTORS FORCE AND TORQUE
C     WHICH = DPOS(I) =D(POS(I))/DT
C
      DIMENSION POS(18),DPOS(18)
      COMMON/FT/FORCE(3,3),TORQUE(3,3)
      DIMENSION FI(3),TI(3),FJ(3),TJ(3),RMUF(3),RMUT(3)
      DIMENSION RMUTT(3,3),RMUTR(3,3),RMURR(3,3)
C
C     INITIALIZE DPOS = DPOS/DT TO ZERO
C
      DO 1 I=1, 18
      DPOS(I)=0.0D0
 1    CONTINUE
      DO 2 I=1,3
      CALL GETFORC(I,TIME,FI)
      CALL GETTORC(I,TIME,TI)
C
C     DO DIAGONAL TERMS
C
      CALL GETRMU(I,I,POS,RMUTT,RMUTR,RMURR)
      CALL RMV(RMUTT,FI,RMUF)
      CALL ADDPOS(DPOS,RMUF,I)
      CALL RMV(RMURR,TI,RMUT)
      CALL ADDANG(DPOS,RMUT,I)
C
C     NOW THE OFF-DIAGONAL TERMS
C
      DO 3 J=1,3
```

```
      CALL GETFORC(J,TIME,FJ)
      CALL GETTORC(J,TIME,TJ)
      CALL GETRMU(I,J,POS,RMUTT,RMUTR,RMURR)
      IF(I.LE.J) GO TO 4
C
C     TRANSLATIONAL CONTRIBUTION
C
      CALL RMV(RMUTT,FJ,RMUF)
      CALL ADDPOS(DPOS,RMUF,I)
C
C     NOW RMUTT = RMUTT(J,I)
C
      CALL TRANPOS(RMUTT,RMUTT)
      CALL RMV(RMUTT,FI,RMUF)
      CALL ADDPOS(DPOS,RMUF,J)
C
C     NOW DO ROTATIONAL CONTRIBUTION
C
      CALL RMV(RMURR,TJ,RMUT)
      CALL ADDANG(DPOS,RMUT,I)
C
C     NOW RMURR = RMURR(J,I)
C
      CALL TRANPOS(RMURR,RMURR)
      CALL RMV(RMUTT,TI,RMUT)
      CALL ADDANG(DPOS,RMUT,J)
 4    CONTINUE
C
C     TRANSLATION-ROTATION CONTRIBUTION
C
      CALL RMV(RMUTR,TJ,RMUT)
      CALL ADDPOS(DPOS,RMUT,I)
C
C     NOW RMUTR = RMUTR(J,I)
C
      CALL TRANPOS(RMUTR,RMUTR)
      CALL RMV(RMUTR,FJ,RMUF)
      CALL ADDANG(DPOS,RMUF,I)
 3    CONTINUE
 2    CONTINUE
      RETURN
      END
C
```

```
C###############################################################
C
C     MOBILITY MATRIX ROUTINE THIS VERSION IS KEITH BRIGGS REALIZA-
TION OF
C         THE MAZUR AND VAN SAARLOOS SERIES EXPANSION FOR THREE
SPHERES UP TO R-4
C     I.E. P 43 OF MAZUR AND VAN SAARLOOS , PHYSICA, 115A 21-57 (1982).
C
C     FOR THIS PROBLEM AS TORQUE = 0.0 ONLY RMUTT AND RMUTR
C     ARE NEEDED AND THUS RMURR IS ONLY FOR THE TORQUE
C     FREE CASE
C
      SUBROUTINE GETRMU(I,J,POS,RMUTT,RMUTR,RMURR)
C
C     GET THE IJ BLOCK OF THE MOBILITY TENSOR  MAZUR AND VAN SAAR-
LOOS P43
C
      DIMENSION POS(18),RMUTT(3,3),RMUTR(3,3),RMURR(3,3)
      COMMON/DATAIN/A(3)
C
C     INITIALIZE MOBILITY ARRAYS TO ZERO
C
      DO 1 K=1,3
      DO 2 L=1,3
      RMUTT(K,L)=0.0
      RMUTR(K,L)=0.0
      RMURR(K,L)=0.0
 2    CONTINUE
 1    CONTINUE
      IF(I.NE.J) GO TO 3
C
C     I = J
C
      B=1.0/A(I)
      C=0.75*B/SQRT(A(I))
      RMUTT(1,1)=B
      RMUTT(2,2)=B
      RMUTT(3,3)=B
      RMURR(1,1)=C
      RMURR(2,2)=C
      RMURR(3,3)=C
      GO TO 4
 3    CONTINUE
```

```
C
C     O(A/R) TO O((A/R)**3) TERMS IF I NE J
C
      IT=3*I
      JT=3*J
      R1=POS(IT-2)-POS(JT-2)
      R2=POS(IT-1)-POS(JT-1)
      R3=POS(IT  )-POS(JT  )
      RS=R1*R1+R2*R2+R3*R3
      R=SQRT(RS)
      R1=R1/R
      R2=R2/R
      R3=R3/R
      B=0.75/R
      C=0.25*(A(I)**2+A(J)**2)/RS/R
      D=B+C
      E=B-3.0*C
      RMUTT(1,1)=D+E*R1*R1
      RMUTT(2,2)=D+E*R2*R2
      RMUTT(3,3)=D+E*R3*R3
C
      RMUTT(1,2)=E*R1*R2
      RMUTT(1,3)=E*R1*R3
      RMUTT(2,3)=E*R2*R3
C
      RMUTT(2,1)=RMUTT(1,2)
      RMUTT(3,1)=RMUTT(1,3)
      RMUTT(3,2)=RMUTT(2,3)
C
      B=-0.75/RS
      RMUTR(1,2)= B*R3
      RMUTR(1,3)=-B*R2
      RMUTR(2,3)= B*R1
      RMUTR(2,1)=-RMUTR(1,2)
      RMUTR(3,1)=-RMUTR(1,3)
      RMUTR(3,2)=-RMUTR(2,3)
    4 CONTINUE
C
C     ADD THREE-BODY TERMS
C
      D=15.0/8.0
      DO 5 K=1,3
      IF(I.EQ.K.OR.J.EQ.K) GO TO 6
```

```
C
C     I NE K AND J NE K
C
      IT=3*I
      JT=3*J
      KT=3*K
C
C
      RIK1=POS(IT-2)-POS(KT-2)
      RIK2=POS(IT-1)-POS(KT-1)
      RIK3=POS(IT  )-POS(KT  )
      RIKS=RIK1*RIK1+RIK2*RIK2+RIK3*RIK3
      RIK=SQRT(RIKS)
      RIK1=RIK1/RIK
      RIK2=RIK2/RIK
      RIK3=RIK3/RIK
C
      RKJ1=POS(KT-2)-POS(JT-2)
      RKJ2=POS(KT-1)-POS(JT-1)
      RKJ3=POS(KT  )-POS(JT  )
      RKJS=RKJ1*RKJ1+RKJ2*RKJ2+RKJ3*RKJ3
      RKJ=SQRT(RKJS)
      RKJ1=RKJ1/RKJ
      RKJ2=RKJ2/RKJ
      RKJ3=RKJ3/RKJ
C
      R11=RIK1*RKJ1
      R22=RIK2*RKJ2
      R33=RIK3*RKJ3
      B=(R11+R22+R33)**2
      C=D*A(K)*(A(K)**2)/RIKS/RKJS*(3.0*B-1.0)
C
      RMUTT(1,1)=RMUTT(1,1)+C*R11
      RMUTT(2,2)=RMUTT(2,2)+C*R22
      RMUTT(3,3)=RMUTT(3,3)+C*R33
      RMUTT(1,2)=RMUTT(1,2)+C*RIK1*RKJ2
      RMUTT(1,3)=RMUTT(1,3)+C*RIK1*RKJ3
      RMUTT(2,3)=RMUTT(2,3)+C*RIK2*RKJ3
      RMUTT(2,1)=RMUTT(2,1)+C*RIK2*RKJ1
      RMUTT(3,1)=RMUTT(3,1)+C*RIK3*RKJ1
      RMUTT(3,2)=RMUTT(3,2)+C*RIK3*RKJ2
C
C     LET RIJ = C*(RIK CROSS RKJ)
```

```
C
      C=45.0/8.0*A(K)*(A(K)**2)*B/RKJS/RIKS/RIK
      RIJ1=C*(RIK2*RKJ3-RIK3*RKJ2)
      RIJ2=C*(RIK3*RKJ1-RIK1*RKJ3)
      RIJ3=C*(RIK1*RKJ2-RIK2*RKJ1)
      RMUTR(1,1)=RMUTR(1,1)+RIJ1*RKJ1
      RMUTR(2,2)=RMUTR(2,2)+RIJ2*RKJ2
      RMUTR(3,3)=RMUTR(3,3)+RIJ3*RKJ3
      RMUTR(1,2)=RMUTR(1,2)+RIJ1*RKJ2
      RMUTR(1,3)=RMUTR(1,3)+RIJ1*RKJ3
      RMUTR(2,1)=RMUTR(2,1)+RIJ2*RKJ1
      RMUTR(3,1)=RMUTR(3,1)+RIJ3*RKJ1
      RMUTR(3,2)=RMUTR(3,2)+RIJ3*RKJ2
      RMUTR(2,3)=RMUTR(3,2)+RIJ2*RKJ3
 6    CONTINUE
 5    CONTINUE
      RETURN
      END
C
C#########################################################
C
C      RUNGE-KUTTA D.E. SOLVER
C
C      SET UP CONSTANTS FOR DOPRI5
C
      SUBROUTINE DOPCON(I)
C
C     SETS UP THE CONSTANTS FOR THE DORMAND-PRINCE
C     (4)5 ORDER RUNGE-KUTTA INTEGRATION ROUTINE
C      SEE E.HAIRE, "SOLVING ORDINARY DIFFERENTIAL EQUATIONS  I",
C      APPENDIX. THIS IS A VERSION BY I.K.S. ,25/9/1990
C
      COMMON/RUNGE/C1,C2,C3,C4,C5,C6,C7,C8,C9,C10,
     1C11,C12,C13,C14,C15,C16,C17,C18,C19,C20,
     2C21,C22,C23,C24,C25,C26
C
      C1=3.0/40.0
      C2=9.0/40.0
C
      C3=44.0/45.0
      C4=56.0/15.0
      C5=32.0/9.0
C
```

```
      C6=19372.0/6561.0
      C7=25360.0/2187.0
      C8=64448.0/6561.0
      C9=212.0/729.0
C
      C10=8.0/9.0
C
      C11=9017.0/3168.0
      C12=355.0/33.0
C13=46732.0/5247.0
      C14=49.0/176.0
      C15=5103.0/18656.0
C
      C16=35.0/384.0
      C17=500.0/1113.0
      C18=125.0/192.0
      C19=2187.0/6784.0
      C20=11.0/84.0
C
      C21=71.0/57600.0
      C22=71.0/16695.0
      C23=71.0/1920.0
      C24=17253.0/339200.0
      C25=22.0/525.0
C
      C26=1.0/40.0
      RETURN
      END
C
C##################################################################
C
      SUBROUTINE DOPRI5(X,Y,XEND,EPSIN,HMAX,HGUESS)
C
C     DORMAND-PRINCE RUNGE-KUTTA (4)5 ORDER METHOD
C     SOLVES Y ' = FCN(X,Y) OVER X TO XEND TO ACCURACY EPS.
C     SEE E.HAIRE," SOLVING ORDINARY DIFFERENTIAL EQUATIONS  I",
C     APPENDIX. THIS IS A VERSION BY I.K.S. ,25/9/1990
C
      DIMENSION RK1(18),RK2(18),RK3(18),RK4(18),RK5(18)
      DIMENSION Y1(18),Y(18)
      COMMON/RUNGE/C1,C2,C3,C4,C5,C6,C7,C8,C9,C10,
     1C11,C12,C13,C14,C15,C16,C17,C18,C19,C20,
     2C21,C22,C23,C24,C25,C26
```

```
      LOGICAL REJECT
      COMMON/STAT/NFCN,NSTEP,NACCEPT,NREJECT
      MAXSTEP=3000
      ZERO=0.0
      ONE=1.0
      UROUND=1.0E-10
 1    CONTINUE
      UROUND=UROUND/2.0
      IF((ONE+UROUND).GT.ONE) GO TO 1
C     PRINT 111,UROUND
 111  FORMAT(' UROUND =',E12.6)
      POSNEG=SIGN(ONE,XEND-X)
C
C      INITIAL PRERARATION
C
      H=MIN1(HGUESS,ABS(HMAX))
      H=SIGN(H,POSNEG)
      EPS=MAX1(ABS(EPSIN),7.0*UROUND)
      REJECT=.FALSE.
      NACCEPT=0
      NREJECT=0
      NFCN=1
      NSTEP=0
C     CALL SOLOUT(X,Y)
      CALL F(X,Y,RK1)
C
C      BASIC INTEGRATION STEP
C
 2    CONTINUE
      XH=X+H/10.0
       IF(NSTEP.GT.MAXSTEP.OR.XH.EQ.X) GO TO 79
      IF(((X-XEND)*POSNEG+UROUND).GT.ZERO) RETURN
      IF((X+H-XEND)*POSNEG.GT.ZERO) H=XEND-X
      NSTEP=NSTEP+1
      CH=H/5.0
C
C     THE FIRST 6 STAGES
C
      DO 22 I=1,18
      Y1(I)=Y(I)+CH*RK1(I)
 22   CONTINUE
      CALL F( X+CH,Y1,RK2)
      DO 23 I=1,18
```

```
      Y1(I)=Y(I)+H*(C1*RK1(I)+C2*RK2(I))
 23   CONTINUE
      CALL F(X+0.3*H,Y1,RK3)
      DO 24 I=1,18
      Y1(I)=Y(I)+H*(
     1C3*RK1(I)
     2-C4*RK2(I)
     3+C5*RK3(I))
 24   CONTINUE
      CALL F(X+0.8*H,Y1,RK4)
      DO 25 I=1,18
      Y1(I)=Y(I)+H*(
     1C6*RK1(I)
     2-C7*RK2(I)
     3+C8*RK3(I)
     4-C9*RK4(I))
 25   CONTINUE
      CALL F(X+C10*H,Y1,RK5)
      DO 26 I=1,18
      Y1(I)=Y(I)+H*(
     1C11*RK1(I)
     2-C12*RK2(I)
     3+C13*RK3(I)
     4+C14*RK4(I)
     5-C15*RK5(I))
 26   CONTINUE
      XPH=X+H
      CALL F(XPH,Y1,RK2)
      DO 27 I=1,18
      Y1(I)=Y(I)+H*(
     1C16*RK1(I)
     2+C17*RK3(I)
     3+C18*RK4(I)
     4-C19*RK5(I)
     5+C20*RK2(I))
      RK2(I)=(C21*RK1(I)
     1-C22*RK3(I)
     2+C23*RK4(I)
     3-C24*RK5(I)
     4+C25*RK2(I))
 27   CONTINUE
C     THE LAST STAGE
C
```

```
      CALL F(XPH,Y1,RK3)
      DO 28 I=1,18
      RK4(I)=(RK2(I)-C26*RK3(I))*H
 28   CONTINUE
      NFCN=NFCN+6
C
C     ERROR ESTIMATION
C
      ERR=ZERO
      CON=MAX1(1.0E-5,2.0*UROUND/EPS)
      DO 41 I=1,18
      DENOM=MAX1(ABS(Y1(I)),ABS(Y(I)),CON)
      ERR=ERR+(RK4(I)/DENOM)**2
 41   CONTINUE
      ERR=DSQRT(ERR/18.0)
C
C     COMPUTATION OF HNEW
C
C     WE  REQUIRE 0.2 < = HNEW/H < = 10.0
C
      FAC=MAX1(0.1,MIN1(5.0,((ERR/EPS)**0.2)/0.9))
      HNEW=H/FAC
      IF(ERR.LE.EPS)THEN
C
C     STEP ACCEPTED
C
      NACCEPT=NACCEPT+1
      DO 44 I=1,18
      RK1(I)=RK3(I)
      Y(I)=Y1(I)
 44   CONTINUE
      X=XPH
      IF(ABS(HNEW).GT.HMAX)HNEW=POSNEG*HMAX
      IF(REJECT)HNEW=POSNEG *MIN1(ABS(HNEW),ABS(H))
      REJECT=.FALSE.
      ELSE
      REJECT=.TRUE.
C
C     STEP IS REJECTED
C
      IF(NACCEPT.GT.1) NREJECT=NREJECT+1
      ENDIF
      H=HNEW
```

```
      GO TO 2
C
C     FAIL EXIT
C
 79   PRINT 80,X
 80   FORMAT(' DOPRI5 FAILED TO SATISFY TOLERANCE AT X = ',E11.4)
      PRINT 90,NACCEPT
      PRINT 91,NREJECT
 90   FORMAT(' ACCEPTED STEPS ',I9)
 91   FORMAT(' REJECTED STEPS ',I9)
      RETURN
      END
C
C########################################################################
C
      SUBROUTINE SOLOUT(X,Y)
C
      DIMENSION Y(18)
      X=X
      Y(1)=Y(1)
      RETURN
      END
C
C########################################################################
C
C     ROUTINE TO CHANGE POSITION ARRAY POS FROM OLD NOTATION TO
NEW,THE RESULT OF
C     WHICH IS STORED IN ARRAY SOP
C
      SUBROUTINE CHANOT(POS,SOP)
C
C     CHANGE NOTATION FOR POS
C
      DIMENSION POS(18),SOP(18)
      SOP(1)=POS(1)
      SOP(2)=POS(4)
      SOP(3)=POS(7)
      SOP(4)=POS(5)
      SOP(5)=POS(5)
      SOP(6)=POS(8)
      SOP(7)=POS(3)
      SOP(8)=POS(6)
      SOP(9)=POS(9)
```

```
SOP(10)=POS(10)
SOP(11)=POS(13)
SOP(12)=POS(16)
SOP(13)=POS(11)
SOP(14)=POS(14)
SOP(15)=POS(17)
SOP(16)=POS(12)
SOP(17)=POS(15)
SOP(18)=POS(18)
RETURN
END
```

Index

Printed and bound by CPI Group (UK) Ltd, Croydon, CR0 4YY
08/05/2025
01864806-0002